IFIP Advances in Information and Communication Technology 741

IFIP Advances in Information and Communication Technology

The IFIP AICT series publishes state-of-the-art results in the sciences and technologies of information and communication. The scope of the series includes: foundations of computer science; software theory and practice; education; computer applications in technology; communication systems; systems modeling and optimization; information systems; ICT and society; computer systems technology; security and protection in information processing systems; artificial intelligence; and human-computer interaction.

Edited volumes and proceedings of refereed international conferences in computer science and interdisciplinary fields are featured. These results often precede journal publication and represent the most current research.

The principal aim of the IFIP AICT series is to encourage education and the dissemination and exchange of information about all aspects of computing.

More information about this series at https://link.springer.com/bookseries/6102

Pradorn Sureephong · Christophe Danjou ·
Abdelaziz Bouras
Editors

Product Lifecycle Management

Leveraging AI, Digital Twins, and Smart Technologies

21st IFIP WG 5.1 International Conference, PLM 2024
Bangkok, Thailand, July 7–10, 2024
Revised Selected Papers Part II

Springer

Editors
Pradorn Sureephong
Chiang Mai University
Chiang Mai, Thailand

Christophe Danjou
Polytechnique Montréal
Montréal, QC, Canada

Abdelaziz Bouras
Qatar University
Doha, Qatar

ISSN 1868-4238 ISSN 1868-422X (electronic)
IFIP Advances in Information and Communication Technology
ISBN 978-3-031-93322-6 ISBN 978-3-031-93323-3 (eBook)
https://doi.org/10.1007/978-3-031-93323-3

This Springer imprint is published by the registered company Springer Nature Switzerland AG
The registered company address is: Gewerbestrasse 11, 6330 Cham, Switzerland

Preface

Since 2003 the International Conference on Product Lifecycle Management (PLM) has brought together researchers, developers and users of Product Lifecycle Management. It aims to integrate business approaches to the collaborative creation, management and dissemination of product and process data throughout the extended enterprises that create, manufacture and operate engineered products and systems. The conference aims to involve all stakeholders of the wide concept of PLM, hoping to shape the future of this field and advance the science and practice of enterprise development.

The 21th International Conference on Product Lifecycle Management (PLM 2024) was hosted in Bangkok, Thailand, from July 7th to July 10th, 2024.

For its 21st anniversary, PLM 2024 registered 188 participants from 16 countries who had the chance to attend nine Special Sessions (Digital Transformation) plus one industrial day. In addition, PLM 2024 featured Academic Keynotes, Industrial Keynotes and round tables.

Keynote Speakers

- Sudarsan Rachuri, Department of Energy, USA
- Chidchanok Lursinsap, Chulalongkorn University, Thailand
- Ram D. Sriram, National Institute of Standards and Technology, USA
- Tiranee Achalakul, Big Data Institute, Thailand
- Sittapong Settapat, SCB Data X Co., Ltd., Thailand
- Santisook Limpeeticharoenchot, STelligence Co., Ltd., Thailand
- Passakon Prathombutr, Digital Economy Promotion Agency, Thailand

Industry day

- Data Driven Government Thailand Used Cases – Tiranee Achalakul, Big Data Institute, Thailand
- Generative Al And Knowledge Graph For Digital Twin In Industrial 4.0 – Sittapong Settapat, SCB Data X Co., Ltd., Thailand
- Change Management In The Era Of Al And Fintech: Strategies For Success – Santisook Limpeeticharoenchot, STelligence Co., Ltd., Thailand
- Smart City Lifecycle – Passakon Prathombutr, Digital Economy Promotion Agency, Thailand

Industrial Round Table

BIM and Digital Twin

- Dconstruct Robotics Pte., Ltd.
- Microsoft Partner
- Altotech Global Co., Ltd.

Digital Twin for Sustainable Smart City

- Pcs Thailand Co., Ltd.
- Ai And Robotics Ventures Company Limited
- Siam Inno City Co., Ltd

All submitted papers were double-blind peer-reviewed by at least two reviewers. In total, 105 contributions were submitted, of which 64 were accepted. Program chairs Christophe Danjou and Pradorn Sureephong organized an amazing scientific conference with presentations grouped in 20 thematic sections.

We would like to thank everyone who directly or indirectly contributed to making the PLM 2024 conference a success, particularly by participating in rich scientific discussions and meeting all of us in Bangkok, Thailand.

Pradorn Sureephong

Organization

Conference Chairs

Abdelaziz Bouras	Qatar University, Qatar
Kitipong Promwong	Office of National Higher Education Science Research and Innovation Policy Council, Thailand
Kriengkrai Thiennukul	Federation of Thailand Industries, Thailand

Program Committee Chairs

Pradorn Sureephong	Chiang Mai University, Thailand
Christophe Danjou	Polytechnique Montréal, Canada

Local Organizing Committee

Krittika Kantawong	University of Payao, Thailand
Suepphong Chernbumroong	Chiang Mai University, Thailand
Tama Duangnamol	Chiang Mai University, Thailand
Santichai Wicha	Mae Fah Luang University, Thailand
Wanus Srimaharaj	Payap University, Thailand
Yootthapong Tongpaeng	Mae Fah Luang University, Thailand
Paweena Suebsombut	Mae Fah Luang University, Thailand
Shisa Singkhamfu	Chiang Mai University, Thailand

Steering Committee Members

Abdelaziz Bouras	Qatar University, Qatar
Anderson Luis Szejka	Pontifical Catholic University of Paraná, Brazil
Balan Gurumoorthy	Indian Institute of Science, India
Benoit Eynard	University of Technology of Compiègne, France
Christophe Danjou	Polytechnique Montréal, Canada
Clement Fortin	Skolkovo Institute of Science and Technology, Russia
Darli Rodrigues Vieira	Université du Québec à Trois-Rivières, Canada

Debashish Dutta	University of Michigan, USA
Frédéric Noël	Grenoble Institute of Technology, France
José Rios	Polytechnic University of Madrid, Spain
Klaus-Dieter Thoben	University of Bremen, Germany
Néjib Moalla	University of Lyon, France
Osiris Canciglieri Junior	Pontifical Catholic University of Paraná, Brazil
Paolo Chiabert	Polytechnic of Turin, Italy
Ramy Harik	University of South Carolina, USA
Romeo Bandinelli	University of Florence, Italy
Sebti Foufou	University of Sharjah, UAE
Sergio Terzi	Politecnico di Milano, Italy

Doctoral Workshop Committee

Monica Rossi	Politecnico di Milano, Italy
Yacine Ouzrout	University of Lyon, France

Scientific Committee

Abdelaziz Bouras	Qatar University, Qatar
Abdelhak Belhi	Qatar University, Qatar
Aicha Sekhari	University of Lyon, France
Alain Bernard	Central School of Nantes, France
Alexander Smirnov	Russian Academy of Sciences, Russia
Alison McKay	University of Leeds, UK
Améziane Aoussat	Arts et Métiers ParisTech, France
Anderson Luis Szejka	Pontifical Catholic University of Paraná, Brazil
Balan Gurumoorthy	Indian Institute of Sciences Bangalore, India
Bas Koomen	CADMES, The Netherlands
Benoit Eynard	University of Technology of Compiègne, France
Bob Young	Loughborough University, UK
Brendan Sullivan	Polytechnic of Milan, Italy
Chris McMahon	Technical University of Denmark, Denmark
Chen Zheng	Northwestern Polytechnical University, China
Christophe Danjou	Polytechnique Montréal, Canada
Claudio Sassanelli	Politecnico di Bari, Italy
Clement Fortin	Skolkovo Institute of Science and Technology, Russia
Daniel Schmid	Zurich University of Applied Sciences, Switzerland

Darli Rodrigues Vieira	Université du Québec à Trois-Rivières, Canada
Debashish Dutta	University of Michigan, USA
Detlef Gerhard	Ruhr University Bochum, Germany
Dimitris Kiritsis	Federal Institute of Technology in Lausanne, Switzerland
Eduardo Zancul	University of São Paulo, Brazil
Erastos Filos	European Commission – Research & Innovation, Belgium
Felix Nyffenegger	Rapperswil University of Applied Sciences, Switzerland
Fernando Mas	University of Seville, Spain
Frédéric Demoly	University of Technology Belfort-Montbéliard, France
Frédéric Noël	Grenoble Institute of Technology, France
Frédéric Segonds	Arts et Métiers ParisTech, France
George Huang	University of Hong Kong, China
Gianluca D'Antonio	Polytechnic of Turin, Italy
Gregor Schweitzer	Fresenius Medical Care, Germany
Guilherme Brittes Benitez	Pontifical Catholic University of Paraná, Brazil
Hannele Lampela	University of Oulu, Finland
Hannu Kärkkäinen	Tampere University of Technology, Finland
Hervé Panetto	University of Lorraine, France
Ilkka Donoghue	LUT University, Finland
Johan Malmqvist	Chalmers University of Technology, Sweden
Jong Gyun Lim	Samsung Advanced Institute of Technology, South Korea
José Rios	Polytechnic University of Madrid, Spain
Julie Jupp	University of Technology Sydney, Australia
Julien Le Duigou	University of Technology of Compiègne, France
Klaus-Dieter Thoben	University of Bremen, Germany
Lionel Roucoules	Arts et Métiers ParisTech, France
Louis Rivest	University of Quebec, Canada
Margherita Peruzzini	University of Modena and Reggio Emilia, Italy
Mariangela Lazoi	University of Salento, Italy
Matthieu Bricogne	University of Technology of Compiègne, France
Michael Schabacker	University of Magdeburg, Germany
Mickael Gardoni	University of Quebec, Canada
Monica Rossi	Polytechnic of Milan, Italy
Néjib Moalla	University of Lyon, France
Nickolas S. Sapidis	University of Western Macedonia, Greece
Nicolas Maranzana	Arts et Métiers ParisTech, France
Nikolaos Bilalis	Technical University of Crete, Greece

Nikolay Shilov	Innopolis University, Russia
Nopasit Chakpitak	Chiang Mai University, Thailand
Osiris Canciglieri Junior	Pontifical Catholic University of Paraná, Brazil
Paolo Chiabert	Polytechnic of Turin, Italy
Paul Hong	University of Toledo, USA
Peter Hehenberger	University of Applied Sciences Upper Austria, Austria
Ramy Harik	University of South Carolina, USA
Rebeca Arista	Airbus, France
Ricardo Jardim-Goncalves	New University of Lisbon, Portugal
Rob Vingerhoeds	ISAE-SUPAERO, France
Romain Pinquié	Grenoble Institute of Technology, France
Romeo Bandinelli	University of Florence, Italy
Sandor Vajna	University of Magdeburg, Germany
Sebti Foufou	Qatar University, Qatar
Sehyun Myung	Youngsan University, South Korea
Sergio Terzi	Politecnico di Milano, Italy
Shuichi Fukuda	Keio University, Japan
Simone Briatore	Skolkovo Institute of Science and Technology, Russia
Sophia Salas Cordero	ISAE-SUPAERO, France
Sudarsan Rachuri	Department of Energy, USA
Thomas Hedberg	University of Maryland, USA
Thomas Vosgien	HILTI Group, Austria
Thorsten Wuest	West Virginia University, USA
Virginia Fani	University of Florence, Italy
Vishal Singh	Indian Institute of Science, India
Wen Feng Lu	National University of Singapore, Singapore
Yacine Ouzrout	University of Lyon, France
Yaroslav Menshenin	Grenoble Institute of Technology, France
Young Won Park	University of Tokyo, Japan

Contents – Part II

Contents – Part I

Discussion of Current Issues in Cost Estimation Decisions for Aircraft Manufacturing

Leonardo Cavalcanti Hernandes[1], Matheus Herman Bernardim Andrade[2], Anderson Luis Szejka[1], and Fernando Mas[3,4](✉)

[1] Industrial and System Engineering Department, Pontifical Catholic University of Parana (PUCPR), Curitiba, Brazil
leonardo.hernandes@pucpr.edu.br, anderson.szejka@pucpr.br
[2] Computer Engineering Department, Pontifical Catholic University of Parana (PUCPR), Curitiba, Brazil
matheus.bernardim@pucpr.edu.br
[3] Mechanical and Manufacturing Engineering Department, University of Sevilla, Sevilla, Spain
fmas@us.es
[4] CT Engineering, Madrid, Spain

Abstract. The development of new products is attracting increasing attention from industry and academia due, among other factors, to the reduced product development process, market differentiation, greater product complexity, rapid changes in technological knowledge and increased customer sophistication. This scenario requires an efficient cost estimation process and quick decision-making to ensure the efficiency of the production process. Consequently, many approaches have been suggested for use in predicting product costs. However, each has its issues and limitations that affect the effectiveness of the final solution. This paper aims to survey approaches related to the cost estimation process in aerospace industry processes and their respective niches, highlighting the gaps and limitations present to identify the appropriate contexts for applying each. It compiles the latest work on cost estimation approaches, artificial neural networks, and the semantic web in the aerospace industry, using the knowledge acquired for better and comprehensive future implementation.

Keywords: Cost Estimation · Semantic Web Technologies · Artificial Neural Networks · Aerospace Industry

1 Introduction

As in any manufacturing sector, there are constant developments in the development and manufacturing of parts, and, along with these improvements, there are several challenges in trying to reduce the time taken for the processes involved and increase precision and assertiveness, generating efficiency and reliability in the process. Due to the complexity of products in the aerospace sector, the development of new technologies and integrated systems with multiple domains is keeping up with demands to reduce costs and time,

Published by Springer Nature Switzerland AG 2025
P. Sureephong et al. (Eds.): PLM 2024, IFIP AICT 741, pp. 1–14, 2025.
https://doi.org/10.1007/978-3-031-93323-3_1

meaning that information is being massively digitized and semantically structured and, consequently, the importance of data and its quality has become something essential and challenging for the interoperability of systems within smart manufacturing [1].

Given the market's competitive nature, industries employ various strategies to differentiate themselves from competitors. Their main objective is to offer customised, high-quality products at the lowest possible price to remain viable and to establish competitive prices. A clear understanding of costs is required. Although detailed information on costs can be obtained after a product has been manufactured through subsequent calculations, it is often necessary to estimate production costs before actual production, even without the exact definition of processes and tools materials and while still in the design phase [2]. To meet this challenge, cost estimation techniques are used to approximate costs within a specified range of accuracy coupled with the knowledge of experts in the field to provide the most accurate estimate possible so as not to rework the entire process. Another solution would be to implement an expert system that includes knowledge about a defined subject and the mechanism to exploit this knowledge, communicating it to the user via an interface [3].

Based on this context, this paper aims to explore the essential guidelines for building an expert cost estimation system for sheet metal complex parts in an Aerospace Industrial. This paper identifies and details the difficulties in designing, developing, and implementing this ontology-based framework. It also highlights the best techniques and tools in the computational field to assist in this process. For this purpose, the research methodology aimed to compile recent updates and new definitions of the major areas involved: costs, knowledge formalisation and Artificial Intelligence (AI) tools, while also returning to concepts defined decades ago that are extremely important for a complete understanding of the subject.

The paper has six sections and is divided as follows. Section 2 defines the problem statement and further details regarding the research main topics. Section 3 presents the methodology used for the research. Section 4 discusses related works and concepts about approaches commonly applied: cost estimation techniques and approaches, Artificial Neural Networks (ANN's), and Semantic Web technologies. Section 5 presents the discussion, issues and gaps found in each approach based on the literature review conducted, and finally, the conclusions of this paper are drawn in Sect. 6.

2 Problem Statement

Due to the difficulty in quantifying the use of each resource in the early stages of the product lifecycle (and especially in the initial phases of product development), given the limited amount of information and the low level of project definition, many approaches to cost estimation are used to overcome this lack of detailed information, aiming to approximate costs within a specific range of accuracy [4]. However, no solutions are not effective, resulting in gaps that generate imprecise results. Therefore, the first stage of the research involves conducting a literature review relevant to the study topic and demonstrating scientific recognition. The paper selection was based on research axes and approaches related to the topic. These research axes are necessary to understand the topic's state of the art and the work's goal. The defined research axes are presented in Fig. 1.

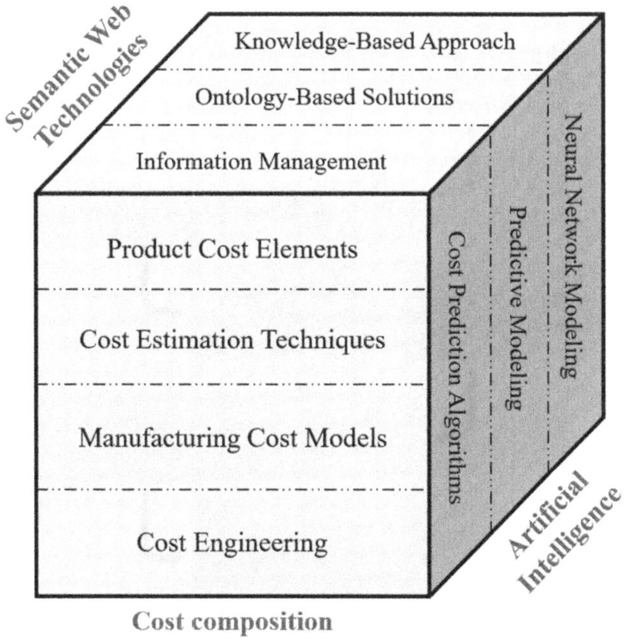

Fig. 1. Research main axes.

In sequence, each research axis gives rise to keywords to characterise the domain of each axis, which can be combined, as shown in Fig. 1. For the analytical cost axis, selected keywords include cost estimation and prediction, product cost estimation, and cost engineering. For the AI, approaches using ANN's were selected. For the third axis related to Web Semantic Technologies, selected keywords include ontology(ies), ontology-based, knowledge-based, intelligent, and expert systems.

3 Research Methodology

The databases used for the research are available at the Coordination for the Improvement of Higher Education Personnel (CAPES)[5], with the main approach in the broad areas of engineering, computer science, aerospace and manufacturing sectors and business and economics. The CAPES database includes several renowned databases such as EBSCO, Emerald, IEEE, Science Direct, Scopus, Springer, Village Engineering, and Web of Science, among others selected for this study. The searches yielded 402 papers, applying the following filters on the data acquired: only English language, date range from 2019 to 2024 and only peer-reviewed papers. By removing redundant keywords and associating the title and relevant keywords, 53 papers remained for the next stage of analysis., as shown in Fig. 2. Considering the scientific relevance of the journals, eleven papers remained that required reading the abstract and main contributions of the papers, and at the end, only six were found aligned with the research theme.

As shown in the Table 1, six papers presented an alignment of concepts and definitions when reading the abstract and are presented in Table 1. Further discussion and details

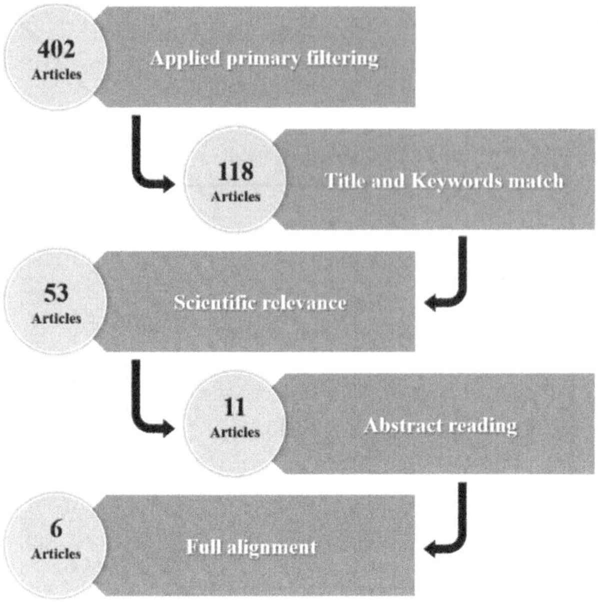

Fig. 2. Filter of papers obtained from the literature review.

of these papers will be presented in the discussion section, along with a summary of all the information obtained during the research.

The search resulted in papers with a publication year between 2020 and 2024 distributed into five different relevant journals, as described in one of the criteria for this literature review. The keywords found in these papers include "artificial intelligence" and "cost", along with others also related to this research, such as ANN's, decision support and cost estimation and analysis.

It is important to note that, during recent research into the subject, it was noted that a large part of the basic concepts was defined outside the defined year limit, and for this reason, the authors also opted to cover consolidated works to update readers on the current stage on the subjects and provide better and more detailed information. Thus, this section will present the combination of papers defining terms and concepts, and some of the citations and those obtained from the literature search process will be described in the discussion section.

Additionally, Connected Papers was used to find relationships between the research findings, using each of the research axes as a trigger and considering the presence of each keyword. The following details the research and contributions found for each of the research axes.

Table 1. Papers description obtained from the literature review

Ref.	Title	Year	Journal
[30]	Manufacturing cost estimation based on a deep- learning method	2020	Journal of manufacturing systems
[31]	Manufacturing cost estimation based on the machining process and deep-learning method	2020	Journal of manufacturing systems
[32]	Explainable artificial intelligence for manufacturing cost estimation and machining feature visualization	2021	Expert systems with applications
[28]	An empirical investigation into intelligent cost analysis in purchasing	2022	Supply chain management
[29]	Digital transformation for cost estimation system via meta-learning and an empirical study in aerospace industry	2023	Computers & industrial engineering
[33]	Big data and machine learning in cost estimation: An automotive case study	2024	International Journal of Production Economics

4 Related Works for Cost Estimation Decisions for Aircraft Manufacturing

This section presents a survey of related works on the cost estimation process based on the different approaches used in each application and their respective domains. Seeking to provide a comprehensive overview of the approaches, methodologies, and challenges faced in the current context, and to identify research gaps and analyse them further. The sought domains for theoretical grounding included Product Cost Estimation (PCE) techniques and their definitions, the utilization of ANN's, and techniques for automatic prediction of production costs for New Products Development (NPD), and finally, a survey of semantic web technologies used for modelling, formalization, and structuring of knowledge in systems within the analysed context.

4.1 Cost Estimation Approach

PCE techniques and methodologies have been widely discussed in the literature over the years, and to start this review, [6] his work presents a hierarchical classification that focuses on grouping techniques with similar characteristics. This characterisation is divided into qualitative, intuitive, analogue, or quantitative techniques and parametric and analytical techniques, each with subsequent divisions, as illustrated in Fig. 3.

Another definition given by [7] argues that a product has new and standardised product elements and cites two pricing approaches. The authors defined Variant-based as the classification that considers the similarity between the current product and previous ones, using the cost records of products already manufactured as a model for the current one. Generative cost estimation is the classification that estimates costs through the

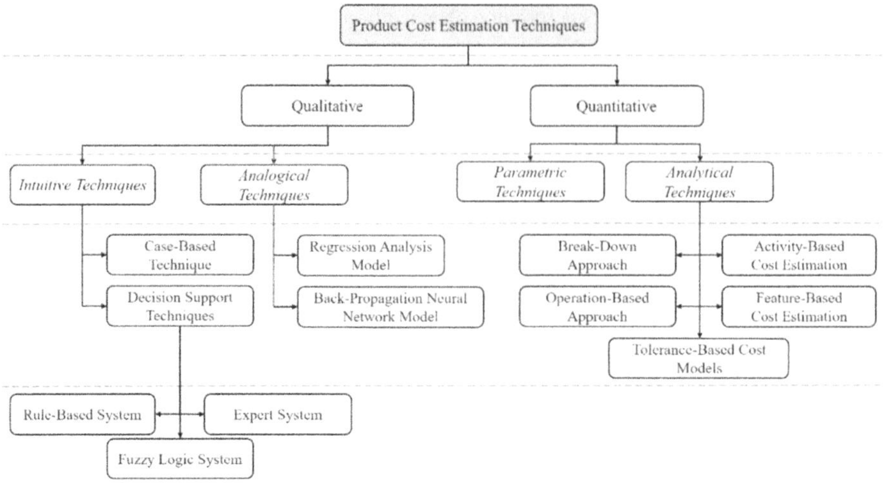

Fig. 3. Classification of PCE techniques adapted from [6].

operations required to produce a product. In addition to the definitions made, the paper also presents a generic framework for applying the concept proposed to a sheet metal example.

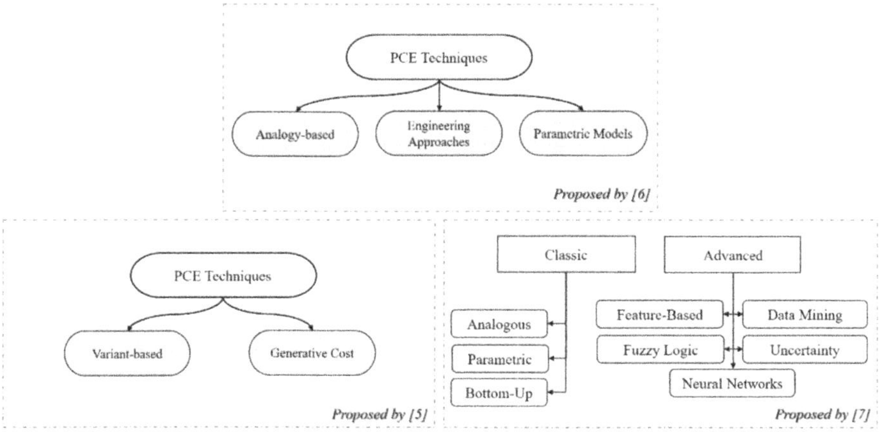

Fig. 4. Overview of the PCE techniques defined in [6–8].

Another approach is proposed by [8] and provides a classification divided into three techniques for cost estimation: Analogy-based, making an analogy between the current product and those already produced in the past; Parametric Models, which apply parametric or analytical functions to analyse the variables involved in the process and the product; and Engineering Approaches, which include more detailed information on the characteristics, processes, and manufacture of the product in question. There is yet another classification made by [9] where the methods are differentiated between Classical (analogous, parametric, and bottom-up) and Advanced (Feature-based, Fuzzy Logic, Neural networks, Uncertainty and Data Mining techniques) and, as well as defining them. The authors demonstrate comparisons between the classical methods and a detailed and referenced explanation of each technique mentioned in the aerospace industry. An overview of all these techniques and their hierarchy is shown in Fig. 4. To provide a more detailed comparison of the techniques, the authors have chosen the definition given in [6] for this comparative analysis as shown in Table 2. To simplify the visualisation of each technique, an identifier has been used to separate them as follows: Analytical Techniques (i), Analogical Techniques (ii), Parametric Techniques (iii) and Intuitive Techniques (iv).

Table 2. Comparison of strengths, limitations and applications on cost estimation techniques.

ID	Strengths	Limitations	Applicability
i	High effectiveness in projects with clear cause and effect relationships and well-defined scope, providing a solid understanding of the main cost factors	Requires significant amount of data and complex calculations. May not be suitable for new or 'unique' projects	Research and development (R&D) or design projects with well-defined processes and established models
ii	Useful for projects similar to past experiences, easy to implementation and good for early estimation processes	May not capture project specifics and accuracy strongly depends on a good analogy finding	Budget planning in the initial product design stages and feasibility studies
iii	Combination of historical data with the project specifications. Provides a structured approach and accurate results	Strong dependence on the quality and relevance of historical data. Not suitable for highly personalised projects	Estimation of similar (or repeatable) projects with historical data and defined parameters
iv	Leverages the knowledge and experience of experts and useful for subjective or non- quantifiable factors	Not easily replicable or verifiable and accuracy relies on experience (tacit knowledge)	Viability studies for new or exclusive projects. Complementing other methods with expert insights or to fill knowledge gaps

4.2 Artificial Neural Networks

Parametric cost estimation approaches are commonly used in situations like analytical approaches, where a predefined cost function is involved, and it operates based on the relationship between product attributes and the manufacturing process. With free parameters, this function is determined based on the engineer's experience and then refined using statistical regression methods applied to previous case data. However, they always require a cost function that accurately represents the process [8]. In contrast, ANN's seek to fit curves to data, identifying hidden functional relationships between product attributes and cost, acting as non-parametric approximators. The advantages of using ANN's include assuming less restrictive underlying distributions, using inherent parallelism for faster calculations, and achieving robustness due to local connections between processing nodes, which do not occur in conventional approaches [10].

In this context, many approaches have been found in the literature relating artificial neural networks to the cost estimation process, as it is a field that has received a lot of attention from manufacturing engineers over the years, as knowing the production cost is a key factor in determining the overall performance of the NPD process [8]. Figure 5 shows a relationship between publications related to both themes.

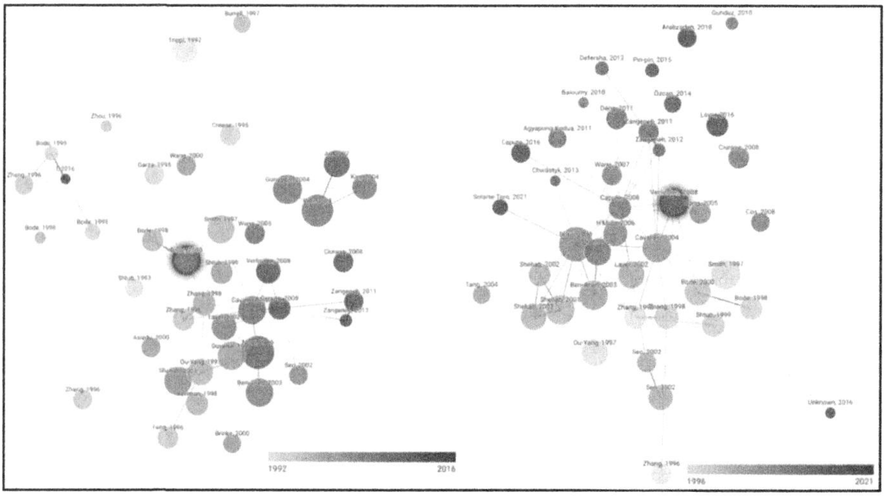

Fig. 5. Related publications.

The image displays connections between various papers in Scopus, all dealing with the same approach. The graph reveals the network of connections among the different studies, highlighting the relationships and similarities of the approaches in cost estimation.

Although ANN's have found applications in various design problems beyond cost estimation, there is still a scarcity of comprehensive evaluations of their performance and comparisons with traditional methods in the literature [11] and [12]. Furthermore, the works of [8, 13, 14] and [15] become important for this research, as they explored this approach with different emphases on model architecture, providing diverse and comprehensive results for performance evaluation.

4.3 Semantic Web Technology

Bringing the focus to where this work is inserted, all the information about the pricing of a product within the aerospace industry must be obtained in the form of data and, with the appropriate treatment, they can be applied to AI techniques to solve a problem. However, given the volume, the integration of different areas, and the complexity of these data, there is a need for proper formalisation so that the information contained in this group of data is applicable. For this purpose, Semantic Web technology links this data set in a structured and intelligent way through specific languages and standards, guaranteeing the interoperability of this knowledge network and allowing Knowledge Management (KM) [16].

One of the research fields for defining knowledge is Knowledge-based Engineering (KBE), which applies methodologies and technologies to capture and reuse product and process engineering knowledge in various areas [17]. The author also conducts a literature review of the various definitions of the subject by different authors. To demonstrate semantic interoperability in the integrated Product Development Process (PDP), the authors in [18] present a systematic literature review highlighting the importance of efficient information exchange and preserving the meaning of information in the different phases of the PDP. Focusing more on the application of AI, the paper [19] discusses the role of KBE in the interaction between AI and CAD, highlighting the need to improve knowledge traceability and some expectations for the area, such as integration with open-source software and programming. Another paper identifies opportunities for integrating AI tools for efficient KM, bringing insights from the industrial sphere into the work [20].

Being defined as "an explicit specification of a conceptualisation" by [21], its use and importance in structuring knowledge and the concept and relationship definitions were also demonstrated in this work, emphasising that ontology is an effective tool for sharing information between individuals for different domains. To associate the application of ontologies in systems engineering, the work [22] provides a systematic framework for Ontology-based Engineering (OBE) systems. It discusses various methodologies and frameworks to support systems engineering practices. The paper [23] attests that OBE systems cover the domains of ontology-based products and applications, increasing the knowledge domain through integrating and transitioning KBE methodologies, thus maintaining continuity between KBE concepts. Another approach is taken in [24], where the authors highlight the difference between evaluating the quality of ontology content and its use on the web, presenting the Ontology Usage Analysis Framework (OUSAF) for this purpose. Still, on the ontological domain, the research by [25] aimed to enhance semantic interoperability in intelligent manufacturing processes by demonstrating the importance of effective information sharing through a knowledge-based expert system.

Finally, some works have an application approach to the mentioned concepts, such as [26], which applies the KM approach with semantic web integration to real industry cases to make knowledge easily accessible and reusable for support. The study [27] also proposes a methodology based on KBE techniques to integrate the design process with cost analysis and demonstrates this integration.

5 Discussion of Current Issues

As an expected result of this work, this section aims to discuss the main points raised in the review and, in addition, add to the set of works already cited by the six recent papers obtained by the review with greater emphasis. This selected group includes one research paper and five other application or solution proposals with different approaches and results as demonstrated by Table 3.

Table 3. Overview of relevant information from the selected papers.

Ref.	Approach	Results
[30]	Convolutional Neural Network (CNN) model training on 3D CAD models, material data, and historical cost data	Captures complex relationships between design and manufacturing costs and achieves accurate estimations highlighting the potential of deep learning application
[31]	Machining process with deep- learning techniques integration to manufacturing cost estimation	Demonstrates the efficiency of deep learning methods in recognising processing resources and estimating the cost of parts
[32]	Explore 3D Deep Learning architectures and Explainable AI (XAI) integration	Provides valuable insights for design optimization and achieves accurate cost estimations on the tests performed by using XAI to highlight machining features impacting cost on the 3D model
[28]	Extensive literature review on cost analysis and intelligent systems focusing on applicability of machine learning (ML) on such tasks	Human judgment and experience are still crucial for interpreting the insights provided by machine learning models but the association between ML and human expertise offers significant potential for cost analysis and estimation
[29]	Proposal for meta-learning model using data from different aerospace projects (designs, materials, costs)	Achieves more accurate and faster cost estimations compared to traditional methods in the aerospace industry using broader expert knowledge range. Enables better decision- making process
[33]	Applicability of big data (using historical data on car models and material costs) and machine learning approaches in automotive industry towards a case study approach	Highlights the benefits of combining big data and machine learning demonstrating complex relationships identification and accuracy on cost. Providing valuable insights for cost control by identifying the most significant cost drivers

The following paragraph describes brief pertinent comments on each work, as shown previously in the table following the sequence of the reference value. First, the research paper [28] demonstrates the empirical research process carried out by experts in the field, with the focus on potential Machine Learning (ML) applications in costing and purchasing within the manufacturing industry. The paper [29] conducts an empirical study seeking to improve cost estimation systems through digital transformation and meta-learning, with the main objective of optimising quotation decisions for companies in the aerospace sector. Addressing the studies that have an application, the works [30] and [31] have similarities in the use of deep-learning techniques and the results since the paper [30] compares the performance of the 2D Convolutional Neural Network (CNN) and the 3D CNN in the cost estimation regression problem. The paper [31] proposes and applies 3D CNN to solve the problem of accurately recognising parts processing features to estimate structurally complex parts. The authors in [32] present an explainable manufacturing cost estimation model for 3D CAD models applied to CNC machines, informing us, for example, which parts of the 3D model need to be modified to reduce manufacturing costs, thus helping us understand cost estimation processes and facilitate decision-making. Finally, the recent work carried out by [33] focuses on the automotive industry using a data-driven approach and evaluating by use cases, issues and benefits of machine learning and big data from the perspective of cost experts.

The relevance of pricing within the industrial and smart manufacturing context is directly related to digital transformation due to the demand for increased productivity and efficiency. As highlighted in [4, 6] and [7], the association between engineering and cost knowledge in the early stages of PDP is relevant to the KBE domain and as shown in [23, 25] and [28], approaches that associate the knowledge-based concepts and OBE have become an effective tool in the context of cost estimation, as they allow the formalisation and representation of knowledge in a structured way and integration with programming languages and their appropriate libraries.

Thus, given the results presented, difficulties and barriers must be overcome in implementing a system based on knowledge and integration with new technologies. Still, the demand from industries requires more responsiveness in decision-making and the need to facilitate processes that are often slow due to a lack of digitisation.

6 Conclusion

This paper provides a comprehensive overview of the different approaches used in the cost estimation process in different niches of aerospace manufacturing, highlighting the problems and shortcomings associated with each. In the context of analytical pricing, the concepts of product cost estimation are already well established in the industry, relying on predefined cost techniques and functions applied by cost engineers. However, solutions aiming to digitise the cost process by applying statistical regression methods suffer from a lack of reliable data, difficulty dealing with uncertainty, and sensitivity to market changes.

However, solutions that aimed at digitizing the cost process through the use of artificial intelligence techniques and the application of statistical regression methods suffered from a lack of reliable data, the difficulty of dealing with uncertainty and sensitivity to market changes, as well as the absence of a formal representation of expert knowledge. Since this is a recurring problem, this study selected the use of semantic technologies, such as ontologies, in association with the use of neural networks in the context of product cost estimation, as a possible solution for capturing and organizing the information contained in the industry manufacturing processes.

There are many approaches of ANN's in cost estimation applications found in the literature. However, each presents a different perspective on the problem at hand, resulting in similar yet different-focused solutions. This implies different limitations and issues related to data, such as lack of standardisation and data inconsistencies.

To conclude, the gaps and problems identified in the literature review predominantly refer to the stages of data processing, digitisation and formalisation, thus leading to a greater dependence on manual work for data collection and analysis. The articles obtained and analysed demonstrates a significant need for data and its absence within the industry due to a lack of structuring and formalization, generating an opportunity to integrate existing knowledge and recent technologies in the context of product cost estimating process. Future work will focus on developing an intelligent system that integrates the various approaches, technologies and solutions covered in this document to provide a more robust, efficient and intelligent cost estimate within the industrial sphere.

Acknowledgments. The authors would like to thank the Pontifical Catholic University of Parana (PUCPR), National Council for Scientific and Technological Development (CNPq), University of Sevilla, and CT Engineering colleagues for their support and contribution during the development of this work.

References

1. Pereira, R.M., Szejka, A.L., Canciglieri Junior, O.: Towards an information semantic interoperability in smart manufacturing systems: contributions, limitations and applications. Int. J. Comput. Integr. Manuf. **34**, 422–439 (2021). https://doi.org/10.1080/0951192X.2021.1891571
2. Corbett, J., Crookall, J.R.: Design for economic manufacture. CIRP Ann. **35**, 93–97 (1986). https://doi.org/10.1016/S0007-8506(07)61846-0
3. Byrd, T.A., Hauser, R.D.: Expert systems in production and operations management: research directions in assessing overall impact. Int. J. Prod. Res. **29**, 2471–2482 (1991). https://doi.org/10.1080/00207549108948097
4. Gómez, A., Mas, F., Menéndez, J.L., Ríos, J.: A knowledge-based application for industrialization design. In: Procedia Engineering, pp. 318–326. Elsevier Ltd. (2013). https://doi.org/10.1016/j.proeng.2013.08.178
5. CAPES: Portal Periódicos. CAPES (2024). https://www-periodicos-capes-gov-br.ezl.periodicos.capes.gov.br/
6. Niazi, A., Dai, J.S., Balabani, S., Seneviratne, L.: Product cost estimation: Technique classification and methodology review. J. Manuf. Sci. Eng. **128**, 563–575 (2006). https://doi.org/10.1115/1.2137750
7. Weustink, I.F., ten Brinke, E., Streppel, A.H., Kals, H.J.J.: A generic framework for cost estimation and cost control in product design. J. Mater. Process. Technol. **103**, 141–148 (2000). https://doi.org/10.1016/S0924-0136(00)00405-2
8. Cavalieri, S., Maccarrone, P., Pinto, R.: Parametric vs. neural network models for the estimation of production costs: A case study in the automotive industry. Int. J. Prod. Econ. **91**, 165–177 (2004). https://doi.org/10.1016/j.ijpe.2003.08.005
9. Curran, R., Raghunathan, S., Price, M.: Review of aerospace engineering cost modelling: the genetic causal approach. Prog. Aerosp. Sci. **40**, 487–534 (2004)
10. Bode, J.: Neural networks for cost estimation: simulations and pilot application. Int. J. Prod. Res. **38**, 1231–1254 (2000). https://doi.org/10.1080/002075400188825
11. Bahrami, A., Dagli, C.H.: Design retrieval by fuzzy neurocomputing. J. Eng. Des. **3**, 339–356 (1992). https://doi.org/10.1080/09544829208914766

12. Zhang, H.C., Huang, S.H.: Applications of neural networks in manufacturing: a state-of-the-art survey. Int. J. Prod. Res. **33**, 705–728 (1995). https://doi.org/10.1080/00207549508930175
13. Verlinden, B., Duflou, J.R., Collin, P., Cattrysse, D.: Cost estimation for sheet metal parts using multiple regression and artificial neural networks: a case study. Int. J. Prod. Econ. **111**, 484–492 (2008). https://doi.org/10.1016/j.ijpe.2007.02.004
14. Bode, J., Hu, Y., Liu, P., Shu, B.: Cost estimation in new product development: Knowledge pluralism, knowledge granularity, and agent-based representation. In: Proceedings of the Conference, pp. 669–678 (1995). https://doi.org/10.1007/978-0-387-34879-7_69
15. Rafiq, M.Y., Bugmann, G., Easterbrook, D.J.: Neural network design for engineering applications. Comput. Struct. **79**, 1541–1552 (2001). https://doi.org/10.1016/S0045-7949(01)00039-6
16. Breslin, J.G., O'Sullivan, D., Passant, A., Vasiliu, L.: Semantic web computing in industry. Comput. Ind. **61**, 729–741 (2010). https://doi.org/10.1016/j.compind.2010.05.002
17. Verhagen, W.J.C., Bermell-Garcia, P., Van Dijk, R.E.C., Curran, R.: A critical review of knowledge-based engineering: an identification of research challenges. Adv. Eng. Inform. **26**, 5–15 (2012). https://doi.org/10.1016/j.aei.2011.06.004
18. Szejka, A.L., et al.: Semantic interoperability for an integrated product development process: a systematic literature review. Int. J. Prod. Res. **55**, 6691–6709 (2017). https://doi.org/10.1080/00207543.2017.1346314
19. Rocca, G.L.: Knowledge based engineering: between AI and CAD. review of a language based technology to support engineering design. Adv. Eng. Inform. **26**, 159–179 (2012). https://doi.org/10.1016/j.aei.2012.02.002
20. Quintana-Amate, S., Bermell-Garcia, P., Tiwari, A.: Transforming expertise into knowledge-based engineering tools: a survey of knowledge sourcing in the context of engineering design. Knowl. Based Syst. **84**, 89–97 (2015). https://doi.org/10.1016/j.knosys.2015.04.002
21. Gruber, T.R.: Toward principles for the design of ontologies used for knowledge sharing. Int. J. Hum. Comput. Stud. **43**, 907–928 (1995). https://doi.org/10.1006/ijhc.1995.1081
22. Yang, L., Cormican, K., Yu, M.: Ontology-based systems engineering: a state-of- the-art review. Comput. Ind. **111**, 148–171 (2019). https://doi.org/10.1016/j.compind.2019.05.003
23. Arista, R., Zheng, X., Lu, J., Mas, F.: An Ontology-based Engineering system to support aircraft manufacturing system design. J. Manuf. Syst. **68**, 270–288 (2023). https://doi.org/10.1016/j.jmsy.2023.02.012
24. Ashraf, J., Chang, E., Hussain, O.K., Hussain, F.K.: Ontology usage analysis in the ontology lifecycle: a state-of-the-art review. Knowl. Based Syst. **80**, 34–47 (2015). https://doi.org/10.1016/j.knosys.2015.02.026
25. Adamczyk, B.S., Szejka, A.L., Canciglieri, O.: Knowledge-based expert system to support the semantic interoperability in smart manufacturing. Comput. Ind. **115** (2020). https://doi.org/10.1016/j.compind.2019.103161
26. Dadzie, A.-S., et al.: Applying semantic web technologies to knowledge sharing in aerospace engineering. J. Intell. Manuf. **20**, 611–623 (2009). https://doi.org/10.1007/s10845-008-0141-1
27. Zhao, X., Verhagen, W.J.C., Curran, R.: Estimation of aircraft component production cost using knowledge-based engineering techniques. Adv. Eng. Inform. **29**, 616–632 (2015). https://doi.org/10.1016/j.aei.2015.05.004
28. Bodendorf, F., Lutz, M., Michelberger, S., Franke, J.: An empirical investigation into intelligent cost analysis in purchasing. Supply Chain Manag. Int. J. **27**, 785–808 (2022). https://doi.org/10.1108/SCM-11-2020-0563
29. Van Nguyen, T.H., Huang, P.M., Chien, C.F., Chang, C.K.: Digital transformation for cost estimation system via meta-learning and an empirical study in aerospace industry. Comput. Ind. Eng. **184** (2023). https://doi.org/10.1016/j.cie.2023.109558

30. Ning, F., et al.: Manufacturing cost estimation based on a deep- learning method. J. Manuf. Syst. **54**, 186–195 (2020). https://doi.org/10.1016/j.jmsy.2019.12.005
31. Ning, F., Shi, Y., Cai, M., et al.: Manufacturing cost estimation based on the machining process and deep-learning method. J. Manuf. Syst. **56**, 11–22 (2020). https://doi.org/10.1016/j.jmsy.2020.04.011
32. Yoo, S., Kang, N.: Explainable artificial intelligence for manufacturing cost estimation and machining feature visualization. Expert Syst. Appl. **183** (2021). https://doi.org/10.1016/j.eswa.2021.115430
33. Hammann, D.: Big data and machine learning in cost estimation: an automotive case study. Int. J. Prod. Econ. **269**, 109137 (2024). https://doi.org/10.1016/j.ijpe.2023.109137

A Computer Vision Approach to Build Manufacturing Simulation Models in the Digital Twin Context

Marcelo Rudek(✉), Ana Paula R. Valle, Ricardo Bertolin, and Osiris Canciglieri Jr.

Pontifical Catholic University of Paraná (PUCPR), Curitiba Paraná, Brazil
marcelo.rudek@pucpr.br

Abstract. The digital transformation with the application of immersive mixed reality has been supporting the digital twin creation within the industrial context. The techniques for 3D reconstruction from the Computer Vision approach make possible the creation of virtual models of real scenery based on a set of digital image frames. In this paper, we explored the problem of building virtual layouts from 3D mapping by photogrammetry to be applied in the corresponding digital twin model. The context of Industry 5.0 is aborded in the research and aims to define a method for preparing humans to be trained using synthetic information, to perform the real operations in the productive processes. From collected RGB images, the layout of a shopfloor can be recreated, and objects can be virtually represented. However, to rebuild a realistic environment is a difficult task. Thus, we studied the main technological requirements to perform an automated method for shopfloor layout digitalization and digital twin creation, conducting an experimental investigation for a manufacturing automation laboratory. The work also discusses the feasibility of the proposed method.

Keywords: Computer Vision · Digital Twin · Simulation · 3D Reconstruction · Immersive Environments · Manufacture · HoloLens

1 Introduction

The Digital Twin concept matches the new trends of digital transformation in the industries performing human integration as a central element of production systems. However, training issues of human operators are demanding analyses in the ergonomic and cognitive domains. Thus, sensors, human-machine interfaces, and immersive systems, such as virtual and augmented reality [1] can be an alternative for the development of the workforce in manufacturing environments. The virtual reality platforms offer an immersive environment for learning through trial and error, allowing the processing of user performance data, such as error rate, time analysis, number of attempts, and task completion rate, among others [2].

Published by Springer Nature Switzerland AG 2025
P. Sureephong et al. (Eds.): PLM 2024, IFIP AICT 741, pp. 15–25, 2025.
https://doi.org/10.1007/978-3-031-93323-3_2

Considering this possibility of interaction, the metaverse can be a complementary tool, representing an interconnected and collaborative digital environment. For the industrial metaverse and Digital Twin application, three-dimensional modeling of the environment for immersive and interactive technology is needed to create a realistic scenario. In addition to this, real-time communication, such as 5G for instance, is essential for data gathering encompassing various parameters. Thus, network connectivity must enable the link between the real and virtual production environments [3].

1.1 Research Proposal

Considering the growth of Digital Twin and the use of immersive realities within industrial environments, the research proposes the integration of these technologies to optimize the training of operators in the manufacturing context. To map the environment, it will be necessary to use computer vision techniques, such as photogrammetry for 3D reconstruction, enabling the development of virtual environments, like a digital twin, with a focus on interaction and training capabilities. In the context of Industry 5.0, the research aims to address a methodology that teaches workers through synthetic information, empowering them to perform real operations in manufacturing processes, as illustrated in Fig. 1.

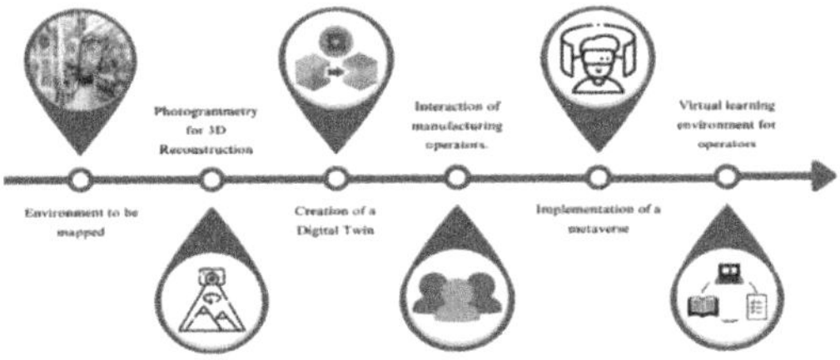

Fig. 1. Proposed diagram with the fundamental elements of simulation environment creation.

Furthermore, it is necessary to assess the accuracy of photogrammetry methods, as the detail of this step is crucial for realistic reconstruction and the automation of mapping, creating a digital twin. The research also aims to integrate the concept of the metaverse, providing interactive and immersive digital environments, potentially expanding the possibilities for training and human interaction in digital twins.

2 3D Mapping for the Creation of Digital Twins

The Digital Twin (DT) is used to dynamically replicate a physical object, achievable through feedback data from sensors and ontological operational records [1]. This system relies on communication channels between virtual/physical controllers, with an execution mechanism, aiming to control both environments. Synchronization between the physical and virtual spaces occurs through industrial communication between software and a Programmable Logic Controller (PLC) hardware [4].

Regarding the application of this technology in industrial training, the virtual environment must resemble the real one, ensuring that the user's experience and learning are not compromised. In this research, it is presented the laboratory of manufacturing automation to be mapped as represented in Fig. 2, and the challenge is to map to virtual all the environment complexity to create its respective digital twin.

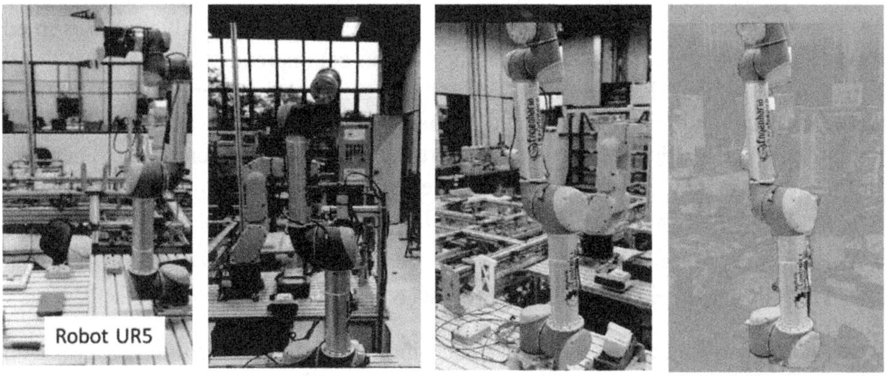

(a) (b) (c) (d)

Fig. 2. Description of the environment to be mapped.

In Figure 2, views (a), (b), and (c) show the position of a manufacturing cell structure, containing the collaborative robot UR5 in the foreground. However, a lot of objects and building details are present inside the environment, such as chairs, small boxes, pens, cables, command buttons, and so on. Thus, the amount of objects increases the complexity of mapping this environment to its virtual model.Also, the different materials, such as windows, white walls, and metallic tables difficult both the reconstruction and rendering procedures. Thus, we need to find a way to segment the robot as in view(d). Different software platforms exist that offer the capability to construct virtual environments, encompassing a diverse array of objectives and applications. These versatile tools

Table 1. Examples of systems commonly applied to academic and industrial simulation.

System\Key Application	Robotic Simulation	AR and VR	Animation	Simulation	Game development	Virtual Prototyping	Architect. Visualization	Training and Education	Digital Manufacturing*	PLM*	Factory Simulation*
Blender	○		●		●	○	●	○	○	○	○
Coppelia SIM	●		●	●				●	○		○
Gazebo	●		●	●		○			○	○	○
PyBullet	●		●	●					○		
Unity 3D	○	●	●	●	●	○	●	●	○	○	○
Unreal Engine	○	●	●	○	●	●	●		○		

Legend (○ partial / ● full). The (*) marked options are fully covered by other systems such as (Siemens Teknomatix, Arena, Flexsim)

find application across various sectors, including architecture, interior design, engineering, industry, and gaming, among others. Fig. 3 lists some examples of such software commonly applied to industrial simulation and academic research, presenting their key applications, advantages, and disadvantages, and Table 1 is the selection made for our research context according to our specific necessity from computer vision.

Table 1 lists some examples of such software commonly applied to industrial simulation and academic research. The key points about advantages and disadvantages described are merely about the most suitable to be employed for study purposes in our research problem, as well as, the availability of non-commercial licenses.

From the system description in Table 1, we selected the Blender. It is an open-source software and has the advantage of many tutorials, plugins, and resources from a community of developers, offering support for animation and rendering. Therefore, for the environment to be reproduced, it is necessary first to employ three-dimensional mapping techniques, which will be discussed in Sect. 2.1.

2.1 The Photogrammetry Approach for Reconstruction

To provide a truly immersive user experience, high-quality visualization is crucial. Considering the landscape of immersive realities, such as Virtual Reality (VR), a valuable concept for the reconstruction of real three-dimensional models stands out: photogrammetry[5]. This technique utilizes 2D images in conjunction with additional sensor data, forming a digitally reconstructed model from real objects, and showcasing potential applications in the digital twin realm [6]. Figure 4 illustrates the application of image acquirement to photogrammetry for the 3D reconstruction.

From the consecutive image frames, the 3D process executes the sequential operations including Image upload, feature extraction, feature matching, structure from motion, dense scene preparation, depth map, mesh filter, and texturing. Thus, a 3D reconstruction software such as Meshroom for instance, generates the digital correspondent image.

As found in the literature the system applies to single-object reconstruction, however here we have a complex scenery with lots of objects in the same field of view.

The DT, in turn, has as its main characteristic the dynamic update of data between the physical and virtual worlds. With the implementation of Artificial Intelligence (AI) in this process, it is expected that the simulation and analysis capabilities will be enhanced. Therefore, 3D virtual modeling represents a vital part of the DT, but it is a complex subject to develop, given the sophisticated and complex information involved [7]. Additionally, it is worth emphasizing the crucial role that photogrammetry techniques play in environment reconstruction, contributing to the accuracy and fidelity of real-world models. By integrating AI into the process, the virtual representation can become enhanced and more immersive for the user, fostering the integration of VR with the DT.

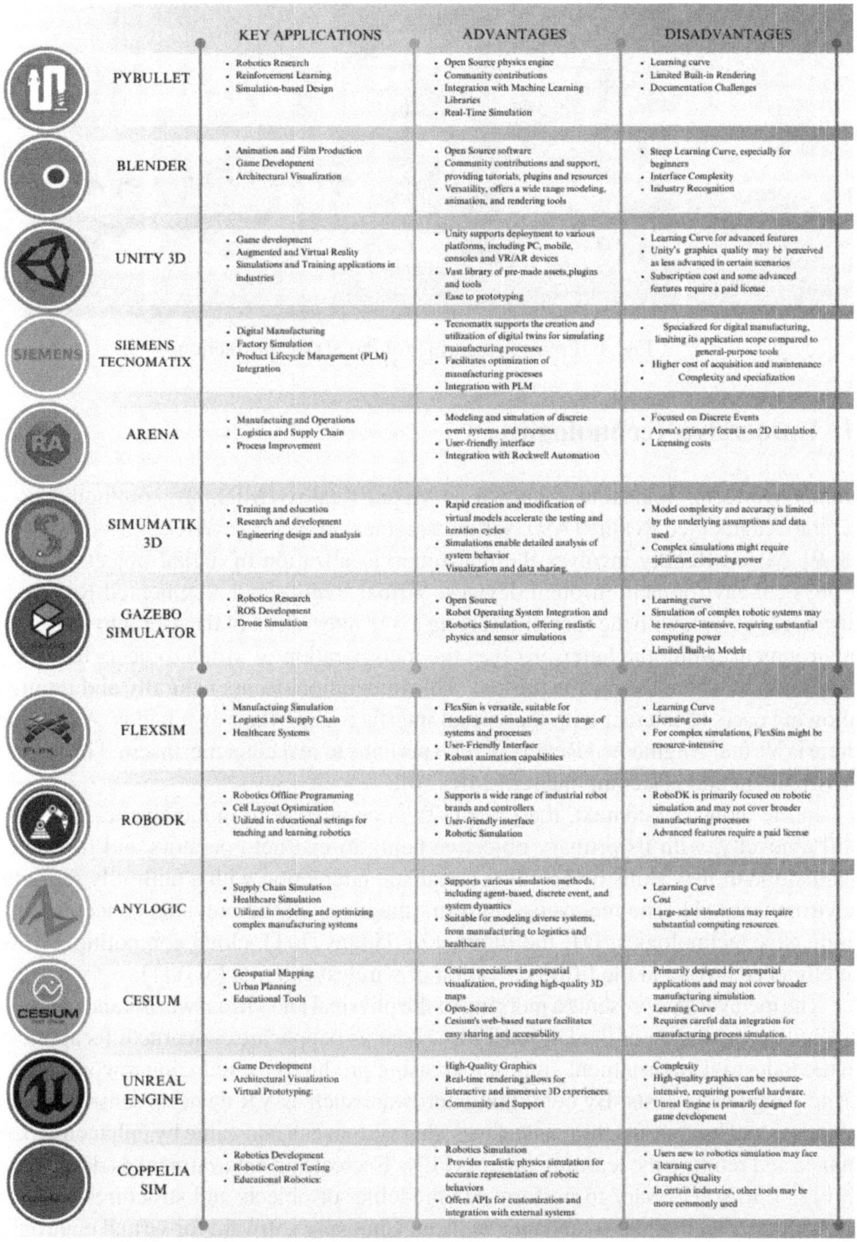

	KEY APPLICATIONS	ADVANTAGES	DISADVANTAGES
PYBULLET	• Robotics Research • Reinforcement Learning • Simulation-based Design	• Open Source physics engine • Community contributions • Integration with Machine Learning Libraries • Real-Time Simulation	• Learning curve • Limited Built-in Rendering • Documentation Challenges
BLENDER	• Animation and Film Production • Game Development • Architectural Visualization	• Open Source software • Community contributions and support, providing tutorials, plugins and resources • Versatility, offers a wide range modeling, animation, and rendering tools	• Steep Learning Curve, especially for beginners • Interface Complexity • Industry Recognition
UNITY 3D	• Game development • Augmented and Virtual Reality • Simulations and Training applications in industries	• Unity supports deployment to various platforms, including PC, mobile, consoles and VR/AR devices • Vast library of pre-made assets,plugins and tools • Ease to prototyping	• Learning Curve for advanced features • Unity's graphics quality may be perceived as less advanced in certain scenarios • Subscription cost and some advanced features require a paid license
SIEMENS TECNOMATIX	• Digital Manufacturing • Factory Simulation • Product Lifecycle Management (PLM) Integration	• Tecnomatix supports the creation and utilization of digital twins for simulating manufacturing processes • Facilitates optimization of manufacturing processes • Integration with PLM	• Specialized for digital manufacturing, limiting its application scope compared to general-purpose tools • Higher cost of acquisition and maintenance • Complexity and specialization
ARENA	• Manufactuing and Operations • Logistics and Supply Chain • Process Improvement	• Modeling and simulation of discrete event systems and processes • User-friendly interface • Integration with Rockwell Automation	• Focused on Discrete Events • Arena's primary focus is on 2D simulation. • Licensing costs
SIMUMATIK 3D	• Training and education • Research and development • Engineering design and analysis	• Rapid creation and modification of virtual models accelerate the testing and iteration cycles • Simulations enable detailed analysis of system behavior • Visualization and data sharing.	• Model complexity and accuracy is limited by the underlying assumptions and data used • Complex simulations might require significant computing power
GAZEBO SIMULATOR	• Robotics Research • ROS Development • Drone Simulation	• Open Source • Robot Operating System Integration and Robotics Simulation, offering realistic physics and sensor simulations	• Learning curve • Simulation of complex robotic systems may be resource-intensive, requiring substantial computing power • Limited Built-in Models
FLEXSIM	• Manufactuing Simulation • Logistics and Supply Chain • Healthcare Systems	• FlexSim is versatile, suitable for modeling and simulating a wide range of systems and processes • User-Friendly Interface • Robust animation capabilities	• Learning Curve • Licensing costs • For complex simulations, FlexSim might be resource-intensive
ROBODK	• Robotics Offline Programming • Cell Layout Optimization • Utilized in educational settings for teaching and learning robotics	• Supports a wide range of industrial robot brands and controllers • User-friendly interface • Robotic Simulation	• RoboDK is primarily focused on robotic simulation and may not cover broader manufacturing processes • Advanced features require a paid license
ANYLOGIC	• Supply Chain Simulation • Healthcare Simulation • Utilized in modeling and optimizing complex manufacturing systems	• Supports various simulation methods, including agent-based, discrete event, and system dynamics • Suitable for modeling diverse systems, from manufacturing to logistics and healthcare	• Learning Curve • Cost • Large-scale simulations may require substantial computing resources.
CESIUM	• Geospatial Mapping • Urban Planning • Educational Tools	• Cesium specializes in geospatial visualization, providing high-quality 3D maps • Open-Source • Cesium's web-based nature facilitates easy sharing and accessibility	• Primarily designed for geospatial applications and may not cover broader manufacturing simulation. • Learning Curve • Requires careful data integration for manufacturing process simulation.
UNREAL ENGINE	• Game Development • Architectural Visualization • Virtual Prototyping:	• High-Quality Graphics • Real-time rendering allows for interactive and immersive 3D experiences • Community and Support	• Complexity • High-quality graphics can be resource-intensive, requiring powerful hardware. • Unreal Engine is primarily designed for game development
COPPELIA SIM	• Robotics Development • Robotic Control Testing • Educational Robotics:	• Robotics Simulation • Provides realistic physics simulation for accurate representation of robotic behaviors • Offers APIs for customization and integration with external systems	• Users new to robotics simulation may face a learning curve • Graphics Quality • In certain industries, other tools may be more commonly used

Fig. 3. Software for virtual environments.

Fig. 4. Frames capturing for the 3D reconstruction.

3 Immersive Technologies

Immersive technologies, also known as Mixed Reality (MR), are commonly used to refer to Augmented Reality (AR), Virtual Augmented Reality (VAR), and Virtual Reality [8, 9]. Mixed Reality involves the real-time visualization of digital objects located in a physical environment through devices. Virtual Reality and Augmented Reality have distinct concepts, with the former referring to the integration of the user into a virtualized environment, while the latter involves the incorporation of virtual objects into the real environment where the user is present. This integration occurs naturally and intuitively, allowing users to interact with the virtual interface using their own hands. Additionally, there is Virtual Augmented Reality, which pertains to real elements inserted into a virtual environment and is a component of MR [10].

In the industrial context, the Digital Twin coupled with immersive technologies is still a novelty, with its primary objective being to instruct operators and facilitate the acquisition of new skills by enabling a realistic interaction with a faithfully reproduced environment [11]. The metaverse, in turn, integrates all the previous concepts, linking immersive technologies, DT, the Internet of Things (IoT), cloud computing, Artificial Intelligence (AI), and the fifth generation of wireless internet (5G) [1].

The metaverse represents a merging of the physical and virtual worlds and can be considered an expansion of the Digital Twin, as far as people are concerned. Its application in the industrial environment aims at increasing productivity, reducing rework, and lowering compliance costs. By combining hardware such as VR devices, sensors, machine learning, and human-machine interfaces, this system can add value by enhancing performance and refining the learning process [12]. To create an environment as demonstrated in Fig. 4, it is necessary to perform 3D modeling of objects and structures using photogrammetry techniques to enhance realism. Choosing software for virtual environment creation, programming interactions, and implementing Artificial Intelligence to enable virtual characters to respond to the real environment are essential steps.

The connections to the real world involve the use of virtual or augmented reality glasses, enabling operators to access the metaverse. Additionally, sensors and tracking devices are utilized to capture user movements, enhancing the immersive experience. Furthermore, the metaverse, coupled with immersive reality, can provide users with work instructions and instructional videos, aiding in the manufacturing process without the need for a tutor to accompany the process.

4 Application Example

According to [13] it presents a computer vision roadmap listing the machine vision artifacts to be applied in the digital transformation of industry. Inside the main artifacts, the 3D reconstruction processes are the basis to link both real and virtual sides. The reconstruction process is based on photogrammetry. The process of virtual model reconstruction was developed using Meshroom software. The steps and technological connections are presented in the Fig. 5.

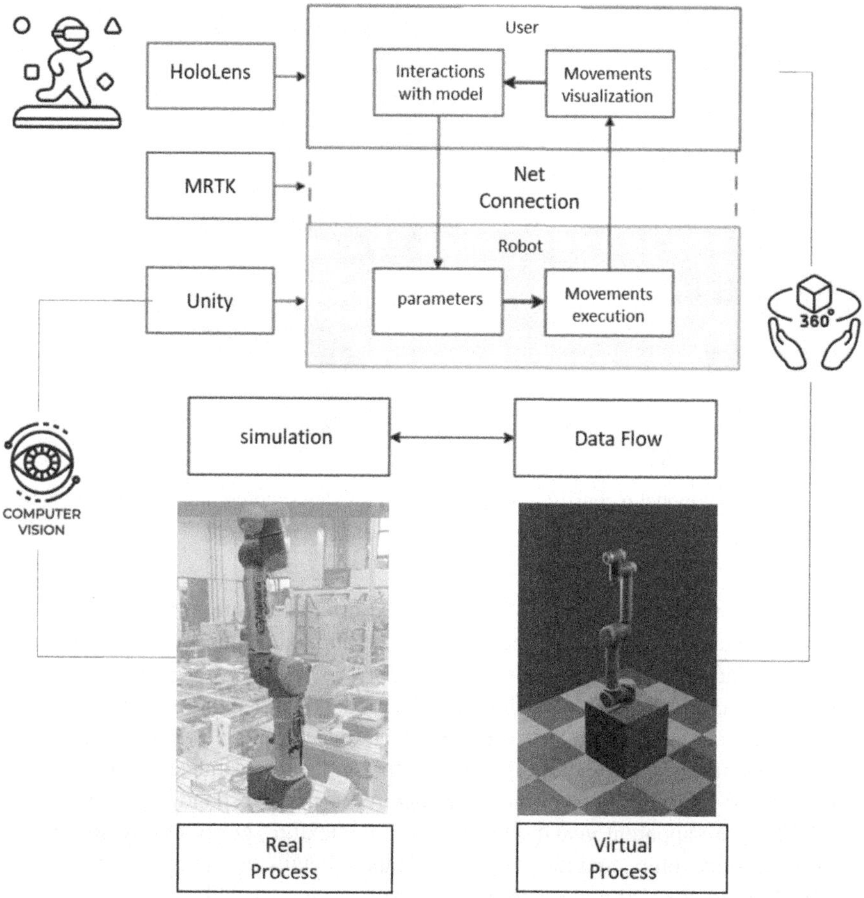

Fig. 5. The conceptual model for the proposed DT scheme.

Figure 5 shows the main components of the digitalization steps. The image capture process is slow, taking around 2 h to perform with less than 200 images, however, considering the amount of sampling, it generated a satisfactory result for using the measurements and basis of the Laboratory model as presented in Figs. 6 and 7.

The image models generated in meshroom were imported to Blender and by manual process "cubes" were modeled in place of the objects, to set the position on the virtual side. After that model was exported to Unity and also it was applied the MRTK settings to the Hololens 2 Emulator. Figure 6 illustrates the virtual model generated.

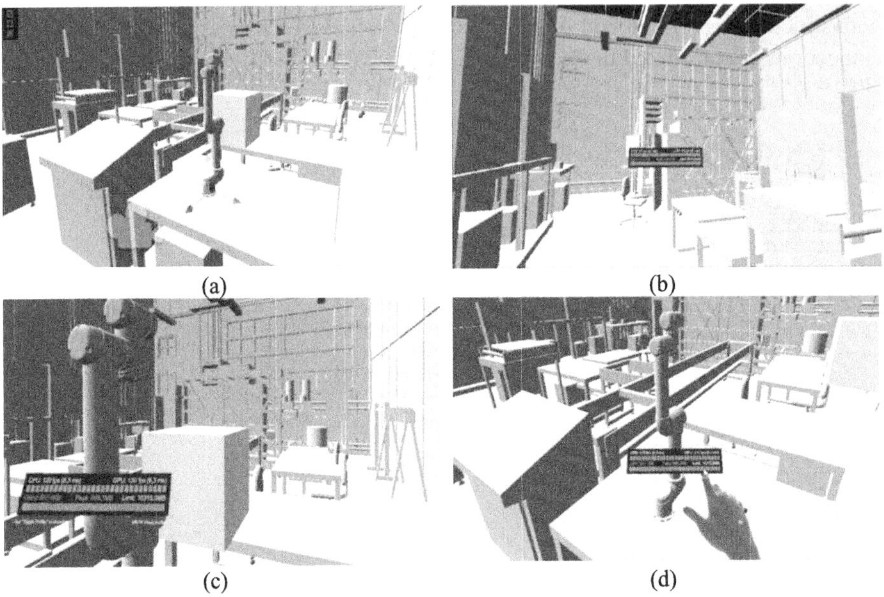

Fig. 6. The virtual model reconstruction with 4 different views representing all directions of the laboratory.

The resulting reconstruction to be visualized after reconstruction is as the Fig. 6. However, some enhancement strategies should be applied to improve the final quality, as presented the Figs. 7 and 8.

Figure 7. (a) presents the three-dimensional model obtained through the photogrammetry process reconstruction, with visible deformities resulting from occlusion, visual imperfections in lines, edges, and angles, and polygonal appearance. Here we applied the RGB color information from the 2D image for each corresponding pixel in the 3D model. There are apparent visual deformities in the texture, generated by non-uniform colors, distortions, blurs, and the absence of graphic details in some parts of the image. Thus, it displays the textured mesh in orthographic view, including improvements by smoothing tools in Blender, as well as additional processing in Unity Fig. 7.(b) presents a hybrid remodeling of the previous three-dimensional model, incorporating components from both photogrammetry and a model molded in Blender to correct imperfections in the original model.

From Fig. 6 we have the virtual model in Unity 3D Environment that permits to create the human interaction. To make it more realistic the textured information was extracted from the collected images and it was built a respective digital twin (in terms of visualization) as presented in Fig. 7.

As can be observed, distorted and mismatched information is presented in the reconstructed textured model. Although the scenery generated image quality lacks precision in comparison with the original, we can add the virtual devices into the 3D image as in Fig. 8(a) by adding the virtual robotic arm in the remodeled scene. In Fig. 8(b) the figure shows the user integration into the virtual reconstructed manufacturing interaction.

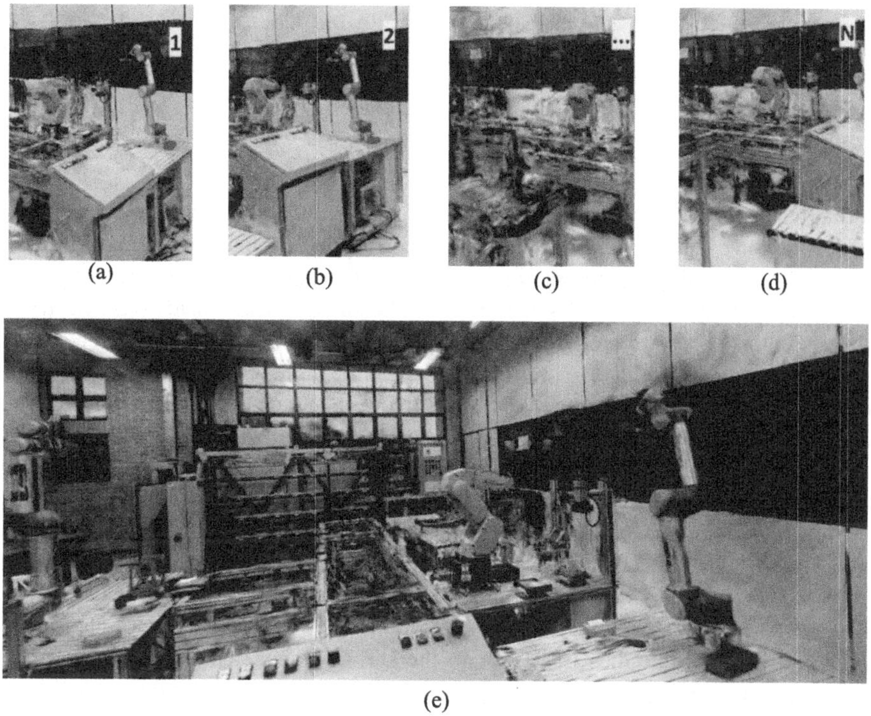

Fig. 7. Example of the digitalization of the environment from the original frames sequence of Fig. 4 and digital model of Fig. 6.

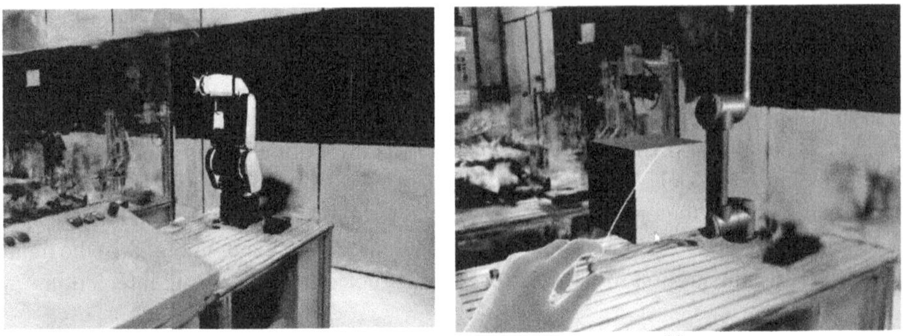

(a) (b)

Fig. 8. The virtual model reconstruction. (a) the virtual robotic arm in the remodeled scene. (b) user integration into the virtual reconstructed manufacturing interaction.

5 Conclusion

In this research we explored the first three phases of Fig. 1. i) environment mapping; reconstruction by photogrammetry and iii) the corresponding digital model as digital twin preparation. A realistic representation is a requirement to create an interactive training virtual environment since the human factors must be considered inside the industry 5.0 context. In this paper, we investigated a way to proceed with the digitalization technique applied to virtual modeling as a preliminary step to the digital twin creation of a complex scenery. We explored in literature the probable tools to create the virtual side of a digital twin of our laboratory, listing the main characteristics with advantages and disadvantages thinking of a low-cost approach. We selected the Blender for an initial investigation and the next step is to evaluate in practice the others available as free or educational licenses. We also mixed the images and respective virtual reconstruction of the environment by using both Blender and Unity 3D platforms. The illustrative result is shown in Fig. 7, where the view of the HoloLens is emulated to see the laboratory components. The results show promisor that strategy to 3D representation. The ongoing step is to integrate the collaborative robot into the virtual side of the DT. Once concluded, the future work is to prepare an experimental course training the operation of the laboratory devices by virtual learning to collect information in terms of realism, interaction, and efficiency.

Acknowledgments. The authors would like to thank the Pontificia Universidade Católica do Paraná-PUCPR for laboratory and equipment availability. Also, the Industrial and System Engineering Graduate Program (PUCPR-PPGEPS) for the financial scholarship to master students.

References

1. Lv, Z., Fridenfalk, M.: Digital twins for building industrial metaverse. J. Adv. Res. (2023). https://doi.org/10.1016/j.jare.2023.11.019
2. Steed, C., Kim, N.: Deep active-learning based model-synchronization pf digital manufacturing stations using human-in-the-loop simulation. J. Manuf. Syst. **70**, 436–450 (2023)

3. Yang, Y., et al.: The use of virtual reality in manufacturing education: state-of-the-art and future directions. Manuf. Lett. **35**, 1214–1221 (2023)
4. Leng, J., et al.: Digital twins-based remote semi-physical commissioning of flow-type smart manufacturing systems. J. Clean. Prod. **306** (2021)
5. Kurka, P.R.G., Rudek, M.: Three-dimensional volume and position recovering using a virtual reference box. IEEE Trans. Image Proc., 573–576 (2007)
6. Tadeja, S.K., et al.: Exploring gestural input engineering surveys of real-life structures in virtual reality using photogrammetric 3D models. Multimed. Tools Appl. **80**, 31039–31058 (2021)
7. Wang, Y., et al. Ontology of 3D virtual modeling in digital twin: a review, analysis and thinking. J. Intell. Manuf. (2023). https://doi.org/10.1007/s10845-023-02246-6
8. Kaluza, A., et al.: Implementing mixed reality in automotive life cycle engineering: a visual analytics based approach. In: (26) CIRP Life Cycle Engineering (LCE) Conference, Procedia CIRP, vol. 80, pp. 717–722 (2019)
9. Silva, O.M., Rudek, M.: Industrial layout mapping by human-centered approach and computer vision. In: Danjou, C., et al. (eds.) IFIP International Federation for Information Processing 2024, Published by Springer Nature Switzerland AG 2024. PLM 2023, IFIP AICT 701, pp. 1–10 (2024)
10. Tori, R., Kirner, C., Siscouto, R.: Fundamentos e tecnologia de Realidade Virtual e Aumentada. Editora SBC, Belém (2006)
11. Martínez-Gutiérrez, A., et al.: Convergence of virtual reality and digital twin technologies to enhance digital operators' training in Ind. 4.0. Int. J. Hum.–Comput. Stud. **180** (2023)
12. Lee, J., Kundu, P.: Integrated cyber-physical systems and industrial metaverse for remote manufacturing. Manuf. Lett. **34**, 12–15 (2022)
13. Silva, R.L., Canciglieri Junior, O., Rudek, M.: A road map for planning-deploying machine vision artifacts in the context of industry 4.0. J. Ind. Prod. Eng. **39**(3), 167–180 (2022)

Anomaly Detection in Supply Chain Business Processes: A Deep Learning Approach

Tahani Abu Musa[1(✉)], Abdelaziz Bouras[2], and Abdelhak Belhi[3]

[1] Community College of Qatar, Doha, Qatar
tahani.abumusa@ccq.edu.qa
[2] Department of Computer Science and Engineering, Qatar University, Doha, Qatar
abdelaziz.bouras@qu.edu.qa
[3] Joaan Bin Jassim Academy for Defense Studies, Al Khor, Qatar
abdelhak.belhi@jbj.edu.qa

Abstract. Business Process Mining is considered one of the emerging fields that focuses on analyzing Business Process Models (BPM), by extracting knowledge from event logs generated by various Supply chain Management information systems, such as ERP and PLM, for the purpose of auditing, monitoring, and analysis of business activities for future improvement and optimization throughout the entire lifecycle of such processes, from creation to conclusion. Such a framework will help enhance the accuracy of anomaly detection in the global Supply Chain, improve the multi-level business processes workflow, and optimize the processes in the Supply Chain in terms of security and automation. In this work, a Bidirectional Long Short-Term Memory (Bi-LSTM)-autoencoder Neural Network was utilized for the prediction of the execution of cases, through training and testing the model on event traces extracted from event logs related to a procurement business process model, which is one of the main components of the supplychain. The approach consisted of three phases: preprocessing the logs, classification, and categorization in addition to all the activities related to implementing the Bi-LSTM model, including network design, training, and model selection. Our results showed that our model was able to predict the next activity in the sequence as well as detect anomalous ones with over 80% accuracy on average.

Keywords: Autoencoder · Bi-LSTM · Business Process Event log · Business Process Mining · Deep Learning · Supply Chain · Procurement Business Process Model

1 Introduction

Business Process Mining or Process Mining is an emerging research area that gained increasing attention in both academic and industrial domains. It aims at analyzing Business Process Models (BPM), by extracting knowledge from event logs generated by various information systems, for the sake of auditing, monitoring, and analysis of business activities for future improvement and optimization. This involves carrying out these

Published by Springer Nature Switzerland AG 2025
P. Sureephong et al. (Eds.): PLM 2024, IFIP AICT 741, pp. 26–40, 2025.
https://doi.org/10.1007/978-3-031-93323-3_3

activities throughout the entire lifecycle of such business processes, from the creation to the conclusion (Dakic & Stefanovic, 2018). Business Process Mining is useful to detect potential exceptions or anomalies in the logged workflow and hence helps identify the source of such exceptions and modify them accordingly.

Typically, a business process may be defined as a group of linked activities produced for a particular purpose (Sarno et al., 2016). A business process has some attributes, a particular goal, needs a certain input, produces a certain output, has a series of events or activities carried out in a predefined order, might need some resources, and involves a number of owners (*Business Process Management Systems: Strategy and Implementation - James F. Chang - Google Books*, n.d.). In this context, an anomaly may be defined as a deviation of that predefined order, which can occur due to different reasons, including system malfunctioning, poor, unexpected, or improvised resource behavior, or even malicious behavior of resources internal or external to an organization (Ko & Comuzzi, 2023). There exist various representations and notations of business processes, including Unified Modeling Language (UML), Business Process Model Notation (BPMN), Event-driven process chain (EPC), and Petri Net Markup Language (PNML)(Sarno et al., 2016).

Due to the availability of event logs produced from different information systems, typical process mining techniques are utilized in process discovery, conformance checking, and model enhancement. Hence, providing valuable business perceptions and insights [5].

In this context, event logs can be considered an important means for extracting knowledge about the sequence of activities, making it possible to discover, monitor, and improve such processes in different application domains of interest. However, those typical systems do not have the ability to make predictions about the next activity in business processes.

In the domain of Supply Chain-based Manufacturing, stakeholders face a number of challenges related to a growing global competitive market, growing product demands, shorter marketing timeframe, and adjustments in production due to variations in demand. This requires incorporating innovative digital technologies including Artificial Intelligence (AI), blockchain, and the Internet of Things (IoT), in order to make sound decisions instantly, for the sake of leveraging product competitiveness [6].

In this paper, a Bidirectional Long Short-Term Memory (Bi-LSTM)-autoencoder Deep learning model is used to predict the next activity of a given normal business process, as well as detect anomalies for the abnormal ones. LSTM is one type of Recurrent Neural Networks (RNNs) that has been used extensively in time series analysis and sequential problems [7–9]. The model presented in this work has been trained on event logs that represent a procurement scenario, allowing for predicting the next activity on a given trace of events that follow another activity or a series of activities as input. To validate our approach, we obtained our preliminary results based on a dataset comprised of 16,000 traces. The tests performed on the trained model showed that it was able to predict the next activity in our business process model with great accuracy. The model was also capable of detecting abnormal behavior recorded in the logs.

The remainder of this paper is structured as follows. In Sect. 2, we present the state of the art of works related to business process mining, and business process anomaly detection. In Sect. 3, we present the methodology of our approach where we present our Bi-LSTM prediction model. Tests and evaluation of our model are presented in Sect. 4. Finally, we draw our conclusion in Sect. 5.

2 Related Work

2.1 Business Process Mining

Process mining can be characterized as the domain situated between data mining and Business Process Modeling (BPM) and analysis. Its primary objective is to scrutinize business processes by deriving insights from event logs produced by various information systems. It aims at increasing the performance of activities such as auditing, monitoring, and analyzing business operations across the entire lifecycle of business processes, spanning from inception to completion.

Several process mining algorithms exist in the literature to support the activities of conformance checking, performance analysis, decision mining, organizational mining, and recommendations [10].

In process mining, data related to events and activities of a given business process are recorded in event logs, which may be defined as a "hierarchically structured file with data on the executions of business processes" [11]. Event logs are considered the input to process mining and typically contain traces of several executions of the same business process. A case or trace is a collection of related events within a single instance of the business process. An event or activity is a single atomic part of the trace, that has several attributes describing it, besides other information related to updating the state of such events, for example (Notify Delivery of the parts). Other attributes of events may include resources allocated and timestamps [12]. The following figure illustrates the relation between an event log, a case, and event (Fig. 1).

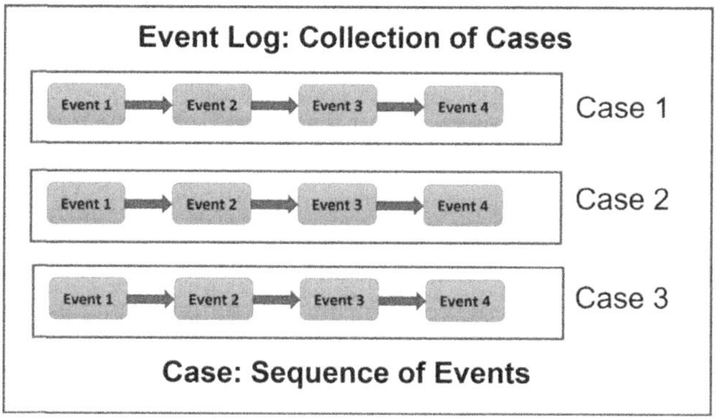

Fig. 1. Relation between a log, cases, and events

According to [11], the three main techniques of process mining are Process Discovery, Process Conformance, and Process Improvement. Process discovery involves the use of an event log as an input and the output will be the corresponding business process model without any previous knowledge. Process Conformance Checking involves the comparison between a business process model and the event log generated via executing the same process model. Such comparison helps evaluate whether the log information and the model are equivalent. Process Improvement involves applying any enhancements to the current model using the current information from the log.

Although event logs can be either structured or unstructured, nowadays, the eXtensible Event Stream (XES) format is the standard format for event log specification considered the 'de facto' standard for storing event logs by the IEEE Task Force back as of September 2010, for the sake of facilitating a unified and seamless way of exchanging data across various systems, with an added value of adding semantics to that data [13]. XML was preceded by the format Mining eXtensible Markup Language (MXML), enabling the exchange of event data between diverse systems, and employing process mining techniques [12].

2.2 Deep Learning for Event Prediction

Recently, the growth and spread of Machine Learning (ML) and Deep Learning (DL) algorithms in the field of process mining, and predictive analysis become a motivation for using them in performing accurate predictions of future events and activities in business processes, by analyzing logged historical activities and events, extracted from event logs. Such methods include Naive Bayes classifier [14], and predictive clustering tree inducer [15].

Besides, and due to the large volumes of data presented by the extracted event logs, deep learning approaches can outperform those typical process mining ones, bringing more accurate predictions about future events in each event log trace, through analyzing data from previous events. Deep learning approaches including deep neural networks have been widely used in predictive analysis and process mining [16].

In [17], the authors proposed an end-to-end Deep Neural Network framework for Supply-demand prediction for Agile Manufacturing, to rapidly respond to customer orders, which relies heavily on multisource data analytics, which provides smart datadriven decision-making strategies to guide manufacturing and supply chain systems. The model was tested with real supply and demand data, gathered from warehouses of a manufacturing company, and the experimental results demonstrated that such a framework was able to eliminate the gap between supply and demand, as well as provide a practical solution to long-term supply-demand prediction.

2.3 LSTM-Based Approaches in Process Mining

Recurrent Neural Networks (RNNs) and their variant the Long-Term-Short Memory (LSTM) have been used extensively in time series analysis and sequential problems. RNNs have two types of input, present and recent past. RNN uses both types of input to verify how they behave in case of new data. In other words, at any given time, the output of an RNN at time step t-1 affects its output at time step t. LSTMs are considered more efficient at obtaining long-term temporal dependencies (Greff et al., 2017)The information contained in LSTMs is beyond the normal flow of the recurrent network in a gated cell. Such Information can be stored, written, or read from a cell, like data in a computer's memory. The cell is able to make decisions about what should be stored and when it should be allowed to read, write, and delete, through gates that open and close. Such gates are implemented with the multiplication of elements by sigmoid, resulting in an output in the range of 0 to 1 [18].

In the area of process mining, there exist several research works related to utilizing LSTM in predicting the next event in business processes. According to [19], the authors proposed the question of how to use Deep Learning techniques to train accurate models of Business process behavior from event logs. The proposed approach trained a neural network with LSTM architecture, in order to predict the sequence of next events, their timestamp, and associated resources. An experimental evaluation on real-life event logs has been performed and showed that the proposed approach outperforms previously other LSTM architectures targeted at this problem, in terms of dimensionality control of the categorical attributes, allowing their proposed approach to achieve consistently good performance even for long sequences [19].

In [20], authors extended the body of research in this area by testing four different variants of Graph Neural Networks (GNN) and a fully connected Multi-layer Perceptron (MLP) with dropout for the tasks of predicting the nature and timestamp of the next process activity.

On the other hand, [21] proposed a model that is composed of a Convolutional Neural Network (CNN) to forecast the next business activity. The use of CNN in process mining was justified by guaranteeing high accuracy in predictions of the next activity, and according to the authors, CNN has achieved faster training and inference than LSTM, even when the processes tend to have longer traces.

2.4 Anomaly Detection in Business Processes

In the field of business process, anomaly detection is concerned with identifying possible deviations that exist in the event logs. Various attempts to utilize Machine Learning and Deep Learning algorithms for anomaly detection have been carried out.

In [22], an anomaly detection model with deep neural networks that acts as a predictive model was proposed. Such an approach is two-fold: training an LSTM-based model to learn the behavior of the business process using the trained model to decide whether a given trace is anomalous or not. This model requires previous knowledge of the Business Process Model.

On the other hand, [23] proposed BINet, an LSTM model of real-time multivariate anomaly detection in business process event logs. Such a model was able to detect both case and attribute levels. This work has also proposed a heuristic to set the anomaly threshold automatically. Unlike [22], such a model requires no domain knowledge about the business process.

Another way to address the anomaly detection challenge is by using autoencoders. Typically, an autoencoder is an unsupervised learning method for neural networks, intended to learn the efficient representation of data. Autoencoders are primarily designed to encode the input into a compressed and yet meaningful representation, then decode it back so the presentation output is as similar as possible to the original input. Hence, they can help learn the 'Normal' patterns in data, and then flag those data points that deviate from such patterns [4].

In a similar context, [24] proposed an anomaly detector model using Graph autoencoders, which relies on the idea of enriching data representation with activity relationships, hence the data will be represented into a graph, then will be fed into the graphautoencoder anomaly detector. Such an approach is emerging in the anomaly detection field and proved to be outperforming the other methods such as t-STIDE OC-SVM, HMM, and Likelihood in terms of accuracy, noise endurance, and generalizability.

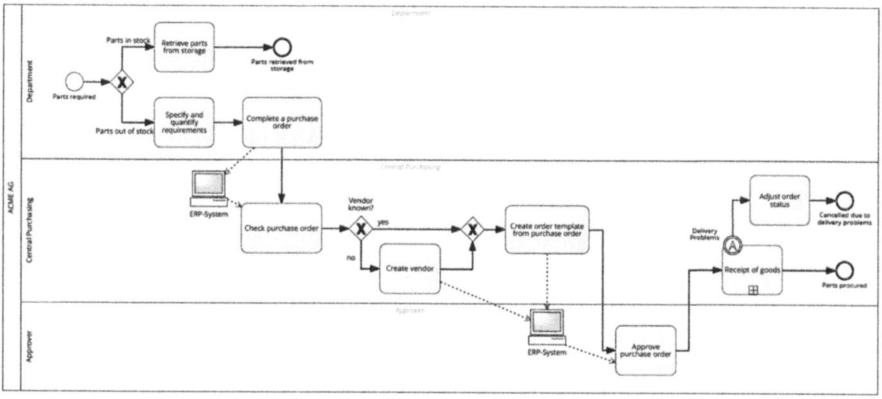

Fig. 2. Procurement Business Process

In this paper, we propose a Bi-LSTM-based autoencoder anomaly detection model, to detect anomalies in event logs of a procurement Business process. Bi-LSTM models outperform conventional LSTM and other state-of-the-art models in terms of accuracy, precision, recall, and F-score values. It also has a much more reduced false alarm rate than the existing models [25].

3 Methodology

3.1 Procurement Business Process Analysis

For our approach, we focus on a generic business process for a procurement scenario illustrated in "Fig. 2". Effective work is also performed on similar scenarios provided by our project industrial stakeholders. For this purpose, we used ProM 6.10, a Process Mining tool developed by Process Mining Group, Math, and CS department, at Eindhoven University of Technology. It has several capabilities related to the simulation of business processes, as well as log generation capabilities. "Table 1". Provides a snapshot of a generated event log resulting from the simulation.

3.2 Proposed Approach

"Figure 3" Illustrates our approach to detecting anomalous activities in the procurement business process using the Bi-LSTM-based autoencoder model. The approach is composed of three phases: 1) Preprocessing of Event log, 2) Classification, and 3) Prediction and detection.

Event Log Preprocessing

This phase is divided into three steps:

Data Extraction:

In which we decide on which attributes we will select from the event log needed for the prediction, in our case the "Activity Name" and "Owner" attributes.

Table 1. Snapshot of a generated Event Log

Process ID	Activity ID	Activity Name	Owner	Timestamp
1	1	Parts Required	Department	2023-09-15 01:59:09.676000 + 03:00
1	2	Specify and Quantify Requirements	Department	2023-09-15 02:02:09.676000 + 03:00
1	3	Complete a Purchase Order	Department	2023-09-15 02:03:09.676000 + 03:00
1	4	Check Purchase Order	Central Purchasing	2023-09-15 02:04:09.676000 + 03:00
1	5	Create Vendor	Central Purchasing	2023-09-15 02:07:09.676000 + 03:00
1	6	Create Order Template from Purchase Order	Central Purchasing	2023-09-15 02:10:09.676000 + 03:00
1	7	Approve Purchase Order	Approver	2023-09-15 02:11:09.676000 + 03:00
1	8	Receipt of Goods	Central Purchasing	2023-09-15 02:12:09.676000 + 03:00
....	......	......	......	......

Trace Identification:
In this step we identify the traces representing different cases in the event log, each case is represented by the corresponding trace. Those traces are extracted and appended to a data frame, preserving their order in the original log.

Segmentation:
After identifying the traces in the previous step, we divide each trace into a set of activities/events. Each event is then converted to a sequence of integers, comprising two sequence lists of "integers", the first contains the list of activities (X) and the second of output activities (Y). Then, the sequence of input activities (X) is converted into a 2-dimensional matrix composed of both the number of sequences and the maximum length of sequences.

Classification
In this phase, we tokenize the sequence of integers of output activities (Y) obtained in the previous phase while making sure that the number of classes will be equal to the size of the vocabulary.

Prediction and Detection
This step is further divided into three sub-steps:

Network Design:
In which we defined the design of the Bi-LSTM network in terms of input, hidden, and output layers, along with the specification of needed parameters.

Network Training:
In this step, we train the Bi-LSTM network using a dataset composed of the sequence list of integers represented by the activities in matrix (X), and the tokenized representation of matrix (Y).

Model Selection:
Proceeding the training phase, in this step the model of Bi-LSTM with the best results will be identified as the final model. Such 'best results' including high accuracy and low Mean Squared Error (MSE), would be the best ones to make accurate predictions. Otherwise, another iteration on training and adjustments of the parameters will be needed.

After these steps, we get the output generated by the Bi-LSTM-autoencoder model, which is the prediction of the next activity in the business process stated earlier, from a single input activity or a series of activities. It uses the trained model (model) to predict the reconstructed (anomalous) data based on the padded encoded data. It calculates the mean squared error (MSE) between the original padded data and the reconstructed data for each sequence. MSE measures the dissimilarity between the original and reconstructed sequences.

3.3 Model Architecture

This section explains in detail the architecture of our Bi-LSTM-autoencoder Model.

Input Layer: It starts by defining an input layer with a shape that matches the max_sequence_length, which represents the length of input sequences.

Embedding Layer: This layer is responsible for converting the input sequences into dense vectors of fixed size (128 in this case).

Bidirectional LSTM Layers (bi_lstm1, bi_lstm2, bi_lstm3, bi_lstm4): our model defines four bidirectional LSTM layers with 16 units each. Bidirectional LSTMs process sequences from both directions (forward and backward), which can help capture complex patterns in the data. The activation functions used are "SoftMax" for the hidden states and "Sigmoid" for the recurrent activations. The first two LSTM layers return sequences, while the last two do not.

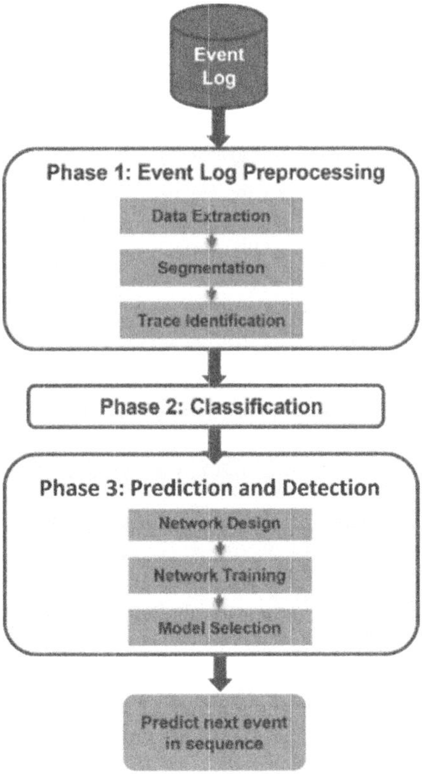

Fig. 3. Proposed approach architecture

Dropout Layers (dropout1, dropout2, dropout3, dropout4): Dropout layers are added after each LSTM layer to prevent overfitting. Dropout randomly sets a fraction of the input units to 0 during training.

RepeatVector Layer: This layer repeats the output of the last dropout layer for each time step in the sequence (max_sequence_length times). It prepares the data for the next set of LSTM layers.

Output Layer: The code adds a TimeDistributed layer followed by a Dense layer with a single output unit. This layer applies a fully connected layer to each time step of the sequence. It's typically used for sequence-to-sequence tasks where each time step produces an output.

3.4 Experimental Setup

"Table 2" lists the hyperparameters configured for our model. For training the network, we used an event log with 16000 traces and 9 different activities in the log. During the training phase, the number of sequences that were identified was 68039. For simplicity, the "generate_abbreviation" function was for the 'Owner' and 'Activity_Name' columns of the DataFrame, effectively abbreviating these columns. It concatenates the 'Activity_Name' and 'Owner' columns with an underscore ('_') in between, creating a new column called 'Activity_Owner' that contains the combined abbreviation. Next, we group the data by 'Case ID' and create sequences of 'Activity_Owner' values based on the specified window size. These sequences are stored as dictionaries with the key 'Input' and appended to a list called sequences. "Table 3" lists the names of the activities captured in the business process, along with their acronyms. As part of the implementation, we used the Keras Python library to build deep-learning models.

Table 2. Model Hyperparameters

Parameter	Value
Epochs	25
Optimizer	Adam
Loss	MSE
Activation	Tanh
Recurrent activation	Sigmoid
Validation Split	0.2

3.5 Evaluation Metrics

In our model, we used the Mean Square Error (MSE) to evaluate the accuracy of our model. It calculates the mean squared error (MSE) between the original padded data and the reconstructed data for each sequence. MSE measures the dissimilarity between the original and reconstructed sequences. Furthermore, we also used the F1 score, which is the harmonic mean of precision and recall. Typically, the precision measures how precisely the anomalies were detected, in other words, how many of those detected anomalies are actual ones. On the other hand, recall measures how many anomalies were detected and how many were not recognized by that model:

$$\text{F1}\ Score = \frac{2 * (precision * recall)}{(precision + recall)} \tag{1}$$

Table 3. Activity Names/Owner Combinations

Activity Name	Owner	Activity_Owner
Parts Required	Department	PR_D
Specify and Quantify Requirements	Department	SQR_D
Complete a Purchase Order	Department	CPO_D
Check Purchase Order	Central Purchasing	CPO_CP
Create Vendor	Central Purchasing	CV_CP
Create Order Template from Purchase Order	Central Purchasing	COTPO_CP
Approve Purchase Order	Approver	APO_A
Receipt of Goods	Central Purchasing	RG_CP
Adjust Order Status	Central Purchasing	AOS_CP
Cancelled Due to Delivery Problems	Central Purchasing	CDDP_CP

Another way of describing the F1 score formula is given by the balanced mean of the true positive rate (tpr) and the true negative rate (tnr):

$$\text{F1}\ Score = \frac{2 * (tpr * tnr)}{(tpr + tnr)} \tag{2}$$

3.6 Threshold-Based Anomaly Detection

In our model, the anomaly threshold value was calculated based on the mean and standard deviation of the MSE values. The threshold is used to classify sequences as normal or not normal (anomalous). It determines which sequences are considered normal (MSE $<=$ threshold) and which are not normal (MSE $> =$ threshold).

4 Results

4.1 Experiment 1

After training the model on normal data, three synthetic anomalous logs of different lengths were used to test it. "Table 4" summarizes the properties of these logs.

Based on the calculated threshold value, which we obtained as explained in section III.F, our model was able to detect the 4 anomalous traces whose MSE values exceeded the threshold calculated with 85% Accuracy for log1.

"Figure 4" depicts the training versus validation loss for the model, which shows how well the model fits training and validation data respectively. The Y-axis shows the trend of loss values for the training and validation phases when building the model, and the X-axis represents the epoch number.

"Figure 5." Depicts the calculated MSE value for all 20 traces of log1, with a threshold value of 1.2993, where traces 6, 8, 11, and 13 are the anomalous ones, comparing their MSE value against the threshold.

Table 4. Anomalous Logs Properties

	No of Traces	No of Events	No. of Anomalous traces
Log 1	20	181	4
Log 2	100	888	25
Log 3	500	4300	130

Table 5. F1-score and Threshold value Comparison between our approach and existing approaches for anomaly detection.

	Bi-LSTM			LSTM			GRU			Random Forest		
	Threshold	Accuracy	F1	Threshold	Accuracy	F1	Threshold	Accuracy	F1	Threshold	Accuracy	F1
Log1	1.2993	0.8125	**0.92**	3.615	.4375	0.61	2.291	0.6875	0.82	1.355	0.75	0.84
Log2	1.6332	**.79**	**0.89**	4.086	0.69	0.57	3.239	.78	0.79	1.89	.765	0.852
Log3	2.053	**0.79**	**0.88**	4.861	.67	0.54	3.782	0.79	0.76	2.63	0.83	0.857

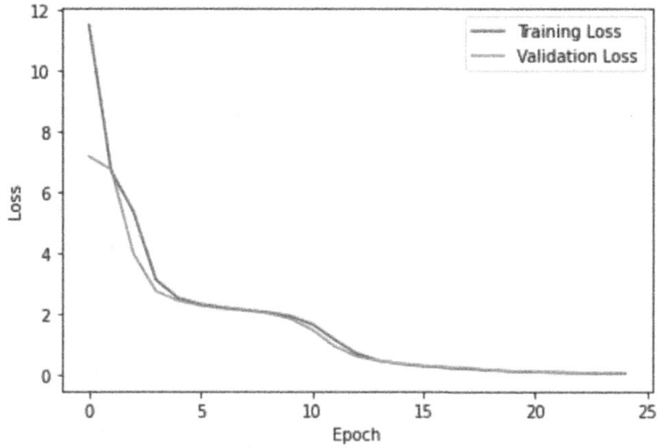

Fig. 4. Training vs. Validation Loss for the Bi-LSTM model

4.2 Experiment 2

In this experiment, we compared the performance of our model against the state-of-the-art machine and deep-learning model in literature, namely: vanilla LSTM, GRU, and Random Forest. "Table 5" shows a comparison of the threshold values computed by each model, and the F1 score resulted from testing the three logs: log1, log2, and Log3 respectively.

"Figure 6" depicts the F1-scores for our approach compared to vanilla LSTM, GRU, and Random Forest.

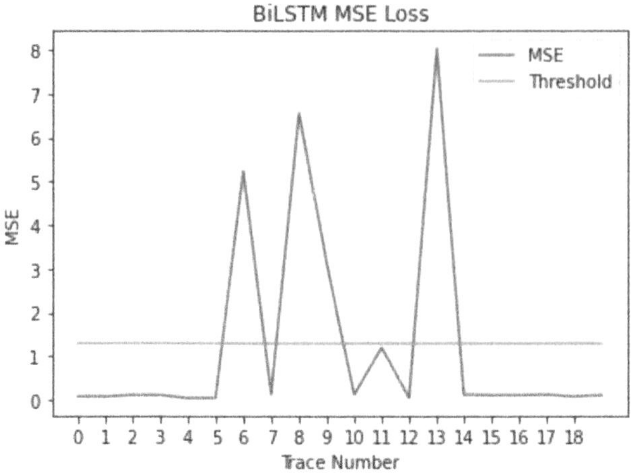

Fig. 5. MSE for testing log

5 Discussion

Comparing the results of "Table 5", BI-LSTM seems to outperform other state-of-the-art models in terms of accuracy and F1 score. For example, when testing all models with log1, our model achieved a 92% F1-score compared to 61%, 82%, and 84% of LSTM, GRU, and Random Forest respectively. As for the accuracy values obtained Bi- LSTM scored 81.25% accuracy compared to 43.75%, 68.75%, and 75% of LSTM, GRU, and Random Forest respectively. For log2 of 100 traces, Bi-LSTM still has the highest value of 89% of the F1-score compared to the other models. These results validate our choice for Bi-LSTM. This is due to the fact that Bi-LSTM is better handling dependencies between features in time-series data.

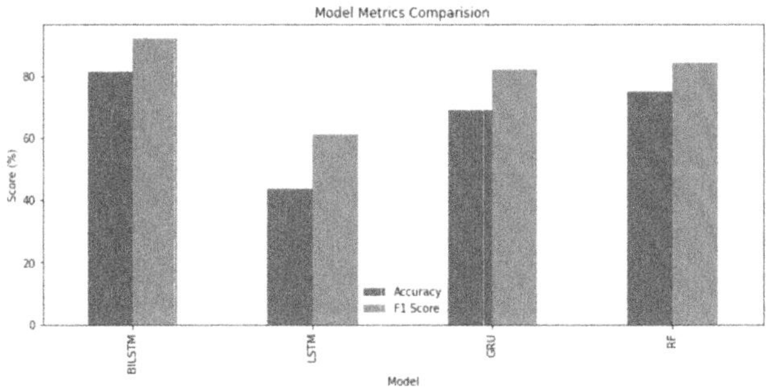

Fig. 6. Accuracy and F1-Score Comparison

6 Conclusion

Deep learning methods including Bi-LSTM-autoencoder can be used in predicting the next activity in a given business process model with 81% accuracy, as well as detecting anomalous transactions. In this work, we utilized a Bi-LSTM-autoencoder neural network model for such a task, by training it on normal traces of events extracted from the event log of a given procurement business process and tested using anomalous logs, with different types of anomalies related to the sequence of the events, or event attributes including their owners. Bi-LSTM-based models achieve better results in terms of in terms of F1-score values. It also has a much more reduced false alarm rate than the conventional LSTM and other state-of-the-art models. In this work, we focused on the Procurement Business Process Model. As a perspective, we intend to extend our work to include other processes of the Supply-Chain.

Acknowledgments. This publication was made possible by NPRP Grant No. NPRP11S-1227-170135 from the Qatar National Research Fund (a member of Qatar Foundation), Qatar.

References

1. Dakic, D., Stefanovic, D.: Business process mining application: a literature review. In: 29th DAAAM Proceedings, pp. 866–875 (2018). https://doi.org/10.2507/29th.daaam.proceedings.125
2. Sarno, R., Wibowo, W.A., Kartini, F.H., Effendi, Y., Sungkono, K.: Determining model using non-linear heuristics miner and control-flow pattern. TELKOMNIKA (Telecommunication Computing Electronics and Control) **14**(1), 349–359 (2016)
3. Chang, J.F.: Business Process Management Systems: Strategy and Implementation. Google Books. https://doi.org/10.1201/9781420031362. Accessed 01 Oct 2023
4. Ko, J., Comuzzi, M.: A systematic review of anomaly detection for business process event logs. Bus. Inf. Syst. Eng., 1–22 (2023)
5. Yurtay, Y.: Process mining in manufacturing: a literature review. Journal **5**(3), 341–355 (2022)
6. Ethirajan, M., Kandasamy, J., Kumaraguru, S.: Connecting engineering technology with enterprise systems for sustainable supply chain management. Smart Sustain. Manuf. Syst. **4**(1), 33–48 (2020)
7. Li, Y., Cao, H.: Prediction for tourism flow based on LSTM neural network. Procedia Comput. Sci. **129**, 277–283 (2018). https://doi.org/10.1016/j.procs.2018.03.076
8. Cortez, B., Carrera, B., Kim, Y.-J., Jung, J.-Y.: An architecture for emergency event prediction using LSTM recurrent neural networks. Expert Syst. Appl. **97**, 315–324 (2018). https://doi.org/10.1016/j.eswa.2017.12.037
9. Liu, F., Chen, Z., Wang, J.: Video image target monitoring based on RNN-LSTM. Multimedia Tools Appl. **78**(4), 4527–4544 (2019). https://doi.org/10.1007/s11042018-6058-6
10. van der Aalst, W.M.P.: Process discovery from event data: Relating models and logs through abstractions. WIREs Data Min. Knowl. Discovery **8**(3), e1244 (2018). https://doi.org/10.1002/widm.1244
11. van der Aalst, W.: Data science in action. In: Process Mining: Data Science in Action, 2nd edn., pp. 3–23. Springer, Berlin, Heidelberg (2016). https://doi.org/10.1007/978-3-66249851-4_1

12. Acampora, G., Vitiello, A., Di Stefano, B., van der Aalst, W., Gunther, C., Verbeek, E.: IEEE 1849: The XES standard: the second IEEE standard sponsored by IEEE computational intelligence society [Society Briefs]. IEEE Comput. Intell. Mag. **12**(2), 4–8 (2017). https://doi.org/10.1109/MCI.2017.2670420
13. Baier, T., Mendling, J., Weske, M.: Bridging abstraction layers in process mining. Inf. Syst. **46**, 123–139 (2014). https://doi.org/10.1016/j.is.2014.04.004
14. Ceci, M., Spagnoletta, M., Lanotte, P.F., Malerba, D.: Distributed learning of process models for next activity prediction. In: IDEAS 2018: 22nd International Database Engineering & Applications Symposium, pp. 278–282. ACM, New York (2018). https://doi.org/10.1145/3216122.3216125
15. Pasquadibisceglie, V., Appice, A., Castellano, G., Malerba, D.: Using convolutional neural networks for predictive process analytics. In: 2019 International Conference on Process Mining (ICPM), pp. 129–136 (2019). https://doi.org/10.1109/ICPM.2019.00028
16. Evermann, J., Rehse, J.-R., Fettke, P.: Predicting process behaviour using deep learning. Decis. Support Syst. **100**, 129–140 (2017). https://doi.org/10.1016/j.dss.2017.04.003
17. Wen, R., Yan, W.: Supply-demand prediction for agile manufacturing with deep neural network. Smart Sustain. Manuf. Syst. **3**(2), 95–105 (2019)
18. Greff, K., Srivastava, R.K., Koutník, J., Steunebrink, B.R., Schmidhuber, J.: LSTM: a search space odyssey. IEEE Trans. Neural Netw. Learn. Syst. **28**(10), 2222–2232 (2017). https://doi.org/10.1109/TNNLS.2016.2582924
19. Camargo, M., Dumas, M., González-Rojas, O.: Learning accurate LSTM models of business processes. In: Hildebrandt, T., van Dongen, B.F., Röglinger, M., Mendling, J. (eds.) Business Process Management, pp. 286–302. Springer, Cham (2019)
20. Venugopal, I., Töllich, J., Fairbank, M., Scherp, A.: A comparison of deep-learning methods for analysing and predicting business processes. In: 2021 International Joint Conference on Neural Networks (IJCNN), pp. 1–8 (2021). https://doi.org/10.1109/IJCNN52387.2021.9533742
21. Obodoekwe, E., Fang, X., Lu, K.: Convolutional neural networks in process mining and data analytics for prediction accuracy. Electronics (Switzerland) **11**(14) (2022). https://doi.org/10.3390/ELECTRONICS11142128
22. Lahann, J., Pfeiffer, P., Fettke, P.: LSTM-based anomaly detection of process instances: benchmark and tweaks. In: International Conference on Process Mining, pp. 229–241. Springer (2022)
23. Nolle, T., Seeliger, A., Mühlhäuser, M.: BINet: multivariate business process anomaly detection using deep learning. In: International Conference on Business Process Management, pp. 271–287. Springer (2018)
24. Huo, S., Völzer, H., Reddy, P., Agarwal, P., Isahagian, V., Muthusamy, V.: Graph autoencoders for business process anomaly detection. In: BPM 2021: 19th International Conference on Business Process Management, pp. 417–433. Springer (2021)
25. Aljbali, S., Roy, K.: Anomaly detection using bidirectional LSTM. In: Proceedings of the 2020 Intelligent Systems Conference (IntelliSys), vol. 1, pp. 612–619. Springer (2021)

Digital Engineering Ecosystem for the Enhancement of Traceability: Industrial Study

Sophia Salas Cordero[1], Brendan Sullivan[2], and Daniel Schmid[3](✉)

[1] ISAE-SUPAERO, Toulouse, France
sophia.salas@isae-supaero.fr
[2] Department of Management, Economics and Industrial Engineering, Politecnico di Milano, Milano, Italy
brendan.sullivan@polimi.it
[3] Zurich University of Applied Sciences, Winterthur, Switzerland
daniel.schmid@zhaw.ch

Abstract. Societies have adapted to the highly digital world we are living in. This has yielded the need for new manners to satisfy their market needs. These societal needs can cascade as stakeholder needs. Customers seem to be more and more inclined to fast feedback, quasi-immediate results, and compromising less with performance-cost trade-offs. Digital Engineering Ecosystems (DEE) are meant to address the complexity and dynamism of modern product development. This paper reviews the elements and characteristics of ecosystems required to achieve an effective digital thread (DTH), presents an industrial example of a DEE and benefits on a discussion of the challenges that arise when setting up an effective DEE and how the goals may be company dependent.

Keywords: Digital Engineering Ecosystem · Digital Thread · Traceability

1 Introduction

In the complex and dynamic realm of product development, Digital Engineering (DE) marks a significant departure from traditional paradigms that align with the foresight towards digitally integrated product architectures [1]. The increasing complexity and connectivity of engineering practices are captured in the Department of Defense's (DoD) Instruction 5000.97 [2], highlighting the essential role of digital models in the engineering product life cycle. Central to this pursuit, traceability has become essential, playing a pivotal role in the management of complex supply chains, adherence to design and compliance standards, and achieving the broader goals of digital business transformation. This paradigm shift enhances cross-disciplinary collaboration and enhanced transparency, enabling digital models and datasets to form the backbone to traverse system development, testing, evaluation, and deployment.

Published by Springer Nature Switzerland AG 2025
P. Sureephong et al. (Eds.): PLM 2024, IFIP AICT 741, pp. 41–55, 2025.
https://doi.org/10.1007/978-3-031-93323-3_4

Digital Engineering has been defined in [2] as "a mean of using and integrating digital models and the underlying data to support the development, test and evaluation, and sustainment of a system", and in [3] as "an integrated digital approach that uses authoritative sources of systems' data and models as a continuum across disciplines to support life cycle activities from concept through disposal". Digital Engineering can be understood as an interdisciplinary effort to be utilized throughout the life cycle of a system that stands for the transparent traceability of information no matter the stage of the life cycle. DE enables digital twins (a digital surrogate of the system that incorporates models to emulate the entire system or parts of it) and digital threads (DTH, a set of interconnected, cross-discipline model data that facilitates decision-making) [1].

The Digital Engineering Ecosystem (DEE) enables the deployment of DE approaches. DEE surpasses traditional siloed confines in modelling, tooling, process implementation, and role designation. As a holistic construct, DEE empowers organizations to develop, sustain, and manage systems. DEE is to be treated as a system of systems (SoS) framework; therefore, it necessitates rigorous design practices for optimal functionality, a perspective supported in [4] and substantiated in the Systems Engineering Handbook v5 [1]. The DEE paradigm necessitates a cohesive, interdisciplinary systems engineering strategy adept at addressing the complexity and dynamism of modern product development.

Central to DEE is the digital thread – a flexible, adaptable analytical structure that systematically exchanges technical data and knowledge throughout the system's life cycle. The digital thread exemplifies the efficacy of digital integration, granting stakeholders the ability to access, synthesize, and apply data to yield actionable insights crucial across a broad spectrum of activities, including rigorous requirements analysis and crafting effective training resources. Such capabilities markedly augment the system's value throughout its life cycle [2]. Within a DEE, traceability evolves from a simple documentation tool to a strategic asset in product development, ensuring that each decision and modification is recorded and aligned with project objectives and compliance mandates. The digital thread thus emerges as a critical DEE component, an analytical structure that facilitates the transformation of data into actionable intelligence, supporting activities from requirements analysis to operational testing.

Traceability is the ability to describe and follow requirements, from their origins to development and specification and finally to their subsequent deployment and use [5]. In its essence, this represents a structured approach to documenting and understanding the myriad processes of development, from initial conception to final realization. As an integral part of the digital thread in systems engineering, traceability's role transcends mere record-keeping; it ensures that all facets of product development are systematically chronicled and congruent with project goals and regulatory demands. The assimilation of vertical (VT), horizontal (HT), and end-to-end (E2E) traceability within the DEE equips stakeholders with an all-encompassing perspective of the development trajectory, enhancing cooperation, minimizing errors, and expediting innovation. This paper explores the elements needed for the deployment of a digital engineering ecosystem; as well as the industrial application that led to the enhancement of the digital thread through traceability in a case where the products are characterized by high variation

(7.4 million variants), options (128), and 950 million theoretical combinations. To illustrate this integration's practical implications, we present a use case based on a pressure sensor that demonstrates the utilization of the digital thread, enriched by traceability, in developing and operating a complex system. This use case focuses on the Vegabar 28 product family. It exemplifies how traceability within the digital thread can streamline processes, mitigate risks, and reinforce the integrity of product development from inception to deployment.

2 Traceability and Digital Engineering Ecosystems

The generic approach in modern manufacturing focuses on enhancing efficiency, flexibility, and overall customer satisfaction. The ability to leverage traceability and digital engineering ecosystems has the potential to streamline operations, from initial design to end-of-life, ensuring products to meet or exceed customer expectations. This involves adopting strategies like Assembly-to-Order (ATO) which leverage Make-to-Stock (MTS) modules for rapid customization, thereby aligning with the increasing demand for personalized solutions. The deployment and integration of digital technologies play a pivotal role in achieving these objectives, particularly through the implementation of the digital thread concept, which enhances traceability across the product life cycle. Enabling and facilitating the capacity to track the entire, development, assembly, distribution, and utilization of a product with precision at any point not only ensures customer satisfaction but also quality control, regulatory compliance, and facilitates recalls or improvements.

The digital thread (Fig. 1) traverses the life cycle phases of a system with various digital tools and technologies supporting and providing data for enhanced traceability [6–8]. As shown on the left side of the figure, the Beginning of Life (BoL, design and manufacturing [9]) is marked by the integration of a digital thread, which encapsulates and traces the ideation, design, and development processes. This integration is facilitated through the employment of IoT devices and sensors within the production and compliance and regulatory management software, which collectively lay the foundation for the digital twin concept [10]. The digital twin acts as a central hub for simulation, verification, and validation processes, ensuring adherence to predetermined specifications and regulatory standards. In this context, Requirements Management Tools and CAD systems are critical for the rigorous documentation and version control of design iterations and product specifications, safeguarding the product's integrity throughout its evolution. The Middle of Life (MoL, distribution, use, and support [9]) phase in the center of the figure, the digital thread becomes vital in orchestrating production, quality assurance, and supply chain logistics, integrating with systems like ERP (Enterprise Resource Planning), MES (Manufacturing Execution System), and QMS (Quality Management System). This phase is characterized by the incorporation of artificial intelligence and advanced analytics, which enhance manufacturing processes and ensure continuity of quality and operational efficiency. Lastly, as the product approaches its End of Life (EoL, product retirement, reuse, and recycle [9]) the right side of the figure, the digital thread facilitates a structured decommissioning process, utilizing asset management systems, CRM software, and knowledge management tools to manage the product's termination.

The traceability provided by the digital thread is essential for making informed decisions regarding the product's disposal or repurposing, ensuring environmental sustainability and compliance with regulatory guidelines. The repurposing may encompass retrofitting or re-engineering of the system.

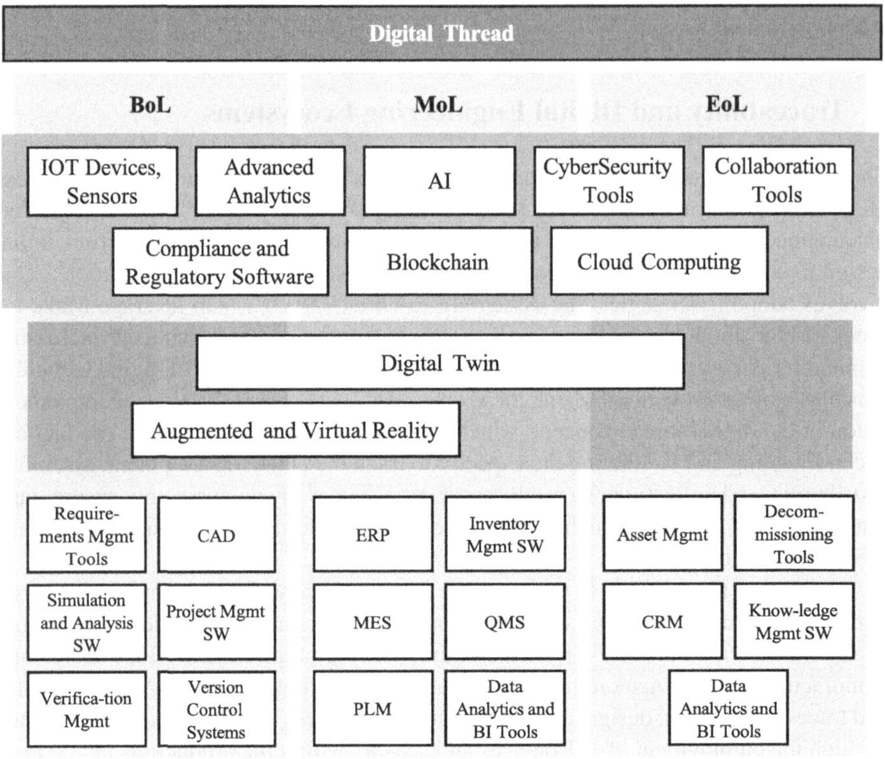

Fig. 1. Generalization of the digital thread per life cycle phase.

Traceability in systems engineering is multifaceted, encompassing various types that collectively contribute to a project's integrity, efficiency, and success. These types are distinguished by their directional focus and scope within the project life cycle, each serving specific functions critical for the systematic management of requirements and facilitating effective communication among project stakeholders.

- Vertical traceability (VT): refers to the ability to trace requirements through the layers of development from high-level needs down to the lowest level of system specifications and back. This type of traceability ensures that each requirement is justified and fulfilled by specific design decisions, implementation strategies, and verification/validation activities. The function of vertical traceability is to maintain coherence and alignment between the stakeholders' needs and the system's delivered capabilities across the system's life cycle (BoL, MoL, and EoL).

- Horizontal traceability (HT): focuses on the relationships and dependencies between requirements at the same level of abstraction. It facilitates the identification of conflicts, dependencies, and overlaps between requirements, ensuring that they are consistently managed throughout the project life cycle. Horizontal traceability is crucial for managing changes and assessing their impacts across the project, thereby enhancing the project's adaptability and resilience to changes, primarily in the MoL.
- End-to-end traceability (E2E): encompasses both vertical and horizontal traceability, providing a comprehensive linkage from the initial requirements through to final verification and validation of the system (BoL, MoL, and EoL). As a product reaches the end of its life cycle, E2E aids in decommissioning, disposal, or recycling, ensuring that these activities comply with regulatory standards and stakeholder expectations. Furthermore, it allows for documenting lessons learned, providing valuable insights for future projects.

The V-Model can serve as a structured approach to systematically address traceability. This model guides designers and engineers to start with the elicitation of stakeholder needs, that follows a set of steps beginning with the Definition of Requirements and then advancing through Functional, Logical, and Physical Design phases; considering all the different technical processes as defined by INCOSE: Business Analysis, Stakeholder Needs and Requirements Definition, System requirements definition, system architecture definition, design definition, system analysis, implementation, integration, verification, validation, operation, maintenance, and disposal. A generic V-model representation can be found in Fig. 2, this figure highlights how the traceability throughout the different lifecycle phases can be digital thread-enabled.

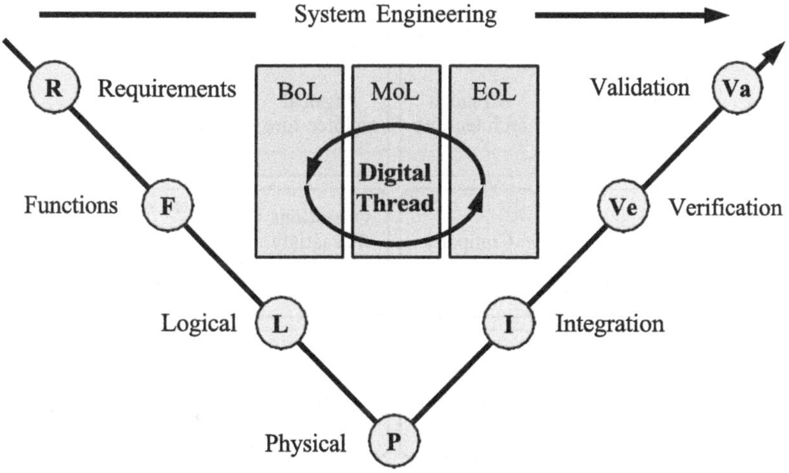

Fig. 2. Generic V-model with digital thread-enabled traceability through the product life cycle phases.

As [5] states, the V-model can be utilized as a base strategy to achieve traceability and, more importantly, to analyze the impact, derivation and coverage of requirements to meet regulations. V-Model considerations allow the early incorporation of traceability, demanding an elaborate system for the detailed monitoring and verification of requirements rationales, components' origins and their progression through the life cycle.

By leveraging and integrating a digital engineering ecosystem that enables a digital thread, the v-model considerations are useful to foresee what data might be available at a point in time during the life cycle phases and to know how it can or needs to be linked with other system data. Ultimately, if E2E traceability is enabled, it is logical to consider that vertical and horizontal traceability are also enabled.

The digital thread, when integrated into traceability efforts, offers a digital footprint that captures and communicates data and information flow from the product to the producer and user for enhanced operation and control purposes. This digital backbone enables stakeholders to access, analyze, and act on information in real-time, thereby significantly enhancing decision-making processes and operational transparency.

Figure 3 presents the customer's viewpoint of information required throughout the life cycle and what types of inputs are expected. For instance, imagine a customer who requires a security system installed in their house. They need to ask themselves: what do they need? Just a security camera or a security company that lends them a security service that includes 24/7 video/audio recordings? Do they want the camera to be able to provide night footage? Do they already have cameras installed in the house, and would they like the new security system to use them? And finally, if they would like to be the only ones available to access the recordings and if any regulations constrain the deployment of the system in question that they need.

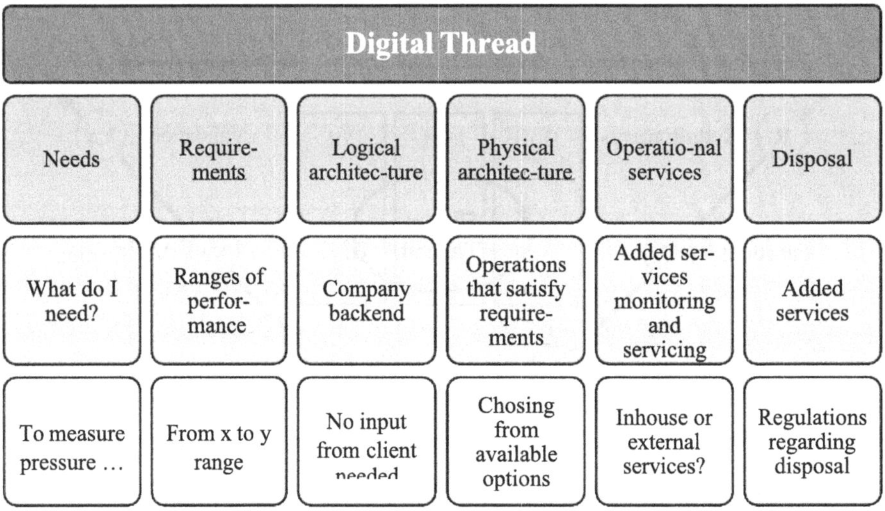

Fig. 3. Customer experience of the digital thread

Efficient traceability for the development of a product encompasses all the relationships between entities of the different technical processes. It is needed to consider different points of view to understand what traceability entails at different levels of abstraction of the product development. Some of the different levels of abstraction to consider: stakeholders; system, sub-sub-system and component level; process level; interfaces, and the recording of the digital assets. The first level, which matches the beginning of the V-model approach, involves tracing the needs, requirements, and expectations of all stakeholders involved in the project, including customers, end-users, project sponsors, regulatory bodies, and others. At this stage the stakeholders and considerations on how they can interact with the digital thread must be identified. Bearing in mind how the customers need or not to experience the digital thread; and if the buyers are different from the end-users.

From the system's viewpoint, traceability is concerned with the overall system requirements that the project aims to fulfill. This includes functional, performance, and usability requirements. The sub-system's traceability viewpoint deals with the requirements and specifications of sub-systems that, together, make up the whole system. It involves tracing the integration and interaction of these sub-systems to ensure they function correctly as part of the larger system. While at the components or modules level traceability tracks the specifications, development, and testing of these components, as they are the individual parts of the product that constitute a sub-system.

In parallel the process level needs to be taken into account in order to determine how the digital thread must be sketched out covering the customer's journey and the functional requirements. From a production point of view, how to handle the product service system within the factory/production venue. The traceability of individual processes or activities within the project, such as design decisions, development tasks, testing procedures, and change management activities. The interface and record-keeping level consider the compatibility and proper interaction between the system elements with the goal of achieving the system's verification and validation (steps in the right side of the generic V-model in Fig. 2).

Several approaches have been attempted to permit digital continuity. Model-based systems engineering (MBSE) has been identified as a key enabler of digital threads and digital twins [1]. MBSE is the formalized application of modeling to support system requirements, design, analysis, verification, and validation activities beginning in the concept stage and continuing throughout development and later life cycle stages [1]. There are as well initiatives that aim to aid the deployment of digital engineering ecosystems such as Open Services for Lifecycle Collaboration (OSLC) and Open Model Based Engineering Environment (OpenMBEE). OSLC's standard is currently mainly adopted in software and systems engineering domains, but with the potential to provide value to any domain with data integration challenges [11]. OpenMBEE proposes and discusses models created in an open-sourced model-based engineering environment, and the components required to achieve such an environment [12]. Other initiatives on tooling and environments with a digital continuity mindset can be found on [13]. The following section of this paper focuses on a specific industrial case study and their approach for the deployment of a digital engineering ecosystem.

3 Use Case

In this section, an industrial example regarding traceability with a digital thread is described and analyzed. This example refers to the pressure sensor families Vegabar 28 [14], 29, 38, and 39 by the company Vega [15]. The primary differentiation between the product families is the sensing element's technology: capacity-ceramic-cell (product families 28/38, up to 60 bar measuring range) and piezo-resistive cell or strain gauge (product families 29/39, up to 600 bar measuring range), and without display (product families 28/29) and with display (product families 38/39). Each family is an assemble-to-order (AtO) product with prefabricated make-to-stock (MtS) modules, as it favours a large number of potential configurations, fast production, and seamless fulfilment of requirements. For example, the Vegabar 28 product family [14] is offered with 7.4 theoretical million variants when not considering the rules of exceptions of modules that have no developed interface. There are also 128 additional configuration options, resulting in a total of 952 million possible theoretical combinations. Vega also possesses digital supporting tools, such as the myVEGA customer platform [16], the Vega Tools App [17] for smartphones, and the DTM Collection and PACTware [18] for computers. They are free of charge, while only the full version of the DTM Collection is charged separately.

Vega offers a digital ecosystem that enables the abovementioned variety of products and additional tools. The aim is to deliver information to the customers (website), enhance the customer relationship (myVEGA for customer binding), enrich the product's functionality by add-on tools (Vega Tools, DTM Collection), and extend the offering to a product service system, e.g. by the Inventory System [19].

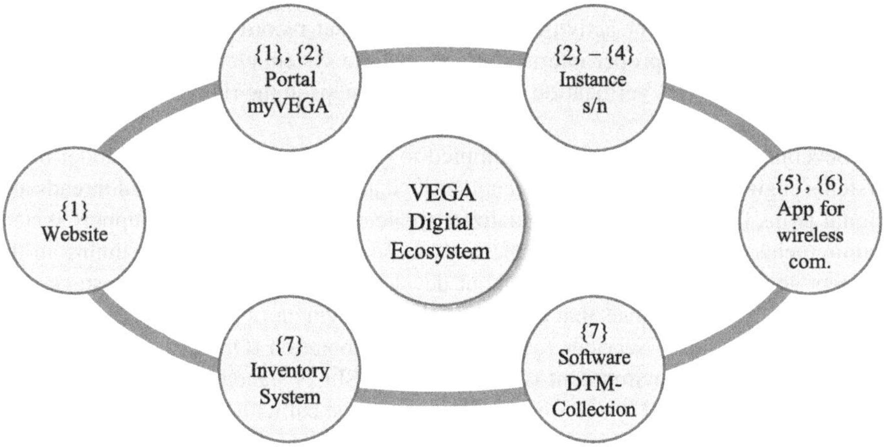

Fig. 4. Vega's digital ecosystem

The starting point of a product customization starts with the configuration by the customer through the digital ecosystem. The different elements of the digital ecosystem are described as follows (the numbers in curly brackets refer to the elements in Fig. 4 that enable Vega's digital engineering ecosystem):

{1} **Connect the instance with the customer.** Via the website from Vega or directly under myVEGA, the customer selects the application first, then the product family (type), and finally, the bespoke configuration. The product connects directly with the customer information, e.g. the company, the delivery address, and the contact person.

Later, the production line adds the related instance's serial number (s/n) [20] and article number. That is the starting point of the digital thread, and the s/n is the reference for traceability.

{2} **Modules with own s/n.** One of the product features is the fast availability and, despite the high number of configuration possibilities, a product consisting of prefabricated make-to-stock modules. The two significant modules are the transducer (consisting of the pressure cell and the process connection) and the electronics with the firmware.

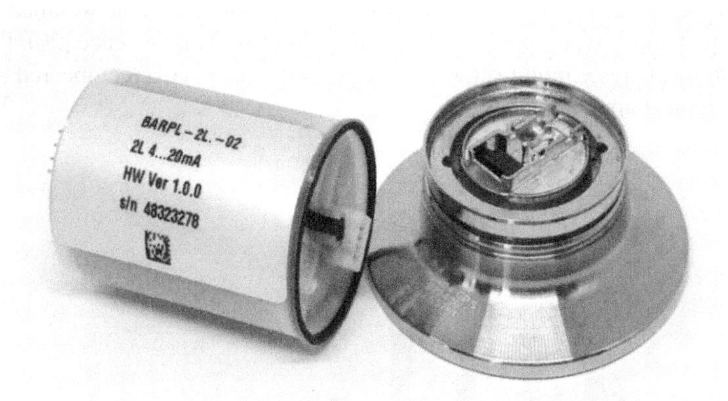

Fig. 5. Electronic module (s/n 48323278) and transducer

The electronic unit's s/n is well visible by a sticker. The one from the transducer is visible only during production (add-on element) and as data available on the electronic part. In the end, the different s/n are combined by reading. Therefore, the final sensor has its s/n and combines digitally all related s/n of its components. Such information is, e.g., the basis for establishing the material certificate 3.1.

Furthermore, the transducer has engraved its article number (2.55791), the batch number of the raw material (972438), and the material information (1.4404 GP 5). The engravements are visible in Fig. 5.

{3} **Device-specific data**, e.g., the mentioned 3.1 material certificate, the Bluetooth PIN, and the 3-point calibration protocol, are connected to the sensor's s/n via the Enterprise Resource Planning (ERP) or the Manufacturing Execution System (MES). The transducer is capable of storing a small amount of data, but VEGA has not yet introduced TEDS (Transducer Electronic Data Sheet [21, 22]). Such TEDS units are one possibility to create the opportunity to store all related device-specific elements and make them machine-readable. Connecting the transducer with the final serial number is automated via the production system at VEGA.

{4} **Load profiles** are interesting to know how a product is used and to support the customer in case of challenges during the sensor's application. Accessing those profiles requires related memory capacity on the sensor. Related histograms help to reduce the amount of data. For example, each minute, one measuring value is taken and sorted into equally distributed ranges of the sensor's full-scale range (FSR) and above and below the specified FSR (see Fig. 5).

In case of problems, those profiles indicate where the problem might be. For example, if the measured pressures are relatively low, the precision might not fulfil the customer's expectations. It might lie in the particular precision definition, which is typically proportional to the sensor's FSR. The sensor supplier can propose a sensor with a lower measuring range. In Fig. 6, one with an FSR of 0 … 60 bar would be optimal. Furthermore, load profiles indicate if a sensor is used beyond its specified measuring range. Again, the sensor supplier can propose a better product that suits the customer's requirements.

If the sensor producer collects information about the whole installed base of a product family (type), portfolio adaptations are based on evidence [20]. For example, those analyses might show that the present sensor is overengineered according to its typical applications.

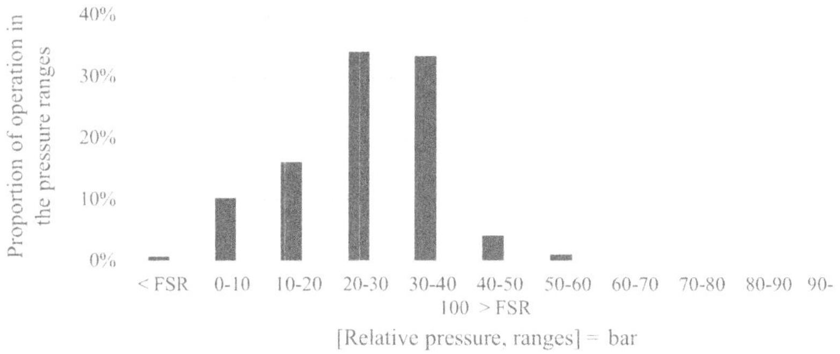

Fig. 6. Example of a load profile as a histogram

VEGA does not store histograms on the pressure sensors, but they collect the following information:

- Drag indicator value for the reached maximum pressure, including point in time
- Drag indicator value for the reached maximum temperature, including point in time
- How long the sensor has been operated outside its specification
- How often the sensor has been operated outside its specification

VEGA can read the data but not yet remotely. Nevertheless, received data can form the basis for conclusions about what supports the after-sales service business and related customer support.

{5} **Accessibility** of the products that are part of machines or process engineering systems might generally be challenging. Therefore, completing the Digital Thread requires accessing the stored or live data accordingly. If the sensors are returned to the producer for reuse, refurbishment, or recycling, the memory can be read accordingly and connected to the serial number (s/n). A way of accessibility is an application on a mobile device, such as a smartphone. Typically, they are handled less restrictively, and terms and conditions are accepted. Therefore, uploading load profiles as histograms and the s/n is more probable.

{6} **Remote access** for remote support can be established with the same accessibility strategy using a related application on mobile devices. In case of difficulties during the commissioning of the sensor or problems during operation, the operator grants access to the sensor expert. Stored data on the sensor's memory can be downloaded and analyzed. Besides the serial number, it also includes the parameter set. Therefore, suboptimal parameters, such as an unsuitable integration time for the signal damping, can be adjusted accordingly.

{7} The digital ecosystem is completed with powerful **software packages** [18] for computers accessing the pressure sensor during operation and offline. While the application on a mobile device offers occasional access, computer software can offer a higher degree of functionality. For example, the sensor's measurement can be trended over days to determine and trace the process' behavior. Such a function builds the base for analysis with other software (e.g. Excel and Matlab) to draw further conclusions without the requirement to program related data acquisition capabilities via a control system's PLC (Programmable Logic Controller).

The digital ecosystem also allows the connection of distributed sensors in distributed production facilities. This solution for an inventory system [19] is not part of a digital thread baseline.

Vega's digital engineering ecosystem is composed of the abovementioned elements ({1} to {7} found as well in Fig. 4, page 8). These elements create a digital thread that provides traceability from the customer's needs to the sensor's end-of-life. It combines all relevant information to create serial number-oriented traceability for each instance. This holistic approach fulfils Vega's required level of consistency.

4 Observations and Insights of a Theoretical and Practical Juxtaposition

The theoretical context and practical applications of the DEE present a dynamic interplay between ideal methodologies and the tangible execution within industrial contexts. As outlined in Sect. 2, the deployment of DEE is predicated on the seamless integration of the DTH (and associated technologies either in part or in entirety) and the comprehensive traceability facilitated. This theoretical underpinning is crucial for understanding how digital engineering strategies can go beyond traditional barriers in product development, fostering enhanced efficiency and innovation. However, translating these theoretical constructs into practice involves navigating the complexities of real-world applications, business goals, and limitations, as evidenced by the case study presented in Sect. 3, where Vega's approach to leveraging a DEE for its pressure sensor families demonstrates both the potential and the limitations of implementation. The analysis revealed the intricacies of designing and developing products with an emphasis on the RFLP (Requirements,

Functional, Logical, and Physical) elements, the debate between utilizing a unified platform versus single tools for enhancing the DTH and Traceability, and the crucial role of interface management within the broader digital ecosystem.

Table 1. Comparison of capabilities to support traceability (*No, Partial, Yes*)

Aspect	Theory (Sect. 2)	Use case (Sect. 3)
Platform-centric	Yes	No
Seamless Data Integration	Yes	No
Lifecycle Data Management	Yes	Partial
Real-Time Data Access and Analysis	Yes	No
Collaboration Tools	Yes	Yes
Standardization of Data and Processes	Yes	No
Cybersecurity Measures	Yes	Yes
Scalability and Flexibility	Yes	Partial
Predictive Analytics and Machine Learning	Yes	No
Digital Twins	Yes	No
Sustainability and Compliance Tracking	Yes	Partial
Customer and Supply Chain Integration	Yes	Yes
Continuous Improvement and Innovation	Yes	Yes
Documentation and Knowledge Management	Yes	Partial

However, as shown in Table 1, the practical application of these principles, as detailed in Sect. 3, reveals a more nuanced approach to the adoption of the V-model presented in Sect. 2. The case study presents a digital ecosystem that, while supporting traceability and the digital thread, diverges from the theoretical model by leveraging an assemble-to-order approach utilizing prefabricated modules. This strategy necessitates the meticulous integration of secondary (often disconnected, purpose-built) tools and technologies within existing digital ecosystems, highlighting an ambition and desire for high flexibility and specificity in product assembly and management. Specifically, Vega's employment of stand-alone tools for customer configuration, module management, and device-specific data handling demonstrates a deviation from the theoretical ideal of a fully integrated, platform-centric approach. Instead, it underscores the practical challenges and adaptations required to translate DTH and DEE theoretical concepts into operational realities.

This deviation between theory and practice is not entirely indicative of a theoretical deficiency in the company but rather reflects the complex realities of implementing DTH in a corporate setting. While theory puts forward an idealized, integrated platform for data and process management across the product life cycle, Vega's case illustrates the practical limitations and strategic decisions companies must navigate to align with traceability ambitions. The use of specific tools for targeted aspects of the product life cycle – ranging from customer interaction to module management – highlights a pragmatic approach to achieving the benefits of digital engineering within the constraints of current technologies and organizational structures.

Moreover, the integration of DTH elements within Vega's ecosystem, although partial, signifies an important step towards realizing the benefits of DEE. Embodying a practical translation of theoretical concepts into tangible operational actions, albeit with adjustments to fit the company's specific context and technological landscape. This approach allows Vega to leverage the advantages of traceability and digital threading while accommodating the need for flexibility in product design and customization, illustrating a balance between theoretical aspirations and practical capabilities.

5 Conclusion and Recommendations/Outlook

This paper provided an overview of the context of the development and deployment of digital engineering ecosystems as digital models have acquired an essential role in the engineering of products' life cycles. This paper presented a background on digital engineering and its enabling ecosystems, the generalization of the digital thread per life cycle phase, how the v-model approach can be used to assure horizontal, vertical and end-to-end traceability and how DEEs can streamline product life cycle processes. This work accentuates on how DEE surpasses traditional siloed confines in modelling, tooling, process implementation, and role designation. DEE is a holistic construct that shall empower organizations to develop, sustain, and manage complex systems with an end-to-end traceability.

An industrial study on a digital engineering ecosystem introduced in this paper aims to aid in conceptualizing its deployment and operations, which can depend on the company. The elements of a digital ecosystem were identified overall to leverage a better traceability, ensuring the companies to meet or exceed customer expectations at a lesser effort. In this case, Vega presents their own community of practice know-how and exploits an assemble-to-order approach utilizing prefabricated modules to achieve what the customer perceives as a seamless engine of traceability linked to their needs. This company does not follow a DEE that is fully integrated or platform-centric; its DEE is constituted of modules that need to interface with each other when a new query is inputted. This is a reminder that the implementation of a sort of DEE deployment is company-dependent. This aspect is also highlighted in NASA's latest report on digital engineering expectations and challenges: "Professional organizations, associations, and individual departments and agencies recognize the potential for using models combined with computing, or digital, environments. However, adopting a new method or practice is difficult for many implementing organizations. They have successfully executed current methods and practices over many years, resulting in capabilities that have met or exceeded expectations. Disrupting the current practices with different methods could appear to put that success at risk, with potentially negative consequences for the taxpayer, system user, and recipient of the engineered item." [23].

As for future work the authors acknowledge that the industrial case presented in this paper constitutes a singular case, and it would be of interest to further a broader study of the deployment of DEE by other companies in order to evaluate a larger sample and explore the percentage of effort required to deploy a DEE versus its added value to the company.

Acknowledgements. This study was carried out within the MICS (Made in Italy – Circular and Sustainable) Extended Partnership and received funding from the European Union Next-Generation EU (PIANO NAZIONALE DI RIPRESA E RESILIENZA (PNRR) – MISSIONE 4 COMPONENTE 2, INVESTIMENTO 1.3 – DD 1551.11-10-2022, PE00000004). This manuscript reflects only the authors' views and opinions, neither the European Union nor the European Commission can be considered responsible for them.

References

1. Walden, D., et al. (eds.): INCOSE Systems Engineering Handbook: A Guide for System Life Cycle Processes and Activities, Fifth edition. John Wiley & Sons Ltd., Hoboken (2023)
2. Shyu, H.: DoD Instruction 5000.97: Digital Engineering, vol. 5000.97 (2023). https://www.esd.whs.mil/Portals/54/Docu-ments/DD/issuances/dodi/500097p.PDF?ver=bePIqKXaLUTK_Iu5iTNREw%3D%3D. Accessed 05 Jan 2024
3. DAU, 'Digital Engineering', Defense Acquisition University Glossary. Defense Acquisition University. https://www.dau.edu/glos-sary/digital-engineering. Accessed 03 Dec 2023
4. Schindel, W.D.: Realizing the value promise of digital engineering: planning, implementing, and evolving the ecosystem. Insight **25**(1), 42–49 (2022). https://doi.org/10.1002/inst.12372
5. Berenbach, B., Lo, R.-Y.: An introduction to project and product traceability. Insight **14**(4), 38–42 (2011). https://doi.org/10.1002/inst.201114438
6. Toriello, M.: Creating the digital thread. Insight **25**(1), 34–37 (2022). https://doi.org/10.1002/inst.12370
7. Kinard, D.: Digital thread and industry 4.0 (2017). https://www.nist.gov/system/files/documents/2018/04/09/2p_kinard_digitalthreadi4pt0.pdf
8. Bajaj, M., Hedberg, T.: System lifecycle handler — spinning a digital thread for manufacturing. INCOSE Int. Symp. **28**(1), 1636–1650 (2018). https://doi.org/10.1002/j.2334-5837.2018.00573.x
9. Terzi, S., Bouras, A., Dutta, D., Garetti, M., Kiritsis, D.: Product lifecycle management - from its history to its new role. Int. J. Prod. Lifecycle Manag. **4**, 360–389 (2010). https://doi.org/10.1504/IJPLM.2010.036489
10. Pang, T., Restrepo, J., Cheng, C.-T., Yasin, A., Lim, H., Miletic, M.: Developing a digital twin and digital thread framework for an "Industry 4.0" shipyard. Appl. Sci. **11**, 1097 (2021). https://doi.org/10.3390/app11031097
11. OASIS, 'OSLC', Open Services for Lifecycle Collaboration. https://open-services.net/. Accessed 23 Nov 2023
12. OpenMBEE, 'OpenMBEE', Open Model Based Engineering Environment. https://www.openmbee.org/. Accessed 22 Oct 2023
13. Gerhard, D., et al.: MBSE-PLM integration: initiatives and future outlook. In: Noël, F., Nyffenegger, F., Rivest, L., Bouras, A. (eds.) Product Lifecycle Management. PLM in Transition Times: The Place of Humans and Transformative Technologies, in IFIP Advances in Information and Communication Technology, pp. 165–175. Springer Nature Switzerland, Cham (2023). https://doi.org/10.1007/978-3-031-25182-5_17
14. V. G. KG, 'VEGABAR 28'. https://www.vega.com/de-ch/produkte/produktkatalog/druck/prozessdruck/vegabar-28. Accessed 23 May 2023

15. V. G. KG, 'VEGA Messtechnik AG - Füllstand - Grenzstand - Druck'. https://www.vega.com/de-ch. Accessed 26 Sep 2023
16. V. G. KG, 'myVEGA'. https://www.vega.com/de-ch/myvega/dashboard. Accessed 26 Sep 2023
17. V. G. KG, 'VEGA Tools App'. https://www.vega.com/en-ch/products/product-catalog/signal-conditioning/software/vega-tools-app. Accessed 18 Apr 2023
18. V. G. KG, 'DTM Collection und PACTware'. https://www.vega.com/de-ch/produkte/produktkatalog/signalverarbeitung/soft-ware/dtm-collection-und-pactware. Accessed 18 Apr 2023
19. V. G. KG, 'VEGA Inventory System (VMI software)'.https://www.vega.com/en-ch/products/product-catalog/signal-conditioning/soft-ware/vega-inventory-system---vega-hosting/vega-inventory-system. Accessed 06 Dec 2023
20. Schmid, D., Nyffenegger, F.: "Metro Map" illustrating the digitalisation in industry. In: Noël, F., Nyffenegger, F., Rivest, L., Bouras, A. (eds.) Product Lifecycle Management. PLM in Transition Times: The Place of Humans and Transformative Technologies, in IFIP Advances in Information and Communication Technology, pp. 294–303. Springer Nature Switzerland, Cham (2023). https://doi.org/10.1007/978-3-031-25182-5_29
21. Anonymous, 'TEDS - transducer electronic data sheets', Sens. Rev. **24**(3), 309–310 (2004). https://doi.org/10.1108/sr.2004.08724caf.003
22. '1451.0–2007 - IEEE Standard for a Smart Transducer Interface for Sensors and Actuators - Common Functions, Communication Protocols, and Transducer Electronic Data Sheet (TEDS) Formats | IEEE Standard | IEEE Xplore'. https://ieeexplore.ieee.org/document/4338161. Accessed 05 Dec 2023
23. Hale, J., et al.: Digital model-based engineering: expectations, prerequisites, and challenges of infusion. NASA, NASA/TM—2017–219633 (2017). https://ntrs.nasa.gov/api/citations/20170006995/down-loads/20170006995.pdf. Accessed 15 Nov 2023

A Survey Investigating Skills, Job Profiles, and Education in Circular Economy: A Practice Perspective

Claudio Sassanelli[1](✉), Federica Acerbi[2], Irene Pellucchi[3], Denitsa Dimitrova[4], Sarah Downes[5], Iskra Dukovska-Popovska[6], Giorgio Mossa[1], and Sergio Terzi[2]

[1] Department of Mechanics, Mathematics and Management, Politecnico di Bari, Via Orabona 4, 70125 Bari, Italy
claudio.sassanelli@poliba.it

[2] Department of Management, Economics and Industrial Engineering, Politecnico Di Milano, Via Lambruschini, 4/B, 20156 Milan, Italy

[3] Erion Compliance Organization Scarl, Via Messina 38, 20154 Milan, Italy

[4] Cleantech Bulgaria, Krakra 18 Str, 1504 Sofia, Bulgaria

[5] Repic Limited, REPIC House, Waterfold Business Park, Bury BL9 7BR, UK

[6] Materials and Production, Aalborg University, 9220 Aalborg, Denmark

Abstract. The manufacturing sector requires a highly skilled workforce able to integrate sustainable business models in line with technological advancement. So far, it is still not clear how company deal with the enhancement of the set of skills needed to ease the twin green and digital transition. Without an understanding of the current state of practice, it could be tough for educational institutions to provide educational and training courses capable to meet the actual requests of companies approaching this transition. This research, grounded on an online survey, aims to identify the industrial and practical needs associated with the required skills and professional roles, necessary to address the transition towards a Circular Economy (CE) model. The results of this study (complemented by a literature, empirical interviews, and market investigation) will be exploited to develop courses in the twin green and digital transition at Higher Education (HE) and Vocational Education and Training (VET) levels.

Keywords: Circular Economy · manufacturing · twin transition · skill · job profile · education · training · survey

1 Introduction

The advent and continuous evolution of digital technology is pushing production systems towards the Industry 4.0 (I4.0) transition, making them always more ready to foster the adoption of Circular Economy (CE) practices in their processes [1] and to address sustainability, reduce emissions [2, 3], and achieve Net-Zero [4]. CE can be accelerated by

Published by Springer Nature Switzerland AG 2025
P. Sureephong et al. (Eds.): PLM 2024, IFIP AICT 741, pp. 56–70, 2025.
https://doi.org/10.1007/978-3-031-93323-3_5

seeking solutions around digital innovation and by increasing industry expertise. From this recognition emerges the ambition to intersect key technological disciplines with the CE model with the aim of providing tools and educational paths to implement technical competences and skills [5]. However, the manufacturing sector requires a highly skilled workforce able to integrate sustainable business models in line with technological advancement [6]. This change calls for educational institutions to be more flexible, provide reskilling programmes, and generate new skills [7]. Prior studies unveiled that to enhance CE level not only technical skills (such as supply chain management and machine learning) and green skills are necessary but also social ones (such as critical thinking and adaptability to change) [8].

This set of skills turns out to be critical in the circular business model implementation [9], and their lack can drive up manufacturing costs [10] or even act as roadblocks to the shift [11]. Notwithstanding the growing interest in literature related to education for CE (e.g. [6, 12]), so far it is still not clear how companies deal with the enhancement of the set of skills needed to ease the twin transition. Without an understanding of the current state of practice, it could be tough for educational institutions to provide educational and training courses capable to meet the requests of companies approaching this transition. Indeed, new job profiles with extended or updated skills are needed in companies' organizations to ease their transition from linear to circular business models [9]. Personnel with different competencies could be required to address the different activities related to CE (from organization, business model and research and development, through technology, supply chain and customer involvement, up to performance evaluation and regulations) [13]. To own the right competences related to each of these facets of the CE, a different type of training could be required. In addition, a specific type of skill could be appropriate for different stages of the lifecycle of the circular solution. Furthermore, it is also needed to understand if companies are capable to find and select extant training courses and to reach and exploit them to qualify their workforce, enhance their CE readiness level, and to cope with the obstacles occurring throughout the twin transition. Having strategic approach to complement the need of new CE-related skills, the collaboration with external stakeholders requires to be investigated to unveil which type of actors could better assist companies to become more circular [14]. Finally, models have been recently introduced in literature to help companies to self-assess their maturity in terms of CE [15]. Progression of circular maturity requires companies to be aware of the barriers they could face and which public strategies, action plans, and regulations could be supportive in coping with them [16].

To overcome this challenge, this study, grounded on a survey, aims to identify the industrial and practical needs associated with the required skills and professional roles, necessary to address the transition towards a CE model. The results of this study (complemented by a literature and market investigation) will be exploited to develop courses in the Circular Economy Innovation Ecosystems Redesigning Skills (CERES) project. These courses, dedicated to Higher Education (HE) and Vocational Education and Training (VET) levels, will be made available in the new Circular Economy Digital Innovation Hub (CE-DIH) to be developed in CERES [17]. The paper is structured as follows. Section 2 presents the research method employed to conduct the survey.

Section 3 presents the results, Sect. 4 discusses them, and Sect. 5 concludes the paper also providing limitations and further researches.

2 Research Method

To identify the industrial and practical needs associated with the required professional roles, necessary to address the transition towards a CE model, a questionnaire was developed, and a related online survey was run. In it, respondents were asked to share personal information, data about the organization in which they are currently working for, and their experience about their understanding of CE and about the needs related to CE-related skills. The questionnaire was designed to be filled by employees of companies and organizations interested or involved in the CE domain, to grasp their experience, issues, needs and expectations. The research method is composed of three main phases: survey development, survey execution, and analysis of results. In the first phase, thirty questions were developed in English (and then translated also in Italian and Bulgarian) to ensure accessibility and inclusivity across diverse linguistic audiences. The questionnaire consists of multiple-choice, open-ended, and ranking questions, split in four main sections. The first section (from question 1 to 9) aims at profiling respondents (asking their affiliation, their role within organizations, their level of involvement in the CE-related activities, and their areas of responsibility). The second part, comprising questions from 10 to 13, refers to information purely related to the respondents' organization (i.e., industry, size of the organization, geographical impact of the related business). The third section, from question 14 to question 17, aims to explore the company's understanding of CE, asking their level of maturity concerning CE, and identifying the main barriers in approaching its guiding principles. The fourth and final section includes questions from 18 to 30 and wants to investigate the organizations needs in terms of skills and competences, type of trainings, and the stakeholders capable of better supporting the CE transition. The survey was designed to require around 15 min for its completion. Regarding the survey execution, the questionnaire, in the three different languages, was reported on a professional platform (https://freeonlinesurveys.com/). Each of the versions of the same questionnaire in the three languages corresponded to a different link to the same platform. The English version of the survey was shared for the first time in the first week of November 2023, and then also in Bulgarian and Italian were shared in the following week. Three waves of official dissemination of the survey's links were implemented (beginning of November 2023, of December 2023, and of January 2024) through the CERES project (in which the survey was performed) website and on its LinkedIn page. To enable the widest possible reach, the links were also shared via each project partner private and public social channels. Additionally, the link to the questionnaire in Italian was also shared through Erion's newsletter and the questionnaire in English was pushed by REPIC and Politecnico di Bari through direct e-mail contacts to different companies. At the beginning of February, the survey was closed, and all the data were retrieved and translated in English (when needed). The third phase was the analysis of results. It began with the verification of answers and leading to eliminating 50 not significative (not fully filled) answers from the English-language questionnaire and 24 from the questionnaire in Italian, leading to a total number of 102 useful answers. The answers to the open questions were collected and analysed, grouping similar ones, and grasping recurring themes.

3 Results

This section provides the results coming from the analysis of the survey. Answers are summarized per each question, also reporting the percentages of the respondents who answered the specific question. This number may vary from the total participant count (102), which includes not only individuals who answered to all the questions of the survey but also individuals who submitted only their personal data and/or offered significant answers only to some open-ended or multiple-choice questions. Different types of graphs are inserted to exhaustively conduct the analysis.

The first section of the survey (from question 1 to 9) dealt with personal details of the respondent. Four fields (name, surname, email address, and organization name) were asked to profile participants. The fifth question was related to the respondents' role inside their own company (to be selected among the five available options: *consultant, academic, industrial, policy-maker, other*). The majority (48% of 96% of total participants) indicated *Other*, meaning that none of the alternatives available as predefined answers was descriptive of their role, but no further specifications were provided for this category. Then, in order, *industrial*, *consultant*, and *academic* range from 20% to 10% of respondents, whereas only 2% were *policymakers*. Question 6 was an open question aimed at deepening the profiling of attendees by asking for their job title. Among the replies, *manager* was indicated by several participants, followed by *sustainability manager* and *project manager*. Besides that, *engineer, officer, director, professor, audit, business, sales, and IT representative* were other figures reported. It emerges that a broad spectrum of roles is analysed and that the survey has been shared among employees belonging to different areas and seniority levels. Questions 7 and 8 were the first related to CE, respectively asking respondents if they have been working on CE-related activities, and if yes, how many years. Around 73% answered positively, while the remaining 27% never worked on CE. This fraction is expected to decline in the upcoming years, considering in broad perspective the results from initiatives such as CERES, which aid in creating a new labour force sensible to these topics. Regarding question 8, of the 73% declaring of being enrolled in CE-related activities, the 15% selected less than 1, the 50% 1–4 years, and around the 35% more than 4 years. It means that a minority has only recently started to focus on environmental issues, while a significant percentage has been active for over 4 years. This period marks when the European Union began to tackle important environmental issues (as CO2 emissions) through dedicated regulations and actions linked to CE strategies (e.g., Paris Agreement [18]). Half of the participants began to be interested in these topics 1–4 years ago, following the introduction of such regulations, reflecting a growing interest from both the industry and citizens. This analysis suggests that the minority who declared an experience of less than 1 year is going against the trend, lagging the general trend highlighted by the survey. Question 9 was related to unveil the attendees' area of responsibility (Fig. 1). In this case, multiple answers were allowed (among 14 possible options), and several participants selected three or more options. The majority (39 respondents) selected *Sustainable management*, whereas categories such as *Training & education,* and *End of life – waste management and disposal* recorded 30 votes, followed by *R&D*/Innovation (27). On the contrary, *Logistics, Reverse logistics, After sales services supporting use phase (maintenance, repair, etc.), End of life – remanufacturing, Human resources* were voted by less than 10 respondents.

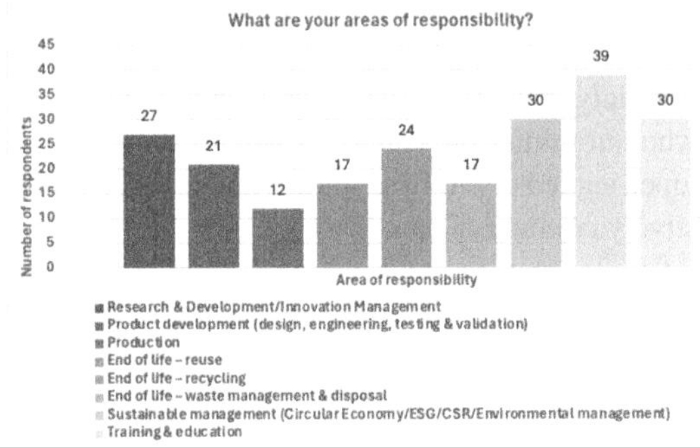

Fig. 1. Areas of responsibility

Questions 10, 11, 12, and 13 were related to the organisation description. Regarding question 10, asking for the industries in which respondents' organisations are involved in, multiple selection was allowed. Most of the respondents selected *Other*, specifying industries such as health care and medical, standardisation and certification, research and teaching, consulting, construction and civil engineering, tourism, architecture, packaging and plastic, hotel, governmental body, food and beverage, NGO, and public administration. Electrical and Electronic Equipment *(EEE)* scored 22 votes, confirming its relevance in the CE domain. Indeed, except the group of several industries gathered under *Other*, the remaining options (in order: textile, ITAD&ITAM, trade association, wind turbine, and automotive) were selected less than 10 times (Fig. 2).

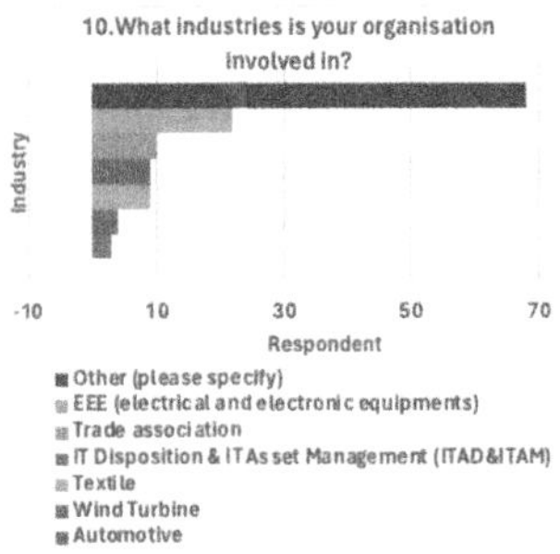

Fig. 2. Industries

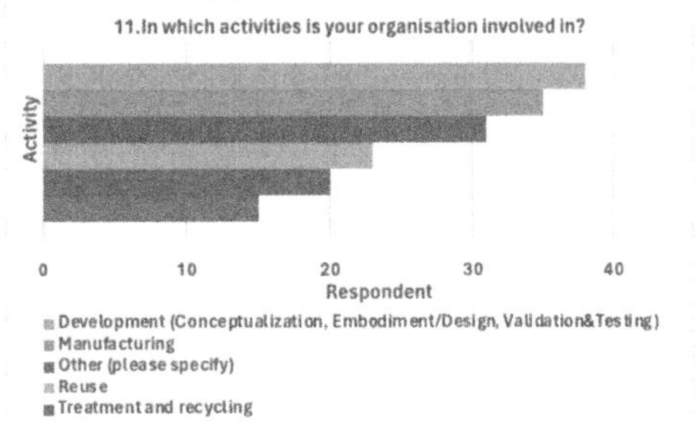

Fig. 3. Activities

In question 11, asking in which activities respondents' organizations are involved in, options *development* and *manufacturing* were the most selected (Fig. 3). Also in this case, the category *Other* was indicated by several respondents, highlighting the heterogeneity of the participants (and including: assessing, certifying, and teaching CE practices; standardization; maintenance of products; retail; audit; education; accounting; development of policies and legislation; and nature conservation). Question 12 aimed at mapping the respondents according to the size of their organization. The survey considered these options: *Large (staff headcount > 250; turnover > 50m), SME (staff headcount < 250; turnover < 50m)*, and *Micro (staff headcount < 10; turnover < 2m)*. Each of the three categories showed percentages around 30%, with *SME* slightly higher (35%). The survey was representative because it encompassed a considerable share of every organization size, even if 5% of 102 participants did not reply to this question. Question 13 investigated the level of granularity of geographical organizations' actions (*Regional/local, national, and international*). Multi-selection of answers was possible. *International* referred to companies working exclusively with foreign partners, *national* with partners within their own country, and *regional/local* if collaborations were confined to the region in which they are located. *National* and *international* categories each received approximately 50 votes, while the regional/local category garnered 24 votes. Despite this, the survey adequately represented all three options.

Questions from 14 to 17 assessed organizations' understanding of the CE, highlighting respectively obstacles (Fig. 4), gaps, maturity, and priorities. Specifically, in question 14, the participants were required to identify the barriers they faced in adopting CE principles. The top three challenges mentioned were (i) a scarcity of financial resources, (ii) a shortage of expertise and sector-specific experience, and (iii) a lack of collaboration with other stakeholders. Businesses encountered financial obstacles to support the deployment of circular strategies, flanked by the absence of benchmarks and prior sector-specific experiences useful to guide the identification and prioritisation of actions. There was a notable challenge in effectively communicating the CE concept within companies, which hindered the comprehension of its entrepreneurial impacts.

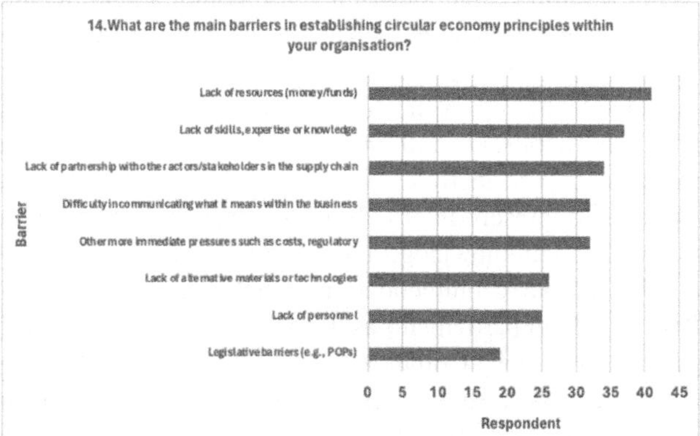

Fig. 4. CE- barriers

In question 15, respondents were asked to select among six options the one which best described their company's level of maturity, from an absence (level 0) up to full adoption (level 6) of CE. Considering that approximately 6% of the participants did not respond, the majority (around 30%) indicated that they have a good understanding of CE and are beginning to adopt related practices. A similar percentage specified that CE is at the core of what they do, whereas 4% only did not know anything about it. These findings are encouraging and significant, underscoring that not only CE is recognized by most of the organizations, but it is also gaining strategic and business relevance.

Approximately 47% of the 102 participants provided answers to question number. The 16.67% of them stated that they did not encounter any gap related to their CE expertise, neither initially nor at any point of their transition. On the other hand, the rest of the participants shared the challenges they faced and/or how they addressed them. In particular, it has been highlighted that there was a widespread absence of regulations and standards, which obstructed the adoption of circular practices. Additionally, there was a cultural hurdle, with some potential customers hesitant to commit to long-term circular solutions, and a prevalent lack of expertise in this area. In practical terms, collecting data that would be functional to CE adoption was challenging since it hadn't been a priority in the past. Consequently, obtaining the necessary information for environmental assessments of materials or products took a considerable amount of time. For example, one company explained its struggle in finding specifications for a suitable equipment to recycle aggregates from buildings and demolition sites. Significant effort was required to develop specialized skills in specific industries (such as construction). Question 16 was strictly related to question 15, asking, only in case of a good understanding and maturity of CE, further details about gaps in organizations related to CE adoption and the way they have been addressed. Companies from different fields and industries described their strategies to build an expertise in CE. As stated by one participant, it is important to notice that it has been (and, to some extent, it still is) a problem to ensure a wide and equal cross-company knowledge. Technical and legislative competences are centralized on very few people who, as part of the business maturity process, must be available to spread further this knowledge within the organization. Consequently, companies have leveraged on their internal talents, promoting commitment at senior level and employment engagement. Figures such as sustainability managers were at the core of the transition to CE. One respondent explained that prevention and reusing practices have been implemented within the office to educate the personnel and promote a transformation of culture and habits. Overall, internal training and communication have been indicated as common tools to increase the awareness about CE. In case the required profiles were not already present internally in the organization, sustainability experts have been hired to bring their knowledge among people in the company, and consultants have been inquired to receive recommendations. On the other hand, external training has been indicated by the 25% of participants as the most efficient strategy to familiarize and gain knowledge about this topic. Collaborations with partners and clients, networking, synergies, market insights on technologies and best practices, scouting and seeking for fundings were also fundamental to embrace the change. The ISO 14001 standard also contributed to set guidelines, and few companies went through the certification process.

Question 17 inquired about primary goals related to CE adoption, and relevant regulations or guidelines followed by organizations to address it. Figure 5 is provided to illustrate this connection, also reporting few cases as examples. Results highlight that the CE topic might be addressed by many regulations due to the interconnectivity of the different concepts. For example, the *lower footprint* is mentioned in *GHG protocol, Net Zero targets, SBTi, Fit4–55, EMAS, ISO 14064, ISO 14067,* and *EU taxonomy*. Moreover, few respondents cited their experiences as virtuous examples of sustainable and circular behaviours. These include *Company A*, a Danish company known as the first one in its field to feature textiles with up to 10% recycled content, as specified in the related reports, and *Company B*, which played a significant role in developing *Recycling City*, recognized as Europe's most eco-friendly scrapyard. Furthermore, *Organization A*, an employer federation from Cyprus and a technical Bulgarian University (*University A*) mentioned their efforts to contribute to climate neutrality and limit their environmental impacts. Among the regulations, the *Green Deal* has been mentioned by the 20% of the respondents who linked it mostly to recycling activities. Indeed, several companies were dedicated to achieving the objectives established in this framework (for instance, sectors like WEEE, textiles, and construction). Beyond recycling, a smaller group of participants underscored the significance of reusing as essential priorities in the textile and WEEE realities. Additionally, the *New European Bauhaus* initiative was mentioned as a novel way to merge the *Green Deal* with the development of sustainable living spaces and experiences. Respondents also mentioned the ISO certification process in 29% of replies, citing standards such as ISO 14001 (for environmental management systems (EMS)), ISO 14064 (for the quantification and reporting of greenhouse gas (GHG) emissions and removals), ISO 14067 (for the quantification and reporting of the carbon footprint of a product (CFP)), ISO 50001 (for establishing, implementing, maintaining and improving an energy management system (EnMS)), and the ISO 59000 series (for guiding on CE approaches and business models). Generally, once addressed these standards, organizations are recognized as capable to guarantee compliance with regulations and to have a sensibility towards sustainability and circularity related topics. These aspects are important also for customers, who prefer companies that respect ISO standards. To conclude, the certification process has been appreciated since it aims to foster the development of circular approaches and an efficient management of energy systems while limiting environmental impact (GHG emissions).

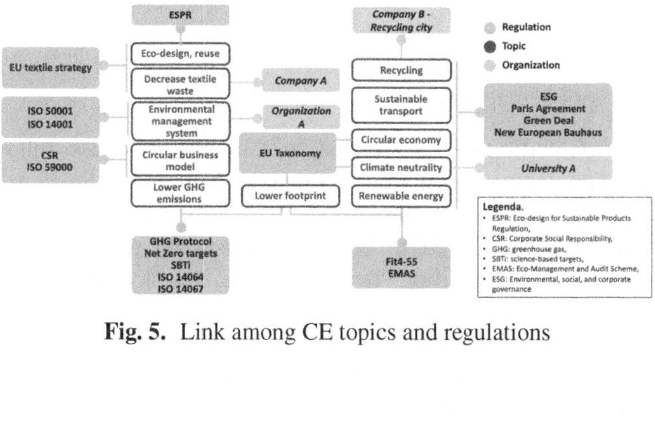

Fig. 5. Link among CE topics and regulations

Questions 18 and 19 asked to rank a list of CE-related skills according to the respondents' preferences. In question 18, concerning advancing CE in their organisation, respondents provided the following priorities for skills development and training:

1. Design of systems, products, and components for durability, reuse, recyclability, and recycled content;
2. Design of circular business models to create added value and incentives including different ways of working with supply chains and customers;
3. Commercial and market awareness to retain profitability and seek funding and investment for change;
4. Procurement and supply chain working to increase closed loop recycling and reuse of components;
5. Digital technologies, information systems and machine learning to manage and use data effectively to manage the performance, use and end of life of products and components;
6. Remanufacturing, refurbishment, and repair of products to extend product life and optimal energy use;
7. Technological advancements to improve design, choice of materials and reuse and recycling of materials;
8. Life cycle assessment to understand impacts of material choice, production, and end of life management of products and materials/whole life costing/net zero;
9. Communication and consumer engagement: to create a sharing economy for products that are used infrequently, wider acceptance of leased product and second hand or refurbishment, etc.;
10. Legislative barriers and opportunities presented by future legislation (e.g., DPP, ESPR);
11. Retail and logistics for more efficient distribution of products, parts and take-back/reverse logistics.

Results underscore the pivotal role of design in the transition from a linear to CE and explains why companies seek for educational opportunities that improve these skills. A transition to CE also needs ad hoc business models and market awareness to create value. Consequently, organizations ranked as second and third these skills, prioritizing education on circular business and marketing.

Related to question 19, respondents ranked the skills needing to be enhanced more to support CE adoption as follows:

1. Resilience Cross-cutting skills (related to Circular Business Innovation) (e.g., Entrepreneurship, Innovation, Circular Business Modelling, Sustainable Competitiveness, Systems, and Design Thinking);
2. Resilience Soft skills (related to communication and teamwork) (e.g., Leadership, Flexibility, Responsibility, Team building, Problem-defining and solving, Selfawareness);
3. Skills for Technology Innovation in the CE (e.g., development, provision and management of technologies as Blockchain, IoT, Artificial Intelligence, Virtual and Augmented reality, Fourth Industrial revolution);

4. Skills for digital transformation of businesses (e.g., Technology Management and Digital Transformation, Smart PSSs, Digital Economy, Platforms in a digital economy, Data-driven approach (GDPR, IPR);
5. Specialised/technical skills for waste management (e.g., Digital Passports, Packaging, Circular Design, Resource flows (collection, sorting, reuse, recycling, etc.), supply chain design, feasibility and sustainability;
6. Specialised/technical skills for Clean Tech and Advanced materials enabling CE (e.g., Composite materials, Zero waste economy, energy efficiency, circular bioeconomy, carbon capture and storage).

From this ranking it is clear that cross-cutting, soft, and digital skills were the most important ones, highlighting the interconnectivity nature of CE. In this context, it is important for companies that every employee involved in CE has a basic knowledge of all the steps of the value chain. This enables an efficient communication with other actors, levering on digital skills to share data and know-how.

Question 20 explored the types of training preferred by respondents (Fig. 4). Three options stood out as favourites: *Continuing Professional Development (CPD) provided by an accredited chartered body, Academic courses (e.g., Master of Science or master's in business administration modules)*, and *Formal certification*. These options were similar in terms of provision of a certificate or a degree upon completion, which is considered a plus for the attendees. Additionally, *Online tutorials* and *MOOCs* came, followed by the option *Other* (*Seminars* and *Business associations* were proposed by respondents). The last choice was *Apprenticeship* (Fig. 6).

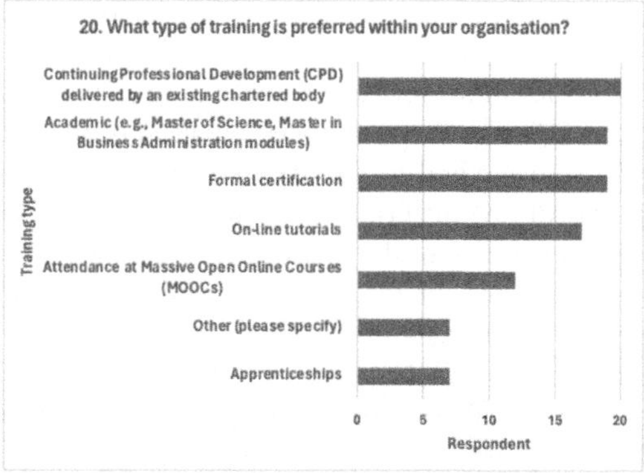

Fig. 6. Types of preferred training

Questions 21, 22, and 23 dealt with the required new technical skills from a CE perspective. Each question focused on a specific life stage of the whole product lifecycle, respectively beginning, middle and end-of-life (BoL, MoL, and Eol).

For question 21, related to BoL, according to most participants (38%), circular design of products was the most required skill. To achieve this competence, materials choice to minimize resource consumed, and Design for X approaches (i.e., disassembly, repair, reuse, and recycling) were the most relevant topics. In this stage, production management systems together with business models should be developed to support the switch to CE. Moreover, LCA has been mentioned as fundamental in circularity assessment and in carbon footprint, environmental, and social impacts evaluation. This method was also important to assess the sustainability of manufacturing processes. Necessary technical skills were related to STEM disciplines (such as engineering), but also interdisciplinary holistic abilities were considered important. Among these, communication, entrepreneurship, and system thinking have been underlined by the attendees. Regarding strategic skills, the following have been specified: good understanding of product life cycle, sensitivity to recognize opportunities for the transition from linear to CE, and stakeholder engagement to share knowledge about it.

For question 22, concerning MoL, understanding of product durability and repairability was fundamental to develop circular strategies. To this aim, planning of regular maintenance and repairing activities allowed to enhance products lifetime. In this regard, the comprehension of the European regulatory framework gave useful guidelines, such as for the implementation of reverse logistic systems. New types of marketing activities and commercial offering were also suggested to improve consumers' and producers' awareness during use. On the other hand, digital skills proved to be at the core of service and products delivering. Indeed, one participant suggested the implementation of a database where manufacturers could upload available components to be sold, and step-by-step manuals for repairing their products.

For question 23, related to EoL, assessing the state of health of goods was the most important skill to implement CE at disposal stage. Specifically, it was fundamental to distinguish between repairable and non-repairable products, and to recognize the actors involved in both circumstances. An efficient separation between these categories made the implementation of reusing strategies easier. To this aim, new competencies in recycling technologies for multi-material and cleaning techniques for post-consumer waste must be developed. It was also mentioned the need of sharing data and labelling materials in a standardized method to facilitate communication among stakeholders. Skills in identifying critical raw materials and valuable chemical substances were also required to value each component of the product. In this context, one participant asked for incentives to build factories for the extensive recovery of materials, and another one suggested the reuse of materials in the same sector from which they come from; for example, aggregates from demolition waste should be utilized in the construction industry. In this context, it was also important to highlight the contribution of consumers that should be facilitated by user-friendly remanufacturing processes. At this purpose, behavioural design was preparatory to understand users and how they can be nudged to dispose products in the correct way.

Question 24 wanted to unveil which are the stakeholders who can play an important role in providing advices about the transition to CE (Fig. 7). Most people believed that *customers* (26%), and *consultants* (23%) were the most useful. Indeed, clients could drive the market by purchasing specific products that were more circular, for example by preferring recycled packaging instead of the conventional one. Consultants can provide knowledge and insights about circular strategies that might be implemented.

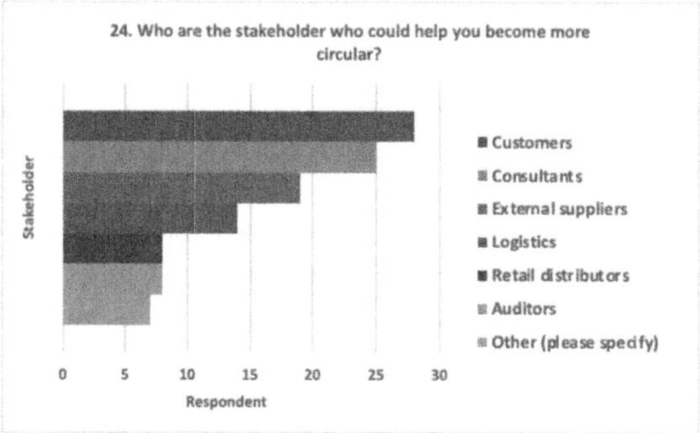

Fig. 7. Stakeholders helping to approach CE

Moving to question 25, the attention moved to the job profiles mostly required to face the economic transition across the value chain (with a 39% of response rate). The majority (90%) believed that there were job profiles required to transit to CE. Personnel knowing how to implement circular business models and to implement sustainable finance have been suggested. Sustainability and circularity managers also were fundamental figures, since they had a holistic view of the supply chain and the chance to interact with several actors. Also technical personnel, such as engineers and repair workers, were necessary to put into effect circular manufacturing lines by developing refurbishments activities, dealing with second-raw materials, and recycling processes. Question 26 asked whether the respondent was aware of training courses available on the market. The majority responded *yes* (61%), and most of them (74%) claimed that these courses were easily available for their companies, as explained in answers to question 27. These outcomes demonstrate that trainings are accessible, but it is difficult to find those addressing specific aspects of the CE, such as for circular design of electronic components or circular business strategies in the textile industry. Question 28 dealt with the preferred types of courses (Fig. 8). *Professional online courses* and *MOOCs* were the most voted categories. This could be because they are often recorded, and thus more easily and flexibly accessible. Furthermore, they are readily available online and focus heavily on professional topics, addressing specific and practical subjects. The 68% of respondents confirmed that they took advantage of the training offers on the market (question 29).

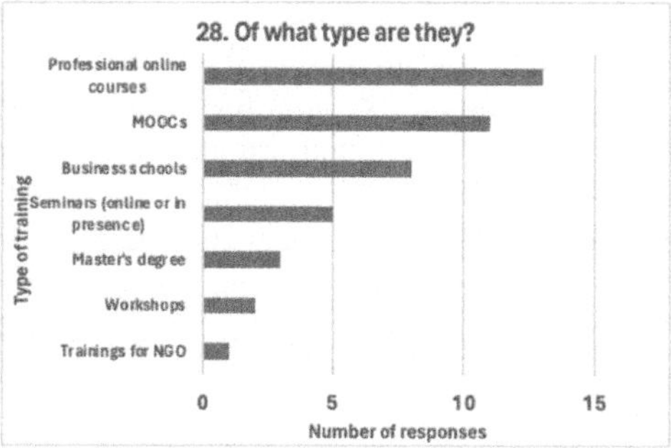

Fig. 8. Types of courses

4 Discussion

The purpose of conducting the survey was to identify the industrial and practical needs associated with the required professional roles, necessary to address the transition towards a CE model as well as training needs and preferred training forms. Even though only about 35% stated that they come from industrial and consultancy sector, and 49% come from *Other* sectors (presumably including services), about 73% have worked with CE, out of which 50% between 1–4 years, and 35% more than 4 years. This indicates that the respondents are knowledgeable about the CE transformation in their own context. Also the years of engagement with CE are in line with general evaluations [18] that the pandemic has made the CE more relevant than ever and that it can create greater resilience to shocks in industry and society (ibid).

Looking at the barriers to CE implementation stated by the respondents, lack of resources (money/funds) is the most frequently stated one. Even though this is not the most frequent barrier identified in the literature [19], the economic barriers have been well identified and recognized as critical ones, embracing high-up front investment costs, long pay-back periods, cheap virgin material, etc. [16, 20]. The second most frequently stated barrier was lack of skills, expertise or knowledge which can also be connected to the fourth barrier related to the difficulty in communicating what CE means to the business. Shortage of wider education in the organizations might prevent internal communication in the companies as there is lack of shared understanding of the concepts, possibilities for solutions and implementation, and of definition of performance metrics and their monitoring. This is in line with the literature that identifies lack of skills and know-how in CE product design (leveraging of Design for X approaches [21]), lack of technical skills, both internally in the focus organization, but also on the supplier and customer side [19]. This might explain the third barrier identified in this survey which relates to lack of partnership with other actors in the supply chain [16]. The lack of skills gained through more formal training and wider knowledge dissemination prevents companies in moving to next levels of CE maturity where projects with external partners start to evolve [22]. This research contributes to both knowledge and practice, also providing managerial ana policy implications. From a theoretical perspective, the survey contributed to unveil which are the skills, job profiles, and types of education more needed to foster education in the CE research domain. Practical gaps and barriers hampering an easy transition have been also collected, together with priorities and regulations adopted by organization to climb their CE maturity levels, unveiling key areas of improvement and of further research in the domain of CE education. From a practical point of view, the results obtained through the survey enabled to clarify the state of practice about the current needs of companies in pursuing a transition towards circularity especially in the field of new skills needed. This can be supportive both for educational institutions in helping them to develop more effective courses and training sessions and *curricula* and for companies willing to approach CE in understanding what could be the key topics to be addressed to upskill and reskill their managers and operators. Finally, policy makers could exploit the results of this research proposing new educationand training-related actions (on the same stream of the CERES project) able to ease the twin transition of manufacturing and of society.

5 Conclusions

This contribution enabled to clarify the state of practice about current practices undertaken by companies, the challenges faced, and the training programmes needed to align the workforce to the twin transition. The research objective has been addressed through a survey developed during the CERES project and covered around 100 respondents.

The answers gathered through this survey will be exploited to better define the structure of the teaching and training courses to be delivered at both HE and VET levels to foster a more effective transition of manufacturing and society towards CE.

Notwithstanding the multiple results obtained, it is worth to mention some limitations such as the restricted number of respondents to really have a complete and generalizable overview of the state of practice. A second iteration could be done to engage more people. Furthermore, a complementary study about the state of the art and of market should be conducted to compare the results obtained in this research.

Acknowledgments. This study was partially funded by the European Education and Culture Agency (EACEA) of the European Commission under grant agreement No 101111684.

References

1. Despeisse, M., Acerbi, F.: Toward eco-efficient and circular industrial systems: ten years of advances in production management systems and a thematic framework. Prod. Manuf. Res. **10**(1), 354–382 (2022). https://doi.org/10.1080/21693277.2022.2088634
2. Colangelo, G., Facchini, F., Ranieri, L., Starace, G., Vitti, M.: Assessment of carbon emissions' effects on the investments in conventional and innovative waste-to-energy treatments. J. Clean. Prod. **388**(135849), 1–15 (2023). https://doi.org/10.1016/j.jclepro.2023.135849
3. Vitti, M., Facchini, F., Mummolo, G.: Assessing the decarbonisation potential of waste- to-hydrogen routes in the energy transition phase: An environmental analytical model. J. Clean. Prod. **457**(142345), 1–16 (2024). https://doi.org/10.1016/j.jclepro.2024.142345
4. Word Economic Forum, The 'No-Excuse' Framework to Accelerate the Path to Net-Zero Manufacturing and Value Chains (2023)
5. Sassanelli, C., Garza-Reyes, J. A., Liu, Y., de Jesus Pacheco, D. A., Luthra, S.: The disruptive action of industry 4.0 technologies cross-fertilizing circular economy throughout society. Comput. Ind. Eng. **183**(109548), 1–5 (2023). https://doi.org/10.1016/j.cie.2023.109548
6. Akyazi, T., Goti, A., Bayón, F., Kohlgrüber, M., Schröder, A.: Identifying the skills requirements related to industrial symbiosis and energy efficiency for the European process industry. Environ. Sci. Eur. **35**(1) (2023). https://doi.org/10.1186/s12302-023-00762-z
7. Whitehill, S., Hayles, C.S., Jenkins, S., Taylour, J.: Engagement with higher education surface pattern design students as a catalyst for circular economy action. Sustainable **14**(3) (2022). https://doi.org/10.3390/su14031146
8. Akyazi, T., Del Val, P., Goti, A., Oyarbide, A.: Identifying future skill requirements of the job profiles for a sustainable European manufacturing industry 4.0. Recycling **7**(3) (2022). https://doi.org/10.3390/recycling7030032
9. Straub, L., Hartley, K., Dyakonov, I., Gupta, H., van Vuuren, D., Kirchherr, J.: Employee skills for circular business model implementation: a taxonomy. J. Clean. Prod. **410**(April), 137027 (2023). https://doi.org/10.1016/j.jclepro.2023.137027

10. Ada, E., Kazancoglu, Y., Mangla, S.K., Aydin, U.: Barriers to cement industry towards circular economy. Int. J. Math. Eng. Manag. Sci. **8**(4), 612–631 (2023). https://doi.org/10.33889/IJMEMS.2023.8.4.035
11. Trevisan, A.H., Lobo, A., Guzzo, D., de V. Gomes, L.A., Mascarenhas, J.: Barriers to employing digital technologies for a circular economy: a multi-level perspective. J. Environ. Manage. **332**(117437) (2023). https://doi.org/10.1016/J.JENVMAN.2023.117437
12. Watkins, M., Casamayor, J.L., Ramirez, M., Moreno, M., Faludi, J., Pigosso, D.C.A.: Sustainable product design education: current practice. She Ji **7**(4), 611–637 (2021). https://doi.org/10.1016/j.sheji.2021.11.003
13. Hofmann Trevisan, A., Acerbi, F., Dukovska-popovska, I., Terzi, S., Sassanelli, C.: Skills for the twin transition in manufacturing: a systematic literature review. J. Clean. Prod. **474**(143603), 1–19 (2024). https://doi.org/10.1016/j.jclepro.2024.143603
14. Pohlmann, C.R., Scavarda, A.J., Alves, M.B., Korzenowski, A.L. : The role of the focal company in sustainable development goals: a Brazilian food poultry supply chain case study. J. Clean. Prod. **245** (2020). https://doi.org/10.1016/j.jclepro.2019.118798
15. Acerbi, F., Sassanelli, C., Taisch, M.: A maturity model enhancing data-driven circular manufacturing. Prod. Plan. Control **2322608**, 1–19 (2024). https://doi.org/10.1080/09537287.2024.2322608
16. Taddei, E., Sassanelli, C., Rosa, P., Terzi, S.: Circular supply chains theoretical gaps and practical barriers: a model to support approaching firms in the era of industry 4.0. Comput. Ind. Eng. **190**(110049), 1–20 (2024). https://doi.org/10.1016/J.CIE.2024.110049
17. Sassanelli, C., Sarbazvatan, S., Demetriou, G., Greyl, L., Mossa, G., Terzi, S.: Coalescing circular and digital servitization transitions of manufacturing companies: the circular economy digital innovation hub. In: IFIP Advances in Information and Communication Technology book series (IFIPAICT), pp. 1–15 (2023). https://doi.org/10.1007/978-3-031-43666-6_11
18. The Ellen MacArthur Foundation, "To fulfil the Paris Agreement we need a circular economy," (2020). https://www.ellenmacarthurfoundation.org/articles/to-fulfil-the-paris-agreement-we-need-a-circular-economy. Accessed 14 Mar 2024
19. Govindan, K., Hasanagic, M.: A systematic review on drivers, barriers, and practices towards circular economy: a supply chain perspective. Int. J. Prod. Res. **7543**, 1–34 (2018). https://doi.org/10.1080/00207543.2017.1402141
20. Kirchherr, J., Hekkert, M., Bour, R., Huibrechtse-Truijens, A., Kostense-Smit, E., Muller, J.: Breaking the barriers to the circular economy (2017)
21. Sassanelli, C., Urbinati, A., Rosa, P., Chiaroni, D., Terzi, S.: Addressing circular economy through design for X approaches: a systematic literature review. Comput. Ind. **120**(103245), 1–23 (2020). https://doi.org/10.1016/j.compind.2020.103245
22. Uhrenholt, J.N., Kristensen, J.H., Rincón, M.C., Adamsen, S., Jensen, S.F., Waehrens, B.V.: Maturity model as a driver for circular economy transformation. Sustain. **14**(12), 1–17 (2022). https://doi.org/10.3390/su14127483

Development of an Augmented Reality Book to Promote the Local Wisdom of Tai Lue Foods in Gastronomy Tourism

Songpon Khanchai[1], Perasuk Worragin[2], Supicha Niemsup[3], and Suepphong Chernbumroong[2](✉)

[1] ChiangMai University, Chiang Mai, Thailand
[2] College of Arts, Media and Technology, Chiang Mai University, Chiang Mai, Thailand
suepphong.c@cmu.ac.th
[3] Department of Library and Information Science, Faculty of Humanities, Chiang Mai University, Chiang Mai, Thailand

Abstract. This study presents the development and evaluation of an augmented reality (AR) book aimed at promoting the local wisdom of Tai Lue foods in gastronomy tourism. In the wake of the COVID-19 pandemic, the global tourism industry seeks innovative strategies for recovery, with a focus on sustainable, inclusive, and culturally rich tourism experiences. The Tai Lue, known for their unique culinary heritage, offer a gastronomic experience that is largely untapped in mainstream tourism. Our project leverages AR technology to provide an immersive experience of Tai Lue cuisine, aiming to foster a deeper appreciation of local cultures and cuisines in a post-pandemic world. Employing a research and development (R&D) methodology, this study follows a pre-post test design to assess the impact of the AR book on participants' knowledge of Tai Lue cuisine. The findings indicate a significant increase in knowledge among participants, demonstrating the effectiveness of AR technology as an educational tool in cultural preservation and gastronomy tourism. This research contributes to the broader efforts to revitalize tourism and underscores the potential of technology in bridging cultural gaps and enhancing tourist experiences.

Keywords: Augmented Reality · Culture Heritage · Gastronomy Tourism · Tai Lue Foods

1 Introduction

In the aftermath of the COVID-19 pandemic, the global tourism industry faces the monumental task of rebuilding and reimagining its future. As countries gradually open their borders and communities seek to recover from economic setbacks, there is an unprecedented opportunity to redefine tourism to be more sustainable, inclusive, and culturally enriching. This period of revitalization calls for innovative approaches to attract tourists and promote cultural heritage, making the development of an Augmented Reality (AR) book to showcase the local wisdom of Tai Lue foods an especially timely endeavor.

Published by Springer Nature Switzerland AG 2025
P. Sureephong et al. (Eds.): PLM 2024, IFIP AICT 741, pp. 71–83, 2025.
https://doi.org/10.1007/978-3-031-93323-3_6

Gastronomy tourism, with its deep ties to cultural identity and local economies, emerges as a vital pathway for post-pandemic recovery and growth.

The Tai Lue, an ethnic group known for their rich culinary traditions and sustainable living practices, offer a unique gastronomic experience that is largely untapped in the mainstream tourism market. The COVID-19 pandemic has underscored the importance of diversifying tourism offerings and enhancing the resilience of local communities through the promotion of intangible cultural heritage.

By integrating AR technology into the exploration of Tai Lue cuisine, this project not only aims to provide an immersive and safe tourism experience in the wake of health concerns but also to foster a deeper appreciation for the complexity and heritage of local cuisines in a post-pandemic world.

Moreover, the pandemic has accelerated digital transformation across sectors, including tourism, making this the opportune moment to leverage technology in connecting tourists with cultural experiences. The AR book dedicated to Tai Lue foods is envisioned not just as a tool for culinary discovery but also as a platform for education and cultural exchange, enabling tourists to engage with the stories, traditions, and values of the Tai Lue people in an interactive format. This approach aligns with the evolving expectations of post-COVID-19 travelers [24, 25], who are increasingly seeking meaningful, authentic, and safe experiences that contribute to the conservation of cultural heritage and the sustainable development of host communities.

In promoting Tai Lue gastronomy through augmented reality, this project contributes to the broader efforts to revitalize and reimagine tourism in a post-pandemic era. It underscores the potential of technology to bridge cultural gaps, enhance tourist experiences, and support the recovery of local economies, thereby offering a model for how gastronomy tourism can adapt and thrive in the new normal.

2 Literature Review

2.1 Gastronomy Tourism

Gastronomy tourism, which includes culinary tourism, gastro tourism, food tourism, wine tourism, and gourmet tourism, is recognized as a vital segment within the tourism industry that provides tourists with authentic and memorable gastronomic experiences [1]. The UNWTO describes it as a type of tourism where individuals participate in activities involving food-related products and services, which may include traditional or innovative culinary activities such as cooking classes, wineries, restaurants, food festivals, and visits to food producers [2]. This form of tourism not only focuses on enjoyment but also on providing tourists with a deep, immersive experience that connects them to the local food culture and community [3]. Rachao et al. add that gastronomy tourism involves active participation and interaction with the local culinary ecosystem, indicating tourists' keenness to engage deeply with the food culture of their destinations [4]. Furthermore, Prayag et al. highlight that such experiences are not only about leisure but also involve elements of skill development, education, socializing, and recreation, classifying them as a form of serious leisure [5]. Manola and Koufadakis analyzed how gastronomy enriches tourism experiences and economic growth, and reviewed literature on its economic impacts and potential. The study found that gastronomy serves as a vital cultural expression that enhances tourism marketing, significantly boosts local economies by drawing tourists who spend on unique culinary experiences and emphasized promoting local gastronomy through events to attract tourists [6]. Similarly, Cabaguing and Lacaba noted that Gastronomy tourism significantly contributes to the local pride, culture, and economic development of Guiuan, Eastern Samar. Therefore, gastronomy tourism can be considered as an influential sector that contributes to the economic and cultural dimensions of tourist destinations, providing tourists with enriched, authentic, and engaging experiences that promote local heritage and foster economic benefits [7].

2.2 AR in Gastronomy Tourism

Augmented Reality (AR) has emerged as a transformative technology in various sectors, notably in tourism and gastronomy. It is defined as a technology that overlays a computer-generated image in the real world in real time, offering a "unified view" that integrates virtual and physical environments [8]. This immersive interaction not only enhances user experience but also facilitates location-based services, cost reduction, task optimization, and experiential marketing [9, 10]. In the context of tourism, AR has significantly improved how visitors engage with their environment, meeting psychological needs more effectively by providing enriched experiential interactions [11]. Park and Stangl highlight AR's role in bridging the information gap for travelers, offering real-time access to travel information which enhances the overall travel experience [12]. AR is understood as one more gastronomic marketing tool, where it is used to attract new customers through innovative methods such as storytelling, virtual games, interactive menus described by chefs, and real-time modifications of food and beverage presentations [13]. This technological integration not only changes how consumers perceive and interact with food but also collects valuable data on navigation, usage, and consumption patterns of the guests, enhancing business intelligence and customer engagement [14]. Loureiro et al. (2020) conducted a comprehensive review of 56 articles and 325 conference proceedings, illustrating that AR and Virtual Reality (VR) significantly increase the immersion of tourists in virtual environments [15].

A practical implementation of AR in tourism is seen in the BRENDA Digital Tours application, which uses gamified AR to enhance gastronomy tourism. This application provides an interactive platform for tourists to explore local cuisine, integrating challenges and rewards that encourage deeper engagement with local food scenes. The application has effectively driven business to local food venues and promoted cultural exchange, leading to increased tourist satisfaction and local economic benefits. Through this technology, BRENDA has enriched the travel experience by making the discovery of local gastronomies both fun and educational [16]. Similarly, a Mobile Augmented Reality (MAR) application developed to enhance tourists' experiences with local food consumption demonstrated significant user engagement. Tourists using the app were highly motivated to interact with local food, accessing information like food names, ingredients, and food miles. The gamification elements within the app, such as collecting badges and earning levels based on local food consumption, further heightened this engagement. Users reported increased pleasure and autonomy in exploring local food options, underscoring the application's effectiveness in delivering an adventurous, competent, and autonomous user experience [17]. Therefore, AR technologies in gastronomy tourism contribute significantly to the enhancement of travel experiences by making them more interactive and enriched. These advancements illustrate the potential of AR to transform industry practices, providing innovative methods for attracting and retaining consumers.

2.3 Tai Lue Foods

Tai Lue cuisine, an integral aspect of the cultural identity of the Tai Lue people living in northern Thailand, exhibits a rich blend of historical influences and culinary traditions. This ethnic group, originating from Xishuangbanna in China's Yunnan province, has managed to retain its distinct cultural characteristics while integrating into Thai society. Their unique culinary culture particularly stands out as a potential attraction for domestic tourists [18, 19]. Local vegetables and herbs grown in kitchen gardens or foraged from nearby forests are commonly used in their dishes, which are generally low in fat and emphasize the freshness of ingredients [20]. This approach to food preparation not only underscores their commitment to sustainability but also contributes to the health and longevity of the community [21, 22]. Culinary practices among the Tai Lue people are a testament to the preservation of their identity, with recipes passed down through generations. These recipes, rooted in the wisdom of their forefathers, are characterized by simplicity and a reliance on locally sourced natural ingredients. The traditional diet includes a variety of dishes that reflect the everyday lifestyle and heritage of the Tai Lue people.

Despite its rich cultural significance, Tai Lue cuisine remains relatively underexplored and underrepresented. Wungrath, along with researchers and local villagers, has worked to classify Tai Lue food into several categories based on preparation methods and traditional recipes. This classification includes eight types of traditional Tai Lue dishes: curry and boiled food: includes diverse varieties like mushroom curry, pork curry, vermicelli curry, and mullet soup. Stir-fried food: features dishes such as stirfried pork with basil leaves, Pad Thai, and stir-fried vegetables. Fried food: encompasses fried pork, fried beef, fried chicken, and other fried items. Grilled food: known locally as Karchi, includes grilled pork, grilled beef, and grilled vegetables. Spicy food: includes bamboo shoot salad, mango salad, and various traditional salads like Ang Thae leaf medicine. Laab dishes: often involves raw or cooked meats, sometimes incorporating fresh animal blood. Chilli paste food: features a variety of chilli pastes such as Fish chilli paste, Crab chilli paste, and Ong chilli paste. Snacks or sweet foods: includes traditional desserts and snacks like pad desserts and ramfuen rice. Other foods: comprises items like dry noodles and steamed rice [23].

2.4 Design and Development of the AR Book

To design and develop the AR book, we employed and adapted a methodology [30, 31] for developing augmented reality for cultural heritage. This methodology comprises several steps: Conceptualization, Content Selection and Scripting, AR Technology and Platform, User Experience (UX) Design, and Development and Testing. Each phase plays a crucial role in ensuring that the AR book not only functions effectively but also engages users meaningfully with cultural heritage content. An overview of this method is illustrated in Fig. 1.

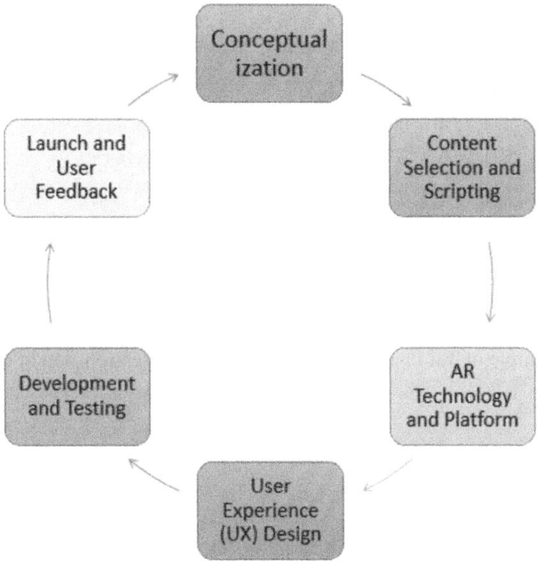

Fig. 1. Overview of the steps in the Design and Development of the AR Book.

2.5 Conceptualization

The initial phase of creating the augmented reality (AR) book commenced with a thorough conceptualization process. This process involved consultations with cultural historians, Tai Lue culinary experts, and AR technology specialists in collaboration with the Faculty of Humanity and the College of Art and Media Technology at Chiang Mai University. The primary goal was to ensure that the book would serve not merely as a technological novelty but as a genuine educational tool that respects and accurately represents the culinary traditions of the Tai Lue people. To achieve this, a multidisciplinary approach was adopted to merge the rich heritage of Tai Lue foods with the latest AR features. The aim was to create an immersive and informative experience for tourists traveling in Chiang Mai.

2.6 Content Selection and Scripting

Selecting content for the AR book necessitated meticulous consideration of which aspects of Tai Lue cuisine would most captivate and educate users. This selection process adhered to the principles of authenticity, educational value, and interactive potential. Dishes were chosen for their cultural significance, uniqueness, and the stories behind their origins and ingredients. Accompanying each selected dish, detailed scripts were crafted, encompassing historical backgrounds, recipes, and the cultural context of their consumption within the Tai Lue community. An example of Tai Lue cuisine is illustrated in Fig. 2.

Fig. 2. Example of Tai Lue cuisine for the content selection step of development

2.7 AR Technology and Platform

The technical foundation of the AR book was anchored in the robust AR platform provided by uniteAR, which allowed for the seamless integration of text, images, and videos within interactive AR experiences. This platform was chosen for its user-friendly interface, wide compatibility with various devices, and its ability to support high-quality, interactive 3D models of Tai Lue dishes. Users are afforded the opportunity to virtually interact with these dishes, enriching their understanding of ingredients and preparation methods through a compelling combination of augmented reality visuals and narratives. Furthermore, the development of 3D models utilized MeshLab for creating photogrammetry-based models, as cited in sources [29] and [29], with subsequent image refinement performed in Adobe Photoshop. The collection of images used to generate Tai Lue cuisine models is presented in Fig. 3, followed by their reconstruction using MeshLab software.

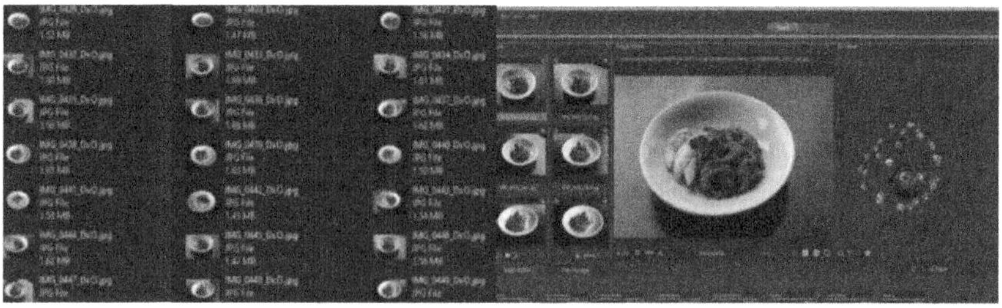

Fig. 3. Collection of images used to generate Tai Lue cuisine models and their reconstruction using MeshLab software.

2.8 User Experience (UX) Design

The user experience (UX) design of the AR book was strategically focused on crafting an intuitive and engaging journey for users. The layout was meticulously organized to facilitate users' culinary exploration of Tai Lue culture, ensuring effortless navigation across various sections and dishes. Interactive features were specifically developed to motivate users to engage more deeply with the stories of each dish, incorporating augmented reality visuals to vividly present the preparation processes and cultural significance of the dishes. Moreover, significant consideration was afforded to making the AR book accessible to a diverse audience, including non-native speakers and individuals with disabilities. The user interface of the AR book is depicted in Fig. 4.

Fig. 4. The user interface of the AR book.

2.9 Development and Testing

The development phase encompassed the iterative creation of the AR book's features, ranging from basic functionality to the incorporation of sophisticated AR elements. Continuous testing with potential users was essential for gathering feedback on usability, engagement, and the educational impact of the book. This feedback proved critical in refining the AR experiences, enhancing navigation, and ensuring that the content was both informative and engaging. Additionally, testing included evaluations of the AR book's performance across various devices and operating systems to ensure widespread accessibility. An example of the book is shown in Fig. 5.

2.10 Launch and User Feedback

Upon completion, the AR book was launched with the aim of reaching both locals and tourists interested in Tai Lue cuisine and culture. An initial feedback loop was established

Fig. 5. Cover of the AR book and an example page from the book.

to collect user impressions and suggestions for improvement. This ongoing process of feedback collection is crucial for the iterative enhancement of the AR book, ensuring it continues to meet the needs of its audience and remains a dynamic tool for cultural education and tourism promotion.

3 Research Methodology

This research adopts an R&D methodology [26, 27], focusing on the iterative development and testing of an augmented reality (AR) book aimed at promoting Tai Lue gastronomy tourism. The core of the research methodology is a pre-post test design, utilized to assess the change in participants' knowledge about Tai Lue cuisine before and after interacting with the AR book. This approach allows for a focused examination of the educational impact of the AR tool.

3.1 Participants

The study engaged a diverse group of participants, specifically tourists with an interest in local culinary culture. The selection criteria included being aged 20 years and above, having an interest in gastronomy tourism, and demonstrating a willingness to participate in the study. A total of 45 participants were successfully recruited. It is important to note that each participant possessed their own mobile phone and had an understanding of augmented reality technology. The demographics are shown in Table 1.

3.2 Instruments and Measurement

To assess the impact of the augmented reality (AR) book on participants' knowledge and engagement with Tai Lue cuisine, several instruments were employed. Notably, the

Table 1. Demographic of participants.

Categories	Data
Sample Size(N)	45
Mean Age (S.D.)	31.15 (5.79)
Male (%)	29 (64.44%)
Female (%)	16 (35.55%)
Has experience using augmented reality (%)	43 (95.55%)
No experience using virtual reality (%)	2(4.44%)

Preand Post-Test Questionnaire emerged as a key tool. This questionnaire consisted of a carefully curated set of structured questions designed to measure participants' comprehension of Tai Lue cuisine, both prior to and following their interaction with the AR book. The format of the questionnaire included 30 items, facilitating a thorough evaluation of the shifts in participants' knowledge. The questionnaire was developed collaboratively by lecturers at Chiang Mai University and experts in Tai Lue cuisine, ensuring both academic rigor and cultural authenticity.

3.3 Procedure

The research methodology unfolded in several distinct phases to comprehensively evaluate the impact of the augmented reality (AR) book on participants' knowledge of and engagement with Tai Lue cuisine. The process began with the recruitment and consent phase, where participants were sought through announcements in local community centers, tourist information boards, and social media platforms associated with Chiang Mai University. Interested individuals received a thorough briefing on the study's objectives and provided informed consent before participating. In the pre-test assessment phase, participants filled out a questionnaire designed to gauge their baseline knowledge of Tai Lue cuisine. During the subsequent AR book interaction phase, they were given access to the AR book along with instructions for its use, enabling them to explore its contents and features independently. This phase aimed to promote engagement with the various aspects of the AR book related to Tai Lue cuisine. The post-test assessment phase then saw participants complete a knowledge questionnaire to measure any changes. Lastly, the data analysis phase involved the use of statistical software to analyze the questionnaires and surveys, identifying significant shifts in the participants' knowledge and patterns of engagement.

4 Results and Data Analysis

The analysis of the intervention group's knowledge improvement on Tai Lue cuisine, facilitated through the use of an augmented reality (AR) book, was assessed via a pretest/post-test approach and analyzed using a paired sample t-test shown in Table 2.

Initially, the pre-test scores among the 45 participants averaged at 8.35 (SD = 4.37), indicating a baseline level of knowledge. Post-intervention, there was a noticeable increase in knowledge, with post-test scores averaging at 13.55 (SD = 5.52). This significant improvement in scores was statistically validated by the paired sample t-test, which yielded a t-value of –5.296 with 44 degrees of freedom and a highly significant p-value of less than .001. This result, marked by a p-value lower than 0.01, unequivocally demonstrates the AR book's effectiveness in augmenting participants' understanding of Tai Lue cuisine. The substantial rise in mean scores from pre-test to post-test, underpinned by the statistical significance of the t-test, attests to the educational potency of the AR book, showcasing its utility in delivering immersive and enriching learning experiences about Tai Lue culinary traditions.

Table 2. Analysis of Results from Pre/Post Tests and Paired sample t-test for intervention group.

	N	Mean	SD	T	df	P-Value
Pre-test	45	8.35	4.37	-5.296	44	< .001*
Post-test	45	13.55	5.52			

Note * P-value less than 0.01

5 Discussion

The utilization of an augmented reality (AR) book to elevate knowledge about Tai Lue cuisine has yielded significant results, indicating a marked improvement in participants' understanding before and after engagement with the AR content. This progression underscores the potential of AR as a powerful educational medium, capable of transforming traditional learning paradigms. The immersive nature of AR, which integrates augmented visuals and narratives, facilitates a deeper connection with the subject matter, suggesting that AR could play a pivotal role in cultural education and preservation, especially in the realm of gastronomy tourism.

Moreover, the study's findings highlight the broader applicability of AR technology in enhancing engagement and learning outcomes in cultural and educational settings. By making learning interactive and engaging, AR technology can significantly contribute to the promotion and preservation of intangible cultural heritage, offering users an innovative way to explore and appreciate diverse cultural traditions.

6 Conclusion, Limitation and Future Work

The study unequivocally showcases the effectiveness of augmented reality (AR) technology in bolstering the educational experience, particularly in the domain of Tai Lue cuisine. The AR book, through its immersive learning environment, has significantly enhanced participants' knowledge, positioning itself as a noteworthy asset in the realm of educational technology. It holds promising potential for contributing to the preservation of cultural heritage and the advancement of gastronomy tourism. However, the study's outcomes are tempered by limitations, including a focus on a specific demographic and geographic area, which may impact the generalizability of the results. Moreover, the investigation into short-term knowledge acquisition does not extend to the evaluation of long-term retention or the influence on participants' culinary habits and tourism behaviors.

Future research endeavors should strive to overcome these constraints by broadening the participant base and examining the enduring effects of AR-facilitated learning on knowledge and behavior. Enhancing the AR book with a wider array of cultural and culinary content could further increase its relevance and appeal. Additionally, ensuring the accessibility and usability of AR technology for diverse age groups and individuals with disabilities is critical, aiming for educational tools that are universally accessible. Ultimately, the AR book marks a significant stride towards employing technology in cultural education and preservation. With the ongoing evolution of AR technology, its capacity to enrich educational experiences and foster global cultural appreciation is vast. Future initiatives ought to capitalize on this technology, exploring new methods for its integration into educational frameworks and broadening its impact to deepen global understanding of rich cultural heritages.

References

1. Pavlidis, G., Markantonatou, S.: Gastronomic tourism in greece and beyond: a thorough review. Int. J. Gastron. Food Sci. **21**, 100229 (2020)
2. World Tourism Organization (UNWTO), Basque Culinary Center eds: Guidelines for the development of gastronomy tourism. World Tourism Organization (UNWTO) (2019)
3. Ellis, A., Park, E., Kim, S., Yeoman, I.: What is food tourism? Tour. Manag. **68**, 250–263 (2018)
4. Rachão, S.A.S., de Jesus Breda, Z., de Oliveira Fernandes, C., Joukes, V.N.P.M.: Drivers of experience co-creation in food-and-wine tourism: an exploratory quantitative analysis. Tour. Manag. Perspect. **37**, 100783 (2021)
5. Prayag, G., Gannon, M.J., Muskat, B., Taheri, B.: A serious leisure perspective of culinary tourism co-creation: the influence of prior knowledge, physical environment and service quality. Int. J. Contemp. Hosp. Manag. (2020). ahead-of-print
6. Manola, M., Koufadakis, S.X.: The gastronomy as an art and its role in the local economic development of a tourism destination: a literature review. SPOUDAI J. Econ. Bus. **70**, 81–92 (2020)
7. Cabaguing, J.M., Lacaba, T.V.G.: Gastronomy and hospitality enterprise image in Guiuan Eastern Samar, Philippines. iRAPA Int. J. Bus. Stud. **2**, 01–06 (2021)
8. McMillan, K., Flood, K., Glaeser, R.: Virtual reality, augmented reality, mixed reality, and the marine conservation movement. Aquat. Conserv. **27**, 162–168 (2017)
9. Blattgerste, J., Renner, P., Pfeiffer, T.: Advantages of eye-gaze over head-gaze-based selection in virtual and augmented reality under varying field of views. In: Proceedings of the Workshop on Communication by Gaze Interaction. ACM, New York, NY, USA (2018)
10. Tsai, S.-P.: Augmented reality enhancing place satisfaction for heritage tourism marketing. Curr. Issues Tourism. **23**, 1078–1083 (2020)
11. Jiang, S., Moyle, B., Yung, R., Tao, L., Scott, N.: Augmented reality and the enhancement of memorable tourism experiences at heritage sites. Curr. Issues Tourism. **26**, 242–257 (2023)

12. Park, S., Stangl, B.: Augmented reality experiences and sensation seeking. Tour. Manag. **77**, 104023 (2020)
13. Spence, C., Okajima, K., Cheok, A.D., Petit, O., Michel, C.: Eating with our eyes: from visual hunger to digital satiation. Brain Cogn. **110**, 53–63 (2016)
14. Batat, W.: How augmented reality (AR) is transforming the restaurant sector: investigating the impact of "Le petit chef" on customers' dining experiences. Technol. Forecast. Soc. Change. **172**, 121013 (2021)
15. Loureiro, S.M.C., Guerreiro, J., Ali, F.: 20 years of research on virtual reality and augmented reality in tourism context: a text-mining approach. Tour. Manag. **77**, 104028 (2020)
16. Deliyannis, I., Poulimenou, S.M., Kaimara, P., Laboura, S.: BRENDA digital tours: designing a gamified augmented reality application to encourage gastronomy tourism and local food exploration. In: Cultural Sustainable Tourism, pp. 101–109. Springer, Cham (2022)
17. Lee, J.: Designing a gamified augmented reality application for tourists to encourage their local food consumption (2019)
18. Boonrueng, L.: Tai Lue community identity and its extension to cultural tourism at ban Luang nuea. Chiang Mai. ABAC J. **42**, 179–201 (2022)
19. Ketsuwan, P., Bejrananda, J., Pramoon, P.: The development of Tai Lue community in north of Thailand with cultural innovation for quality of life and cultural tourism supports. J' Positive School Psychol. **6**, 9595–9601 (2022)
20. Culloty, D.: Food from Northern Laos: The Boat Landing Cookbook. Galangal Press, Wellington (2010)
21. D'Sylva, A., Beagan, B.L.: 'Food is culture, but it's also power': the role of food in ethnic and gender identity construction among Goan Canadian women. J. Gend. Stud. **20**, 279–289 (2011)
22. Quintero-Angel, M., Mendoza, D.M., Quintero-Angel, D.: The cultural transmission of food habits, identity, and social cohesion: a case study in the rural zone of Cali-Colombia. Appetite **139**, 75–83 (2019)
23. Wungrath, J.: Knowledge management for local wisdom of Tai Lue food. Turk. Online J. Qual. Inq. **12**, 1205–1211 (2021)
24. Intawong, K., Olson, D., Chariyalertsak, S.: Application technology to fight the COVID-19 pandemic: lessons learned in Thailand. Biochem. Biophys. Res. Commun. **534**, 830–836 (2021)
25. Intawong, K., Ong-artborirak, P., Boonchieng, W.: Seamless electronic medical record for health management system in emergency patients. In: 2021 Joint International Conference on Digital Arts, Media and Technology with ECTI Northern Section Conference on Electrical, Electronics, Computer and Telecommunication Engineering. IEEE (2021)
26. Gustiani, S.: Research and development (R&D) method as a model design in educational research and its alternatives. Holistics (Hospitality and Linguistics): Jurnal Ilmiah Bahasa Inggris. **11** (2019)
27. Intawong, K., Scuturici, M., Miguet, S.: A new pixel-based quality measure for segmentation algorithms integrating precision, recall and specificity. In: Computer Analysis of Images and Patterns, pp. 188–195. Springer, Heidelberg (2013)
28. Mulcki, R., Rachel, F., Laksana, H.: Preservation of West Java culinary recipe using augmented reality, photogrammetry, and storytelling illustration. IJoCED. **5**, 78–88 (2023)
29. Li, M., Xu, S., Tang, J., Chen, W.: Design and research of digital twin platform for handicraft intangible cultural heritage -Yangxin Cloth Paste. Herit. Sci. **12** (2024)
30. Cranmer, E.E., tom Dieck, M.C., Jung, T.: The role of augmented reality for sustainable development: evidence from cultural heritage tourism. Tour. Manag. Perspect. **49** 101196 (2023)

31. Boboc, R.G., Băutu, E., Gîrbacia, F., Popovici, N., Popovici, D.-M.: Augmented reality in cultural heritage: an overview of the last decade of applications. Appl. Sci. (Basel). **12**, 9859 (2022)

Rail Infrastructure Management Optimization Through Digital Twins: Focus on Data Reliability and Quality

Moussa Issa[1(✉)], Xinyuan Zhai[2], and Guillaume Ducellier[2]

[1] SNCF RESEAU, Saint-Denis, France
ext.moussa.issa@reseau.sncf.fr
[2] Univsity of Technology of Troyes, Troyes, France
{xinyuan.zhai,guillaume.ducellier}@utt.fr

Abstract. The digital twin of the rail system plays a pivotal role in advancing infrastructure management, ushering in greater punctuality and safety for trains. By harnessing data from diverse sources, it facilitates real-time updates throughout the lifecycle of railway assets. This comprehensive digital twin encompasses every aspect of infrastructure management, offering multifaceted functionalities beyond mere asset oversight. These include providing insights into the current state of the rail system and furnishing simulation tools to bolster decision-making processes across all stakeholders.

Given the fundamental reliance on digital data within this framework, our primary focus is on ensuring its usability and quality. Accurate analysis of asset conditions hinges upon the trustworthiness of the data, a requirement shared by maintainers and operators alike. Consequently, we've diligently crafted a set of digital data confidence indicators aimed at instilling trust in the information at hand.

In tandem with the endeavors to develop a digital twin for the rail system, there arises a pressing need to establish digital solutions that capitalize on data pertinent to business objectives. This necessity is underscored by the efforts of SNCF's network description department in modeling both processes and data. However, the efficacy of these digital solutions is contingent upon the relevance, efficiency, and high quality of the data they rely on. To this end, we've embedded the notion of trust within the digital data used to delineate the condition of railway assets.

While these indicators amalgamate various factors impacting user confidence in digital data, further expansion is warranted to ensure the widespread effectiveness and adoption of digital tools by end-users.

Keywords: Digital Twin · Data Processing · Transport · Asset Lifecycle management · Railway Infrastructure · Data Reliability · PLM · BIM

P. Sureephong et al. (Eds.): PLM 2024, IFIP AICT 741, pp. 84–97, 2025.
https://doi.org/10.1007/978-3-031-93323-3_7

1 Introduction

The digital twin of the French railway system promotes better infrastructure management. This will enable more trains to run on time and arrive safely. This digital twin operates thanks to data from heterogeneous sources, allowing it to be updated in realtime during the lifecycle of railway assets. As the data that fuels it is at the core of this digital twin, we have focused on ensuring its quality. To achieve accurate analysis of our assets' condition, we must ensure that the data they contain is trustworthy and of high quality. It is with this objective in mind that we have begun to work on implementing a set of digital data confidence indicators. In the realm of transportation, for more effective management of physical assets without the need for on-site visits and thus saving time, asset managers rely on various sources of information (drone images, LiDAR, BIM, etc.). The emergence of these sources has placed information at the heart of connected assets [1]. The concept of connected assets is particularly highlighted in urban planning with the notion of smart cities, described by [2] as instrumented, interconnected, and living cities. Managing the lifecycle of railway assets requires a thorough understanding of these assets, and therefore high-quality data describing their states. That is why, in this article, we start with a literature review on confidence indicators in a comprehensive manner. Next, we will present the proposed method and the results obtained, and finally, we will discuss the findings in the conclusion.

2 Literature Review

Indicators, in a general sense, are popular for establishing rankings. There are numerous examples today, such as sports competition results, university rankings, rankings of the best cities to live in, top hospitals, highest crime rates, and so on) [3].

One of the most well-known indicators today is that of sustainable development, as it allows for the measurement, in a broad sense, through both quantitative and qualitative techniques among others, of the socio-economic development of a country, thus enabling it to better manage its development strategy [4]. Increasingly, data availability, meaning obtaining the value of sustainability indicators, is not an issue. The main challenges lie in the interpretation and utilization of the indicators.

Researchers have focused on the issue of data trust in general within the industrial world. [5] propose a framework to address challenges in data management, security, and predictive maintenance in what is known as the Industrial Internet of Things or IIOT. [6] tackle the issue of uncertainty and data trust in 3D data and digital models for reverse engineering. On the other hand, have developed a methodology for calculating Key Performance Indicators for industry. These indicators aim to increase confidence in decision-making calculated based on Dempster-Shafer belief function theory [7], which addresses the calculation of uncertainties in data.

Confidence indicators incorporate a notion of uncertainty into their expression. This requires modelling the uncertainty associated with the credibility of the measured information before implementing potentially costly actions.

Outside the industrial domain, in the medical field, the issue of data trust and performance becomes particularly relevant. It is indeed essential to have trustworthy data when ensuring an adequate diagnosis for the patient. For example, [8] have implemented a metrology-based approach that allows for an objective evaluation of data quality, extends data lifespan, enhances their utility, and ensures compliance with FAIR principles (a set of guidelines to make data Findable, Accessible, Interoperable, and Reusable).

The calculation of confidence scores in data is inherently linked to their sources. This is why [9] rely on a data provenance trust model that considers various factors that may affect reliability. Based on these factors, it assigns confidence scores to the data and another confidence score to the data sources. These confidence scores provide key information based on which data users can decide whether to use them and for what purposes.

In the medical domain, according to [10], the concept of data confidence must be based on two fundamental notions: data consistency and truthfulness. Through various techniques that provide an indicator of the truthfulness (rated on a scale of 7 values) of the data, it becomes possible to establish an early warning system based on the patients' health status.

Similarly to the medical and industrial domains, in the economic industry, researchers, for example, [11] have focused on data confidence indicators, such as calculating the Business Confidence Index (BCI) based on economic indicators such as GDP, producer prices, and unemployment rates (see Fig. 1).

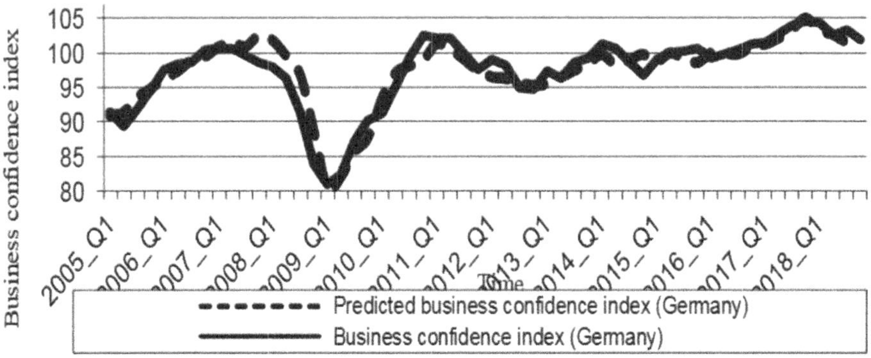

Fig. 1. Business confidence Indicator

The industrial confidence indicator describes the measure of the health status of a country's industry by identifying factors contributing to the formation of this confidence indicator through the development of classical and multivariate discrete regression models [12].

In the railway domain, the question of data confidence depends on the level of digital transformation of each infrastructure manager. The aim of this article is to demonstrate the work done to establish a data confidence indicator for the digital twin of the railway system. In the following sections, we will detail the data confidence indicators that we deemed relevant.

3 Proposed Methodology

In this section we will talk about the basic concept of this methodology, then we will give more details about the formulae's used during the use case and then some results will be shown at the end.

3.1 Basic Concept

For our use case we needed to improve data Reliability and Quality before making decisions. After the studies that we have done, we have been convinced that the data confidence indicators are the key method to measure the degree of the reliability of a piece of data. To make it fully functional, we would first need to define the theories and their corresponding formulas. Secondly, we would need to implement them into an information system ReactJS-Django-PostgreSQL system in our case, that will be provided with more details in Sect. 4.

Hence in the following sub-sections, we will start with the explanation of the indicators that we have defined. Then, several chosen parts of program code will be shown as a demonstration of implementation.

3.2 Formulae

To determine if a given piece of data is reliable and to what extent it is reliable, we need methods to quantify its reliability – thus the individual confidence score. This individual confidence score being out of 10, consists of different indicators that cover all different aspects of the confidence. We have identified 6 indicators in total:

- **Data source consistency:**

  This indicator verifies a given piece of data's consistency of source with other data within its dataset. In the case of SNCF Réseau, if a given piece of data's source differs from others, then the format and degree of precision may vary as well. Therefore, the formula:

$$Consistency = \frac{nsame}{n_{all}} * 10 \tag{1}$$

For example: A dataset has 20 pieces of data in total, among them 19 are of the same data source and 1 is of a different data source. Thus, for the latter, its data source consistency score is 1/20 * 10 = 0.5.

- **Data source reliability:**

  As mentioned in the first indicator, different data sources may simply have different degrees of precision and formats. Hence their reliability differs as well. This indicator reflects hence such difference:

$$Reliability_{source} \tag{2}$$

For example: data source A has a score of 5 and data source B has a score of 7.

- **User Rating:**

Users may as well provide their individual ratings for data confidence. 0 being data unexploitable, and 10 being data completely exploitable.

$$Rating \tag{3}$$

This score can be obtained through questionnaires or direct rating.

- **Data obsolescence:**

$$if t_{current} - t_{update} \leq \Delta\_\max : Obsolescene = 10 - 10 * \frac{t_{current} - t_{update}}{\Delta_{max}} \tag{4}$$

$$\text{else: } Obsolescence = 0 \tag{5}$$

This indicator allows to verify if a piece of data is obsolete, and if so, to what extent it is obsolete. A key variable in this indicator is the maximum range of acceptance (Δ_{max}): if the difference between current time and the time that the data was updated is bigger than the maximum range, then the obsolescence score is 0. Otherwise, it will be proportionate to the amount of time. For example: a piece of data was updated the 2 of April, the date that the piece of data is studied is 20 of April. The maximum range of acceptance is 20 days. Hence Obsolescence = 10 – 10 * 18/20 = 1.

- **Completeness:**

$$Completeness = \frac{n_{filled}}{n_{total}} * 10 \tag{6}$$

This indicator examinates the number of empty fields in a piece of data, if any.

For example: the number of filled fields in a piece of data is 9, and there are 20 fields in total. Therefore Completeness = 9/20 * 10 = 4.5

- **Plausibility:**

$$Plausibility \tag{7}$$

This indicator allows to study to what extent a piece of data is reasonable if it is contradictory with other data and if certain fields of the data are contradictory with others.

Due to its complexity, we are convinced that there is no direct formula that allows its calculation. Its determination is therefore supposed to be done with machine learning.

For example: A train's described maximum speed is 3000 km/h, this data is therefore implausible.

The confidence score can be hence obtained with the following formula:

$$\begin{aligned} Score_{confidence} &= c_1 * Consistence + c_2 + Reliability_{source} + \\ &c_3 * Rating + c_4 + Obsolescence * c_5 * Completeness + c_6 + Plausibility \end{aligned} \tag{8}$$

With:

$$c_1 + c_2 + c_3 + c_4 + c_5 + c_6 = 1 \tag{9}$$

3.3 Program Code

The calculation of all data confidence indicators is performed in the backend environment, thus with Python and Django. Each time the user selects a piece of data available on the interface of frontend environment, it with other data under the same criteria will be transmitted to the backend along with selected indicators. In addition, the calculation of plausibility is not included in program code, as it shall be obtained through data analysis.

- **Data source consistency and data source reliability:**

```
fiabiliteSource = 0.
    coherenceSource = 0
    dataSource = selectedData.get('sourceDonnee', None)
    if dataSource is not None:
        if (dataSource.lower() == 'gaia'):
            fiabiliteSource = 9
        elif (dataSource.lower() == 'armen'):
            fiabiliteSource = 8
        elif dataSource.lower() in ['saisie manuel',
'saisiemanuel', 'user input', 'userinput']:
            fiabiliteSource = 5
        else:
            fiabiliteSource = 1
        countSameSource = 0
        countTotalItem = len(markerData)
        for item in markerData:
            if item.get('sourceDonnee').lower() == data-
Source.lower():
                countSameSource += 1
        coherenceSource = countSameSource/countTotalItem
* 10
```

In our example, we have suggested that the SNCF's database "GAIA" has a confidence score of 9, "ARMEN" has 8, user input is 5 and all other unidentified data sources have a confidence score of 1. The data source consistency is then calculated based on all existing data with same user chosen criteria.

- **User rating:**

```
    note = selectedData.get('noteUtilisateur', None)
```

As note is an existing column in the selected data, we only have to retrieve the rating value of the selected data.

– **Data obsolescence:**

```
obsolescence = 0
    deltaMax = 0
    today = date.today()
    today = datetime(today.year, today.month, today.day)
    equipementType = selectedData.get('equipementType',
None)
    if (equipementType == None):
        obsolescence = 0
    else:

        if equipementType.lower() in ['canton de pose',
'cantondepose']:
            deltaMax = 700

        if equipementType.lower() in ['appareil tendeur',
'appareilTendeur']:
            deltaMax = 1000

        dateMAJdonnees = selectedData.get('dateMAJdon-
nees', None)
        if (dateMAJdonnees == None):
            deltaMax = 0

        if (deltaMax != 0):
            print          ('today:',          today)
            dateMAJdonnees_datetime = datetime.strp-
time(dateMAJdonnees, '%Y-%m-%d')
            if dateMAJdonnees_datetime > today: # if the
update surpasses today then its simply not possible - an
error is present
                deltaMax = 0
            else:
                dateDifference = (today - dateMAJdon-
nees_datetime).days
                print('difference:', dateDifference)
                print('dateMAJdonnees_datetime:',
dateMAJdonnees_datetime)
                obsolescence = 10 - 10*(dateDiffer-
ence/deltaMax)
```

different equipment have different maximum ranges of acceptance, thus the different values for “deltaMax”. The indicator is then calculated based on the number of days between update date and the date of examination(today).

- **Completeness:**

```
totalFields = 0
    emptyFields = 0

    for key, value in selectedData.items():
        totalFields += 1

        if value == "": # true or false values shall not
be counted as empty
            emptyFields += 1

integralite = 10*(totalFields-emptyFields)/totalFields
```

Number of non-empty fields divided by the total number of fields. It is to notice that Boolean values are not considered as empty.

At the end after calculation, the results of calculations will be sent back to frontend and displayed to user:

```
responseData = {'noteUtilisateur': note,
'integralite': integralite,
'fiabiliteSource' : fiabiliteSource,
'obsolescence': obsolescence,
'coherenceSource': coherenceSource,
'plausibilite': plausibilite}
return Response(responseData)
```

4 Results: Application to Heterogeneous Data

Following the definitions of data confidence indicators and their corresponding Python algorithms, we have added more program codes for the frontend visualization (ReactJS, JavaScript), backend management (Django, Python) and the database (PostgreSQL).

Our concrete result consists of two sections: data visualization and data confidence. Data shall be requested and displayed in the data visualization section, and one piece of data selected by user will be analyzed in the data confidence section. Specifically, in the designed mode of operation, all existing databases of SNCF shall have their data synchronized with data visualization section. Such data, available under various forms, can be requested by an agent to be viewed under the same section, or analyzed for reliability inside of the data confidence section. Figure 3 provides an illustration of such functional mechanism (Fig. 2):

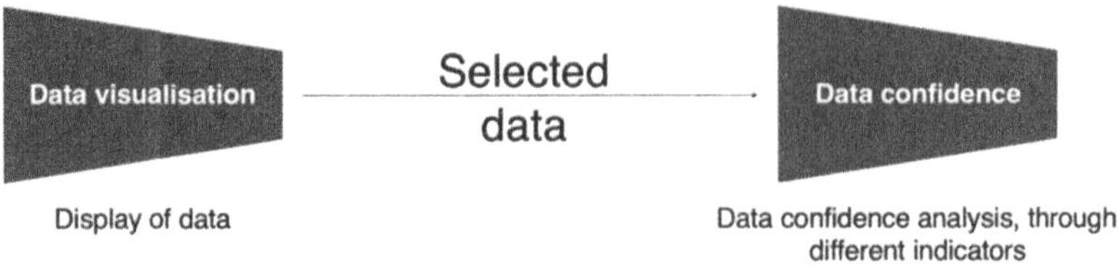

Fig. 2. Architecture of data confidence platform

4.1 Data Visualisation Section

An interface allowing the visualization of existing data can be useful before the evaluation of its confidence level. On demand, relevant equipment's data and its related railway data shall be introduced and displayed in this section:

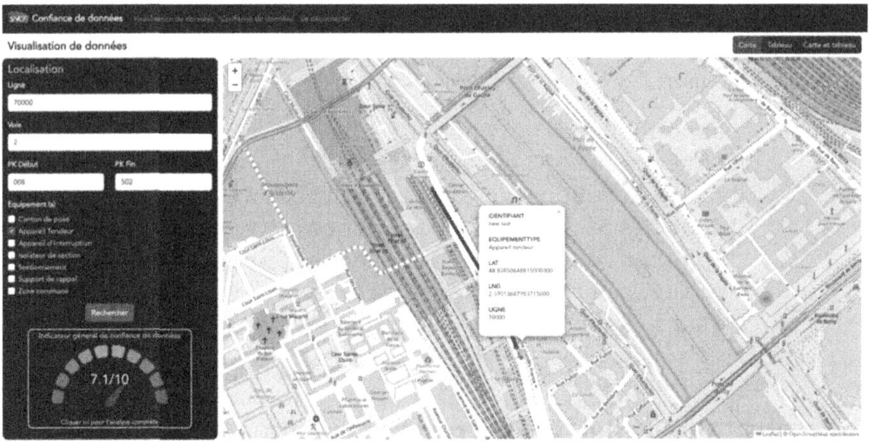

Fig. 3. Data visualisation interface

As seen in Fig. 3, this interface consists of multiple elements (Fig. 4):

- A navigation bar, allowing to pass from one section to another: data visualization, data confidence and connection page.

Fig. 4. Navigation bar

- A lateral bar, allowing to search for one or multiple specific pieces of data, with different fields for precision: line ("ligne"), track ("voie"), start kilometer point ("PK début"), end kilometer point ("PK fin") and the choice of equipment(s). Upon initiating the search function, the platform will launch the search of requested data in the database and retrieve all corresponding results (Fig. 5).

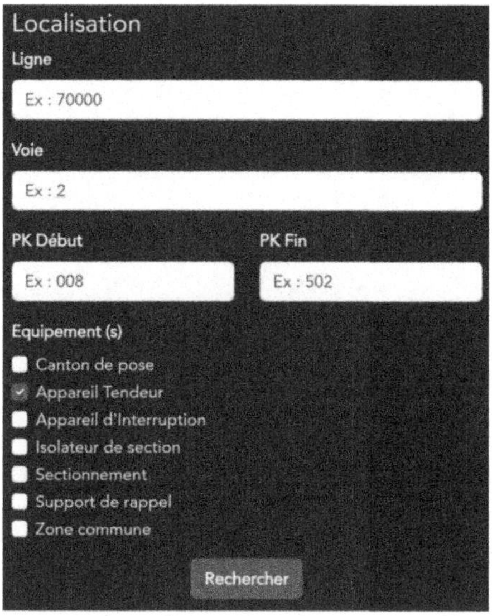

Fig. 5. Lateral bar of data visualization section.

– A general data confidence level indicator, as a result sent back from data confidence section (Fig. 6):

Fig. 6. Data confidence level indicator in data visualization section.

– The data visualization module, allowing the user to view the selected data (Fig. 7):

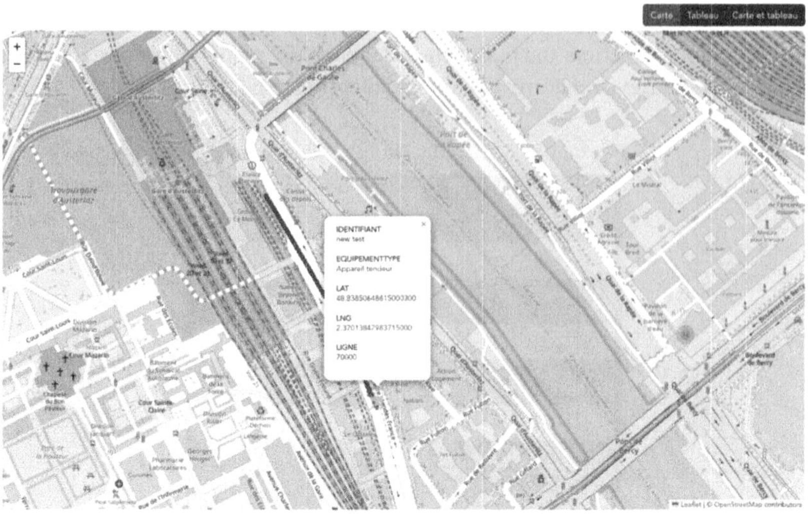

Fig. 7. Data visualization module

As seen in Fig. 8, this module allows the visualization of requested data on a map, along with relevant section(s) of railway. This module also supports visualization in different views:

- Table view:

Importer (CSV) Exporter

Appareil tendeur

Action	ID	Designation	Ligne	Voie	PKD	PKF	Environnement	Localisation	Spécialité	Type appareil tendeur	Type cable ancre	Date de pose ou de dernie
Modifier Supprimer	at-1	tendeur-1	70000	V2	008+665	502+001	environnement-1	ligne 70000	specialite0	type0	type0	2023-10-31

Fig. 8. Table view of data visualization module

This view allows showing all fields of selected data, as well as its modification and deletion.

- Table and map view (Fig. 9):

This view is a combination of map and table view, allowing both its visualization on the map, an overview of its fields and its modification and deletion.

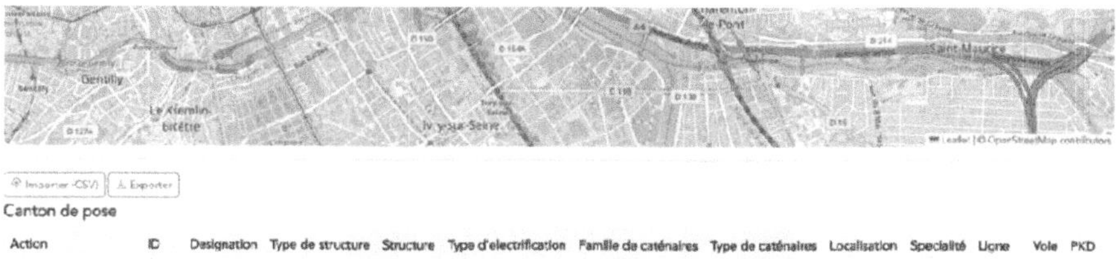

Fig. 9. Table and map view of data visualization section

4.2 Data Confidence Section

The other section of our platform is of data confidence, corresponding to the data confidence analysis of a selected piece of data, based on the indicators that we have previously established (Fig. 10).

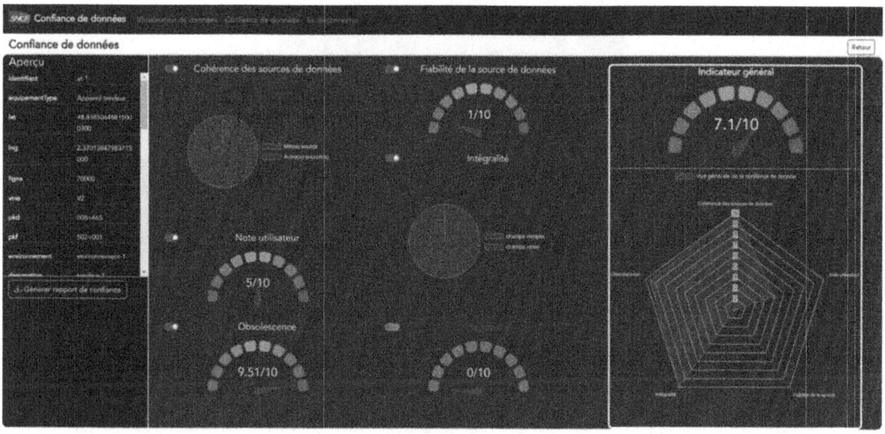

Fig. 10. Data confidence interface

This section consists of multiple subsections:

- A navigation bar, same element as in the data visualization section.
- An overview of selected data, including all its fields (Fig. 11):

- An analysis with various indicators (Fig. 12):

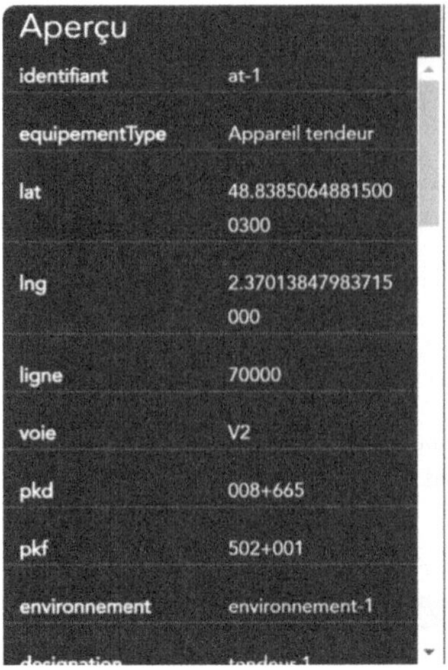

Fig. 11. Data overview of data confidence section.

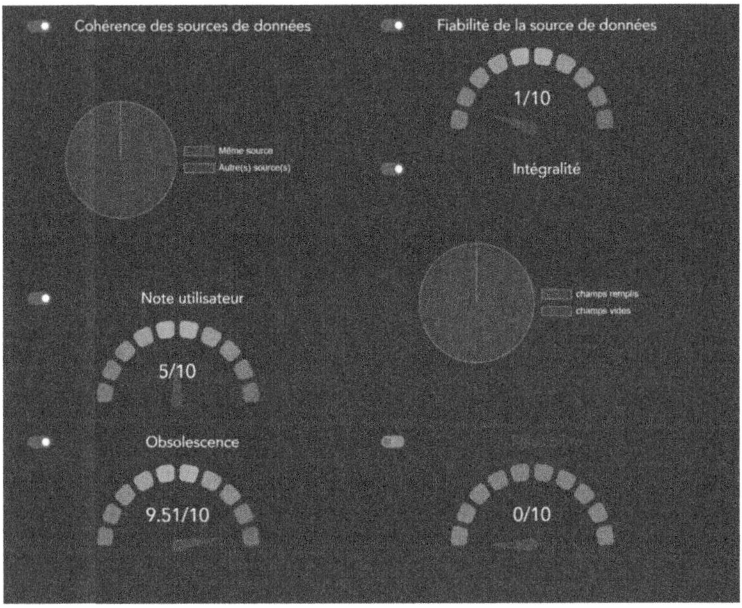

Fig. 12. Analysis with different indicators, in data confidence section.

5 Conclusion

The platform can be used as a prototype to demonstrate possible improvements in data trust assessment in digital twin environments.

In addition, the project has been designed with modularity in mind. If other trust indicators or functionalities are defined, they can be easily integrated into the platform architecture without having to be redone. So, to continue the project, it is entirely possible to develop it further so that it is not just a prototype, but a complete service accessible to all SNCF agents.

Turning to the subject of adding further functionality, several ideas have been conceived but not yet implemented for various reasons. One way forward could be to replace user evaluation measures with user questionnaires. Another possibility is to integrate a role system: the platform's functionality changes according to the user's role by using PLM concept.

References

1. Nylén, D., Holmström, J.: Digital innovation strategy: a framework for diagnosing and improving digital product and service innovation. Bus. Horiz. **58**(1), 57–67 (2015)
2. Harrison, C., et al.: Foundations for smarter cities. IBM J. Res. Dev. **54**(4), 1–16 (2010)
3. Moldan, B., Janoušková, S., Hák, T.: How to understand and measure environmental sustainability: indicators and targets. Ecol. Ind. **17**, 4–13 (2012)
4. Boulanger, P.M.: Les indicateurs de développement durable: un défi scientifique, un enjeu démocratique. Les séminaires de l'Iddri 12, pp. 1–24. IDDRI, Paris (2004)
5. Suhail, S., Hussain, R., Jurdak, R., Hong, C.S.: Trustworthy digital twins in the industrial internet of things with blockchain. IEEE Internet Comput. **26**(3), 58–67 (2021)
6. Ríos, J., Staudter, G., Weber, M., Anderl, R., Bernard, A.: Uncertainty of data and the digital twin: a review. Int. J. Prod. Lifecycle Manage. **12**(4), 329–358 (2020)
7. Gordon, J., Shortliffe, E.H.: The Dempster-Shafer theory of evidence. In: Rule-Based Expert Systems: The MYCIN Experiments of the Stanford Heuristic Programming Project, vol. 3, pp. 832–838. Addison-Wesley, Reading (1984)
8. Smith, N.A.S., Sinden, D., Thomas, S.A., Romanchikova, M., Talbott, J.E., Adeogun, M.: Building confidence in digital health through metrology. Br. J. Radiol. **93**(1109), 20190574 (2020)
9. Dong, W., Yang, B., Wang, K., Yan, J., He, S.: A dual blockchain framework to enhance data trustworthiness in digital twin network. In: 2021 IEEE 1st International Conference on Digital Twins and Parallel Intelligence (DTPI), pp. 144–147. IEEE, Beijing (2021)
10. Götzinger, M., Taherinejad, N., Rahmani, A.M., Liljeberg, P., Jantsch, A., Tenhunen, H.: Enhancing the early warning score system using data confidence. In: Wireless Mobile Communication and Healthcare. LNCS, vol. 192, pp. 91–99. Springer, Milan (2017)
11. Los, V., Ocheretin, D.: Construction of business confidence index based on a system of economic indicators. SHS Web Conf. **65**, 06003 (2019)
12. Mariana, G.: Confidence indicators analysis in the context of Romanian and European economy. Procedia Soc. Behav. Sci.Behav. Sci. **62**, 393–397 (2012)

Survey on Knowledge-Based Engineering Approaches for Complex Product Design: A Focus on Crowd Intelligence

Zhanxi Wang[1], Tengfei Sun[1], Qiongsheng Zheng[2], Kangning Wang[1], Shiqi Gao[1], Benoît Eynard[3], and Chen Zheng[1,2](✉)

[1] School of Mechanical Engineering, Northwestern Polytechnical University, 127 West Youyi Road, Xi'an 710072, Shaanxi, People's Republic of China
chen.zheng@nwpu.edu.cn

[2] Guangdong Provincial Key Laboratory of Advanced Welding Technology for Ships, CSSC Huangpu Wenchong Shipbuilding Company Limited, Guangzhou, People's Republic of China

[3] Department of Mechanical Systems Engineering, Université de Technologie de Compiègne, CNRS UMR7337 Roberval, CS 60319, 60203 Compiègne Cedex, France

Abstract. Modern complex product design is becoming a knowledge-intensive teamwork activity since participants from different disciplines have been involved. Therefore, it is essential to obtain multi-discipline knowledge effectively to enhance the design process. However, considering multimodal, multi-temporal, and multi-spatial characteristics of the design process, it is still challenging to take advantage of crowd intelligence knowledge. Although crowd intelligence knowledge has been studied and applied in many fields, there is still a lack of a systematic analysis of its characteristics, applications, and future directions. This paper presents a state-of-the-art review of the recent advancement in knowledge-based engineering (KBE) approaches for complex product design.

First, this paper discussed the characteristics of KBE approaches and concluded with a systematical definition of product design based on crowd intelligence knowledge. Second, the limitations of existing KBE approaches are pointed out by analyzing current single-discipline and multi-discipline KBE approaches. Third, two knowledge acquisition methods: knowledge query and knowledge push are presented, which are considered potential solutions for crowd intelligence design. Finally, the efficiency, accuracy, and real-time performance of the two methods are discussed, conclusions are drawn, and future research directions of KBE approaches are pointed out.

Keywords: Knowledge-Based Engineering (KBE) · Crowd Intelligence · Complex Product Design

1 Introduction

Modern product design gradually transforms from designer-leading activities into knowledge-intensive activities involving participants from various disciplines. With the development of knowledge acquisition methods such as the Internet of Things, big data,

Published by Springer Nature Switzerland AG 2025
P. Sureephong et al. (Eds.): PLM 2024, IFIP AICT 741, pp. 98–109, 2025.
https://doi.org/10.1007/978-3-031-93323-3_8

and sensing technology, it has been a focus in both academia and industry to enhance complex product design using the advantages of crowd intelligence [1]. Large-scale participants produce a large amount of complex design knowledge. The design knowledge must be accurately organized and obtained to drive the design process effectively and enhance the design activities. However, few useful pieces of knowledge could be extracted due to inadequate organization and acquisition mechanisms for multimodal, multi-temporal, and multi-spatial design knowledge. It has become a critical issue for project delay and failure [2]. Such a problem is particularly prominent for large-scale and complex products, such as advanced manufacturing equipment, and robotic manufacturing systems.

Crowd intelligence is not a simple superposition of individual knowledge. Baeck et al. highlighted that crowd intelligence may not be accurately obtained and utilized if the semantic relationship, knowledge background, and temporal and spatial environment between different kinds of knowledge are not fully considered [3]. Woolley et al. pointed out that the effective utilization of crowd intelligence knowledge needs the integration of various knowledge at modal, temporal, and spatial levels according to the purpose of tasks [4]. Therefore, considering the multimodal, multi-temporal, and multi- spatial characteristics of crowd intelligence knowledge, it has long been regarded as a significant challenge to study further into knowledge organization and acquisition methods to help designers obtain design knowledge effectively.

The structure of this paper is organized as follows. Section 2 summarizes the definition and fundamental characteristics of product design based on crowd intelligence knowledge. Section 3 discusses different organization methods for crowd intelligence knowledge. Section 4 examines current methods for knowledge acquisition. Section 5 concludes with a summary and introduces potential future research directions of complex product design based on crowd intelligence knowledge.

2 Crowd Intelligence-Based Product Design

Product design is a knowledge-driven creative activity. Early in the 1990s, 2-year research in six enterprises in the UK was carried out, which first revealed the critical role of knowledge engineering in improving product design quality and reducing development time and costs [5]. With the increase in product complexity, product design knowledge involves more disciplines. As a result, relying on the knowledge of designers alone is not enough. It has been a main development trend to complete product design using the crowd intelligence knowledge of engineers, enterprise managers, operators, and even relevant personnel in marketing, operation, logistics, sales, etc.

Crowd intelligence has advantages in dealing with complex product design problems [6]. Mecharek et al. proposed a design architecture with product architects, multi-discipline engineers, and users as the source of knowledge. They developed a design platform for knowledge sharing [7]. Kerzner emphasized the importance of project management in the design process and believed that the project managers' experience in cost, schedule, and performance is also an important source of design knowledge [8]. Bahill et al. further confirmed the stakeholders in product design and divided the sources of knowledge into 24 different kinds, such as users, project managers, and operators [9].

The German Aerospace Center and the Netherlands Aerospace Center carried out a complex aerospace product development project with the participation of multiple members covering the whole product supply chain [10]. In addition, research teams in universities also carried out technical research on multi-source knowledge- driven product design, mainly focusing on realizing organization, collaboration, and decision-making of multi-participants in complex product design [11, 12].

However, crowd intelligence represents multimodal, multi-temporal, and multi-spatial characteristics. Take aircraft drilling equipment in the aviation manufacturing industry as an example: new structures, materials, and future significant aircraft development processes change frequently. Designers not only need to master the basic engineering knowledge of electrical, mechanical, and control, follow the design theory and technical standards in various disciplines, but also need to understand that significant aviation components may have machining vibration rebound under the coupling action of large-scale, thin thickness and weak stiffness [1]. Therefore, engineers must provide machining process parameters and technical indicators about drilling, twisting, spot facing, and milling in combination with the deformation rebound mechanism [13]. Besides, operators must provide knowledge of drilling equipment installation, care, maintenance, and recycling throughout the product lifecycle according to actual working conditions.

In general, it can be concluded that the design knowledge provided by the designer group not only changes with the different perspectives of the group members and different stages of the design but also represents a high dependence on time and space. In addition, with the support of the Internet of Things, big data, sensors, and other knowledge acquisition technologies, crowd intelligence knowledge also exhibits multimodal characteristics. The knowledge provided by different members may exist in the knowledge model in a structured or semi-structured form and may be expressed in various modalities such as unstructured data, text, pictures, and natural language. The multimodal, multi-temporal, and multi-spatial characteristics of crowd intelligence knowledge create new demands for the organization and acquisition of design knowledge.

3 Knowledge Organization of Crowd Intelligence

The organization of crowd intelligence knowledge needs to represent the concepts and associations among concepts in a specific discipline to form a knowledge structure system with unified structure, semantics, and granularity so that the group members can acquire and understand the knowledge rapidly and accurately [14]. Recently, research on the organization of crowd intelligence has been carried out mainly from two perspectives: single-discipline and multi-discipline.

3.1 Single-discipline Knowledge Organization

Organization of crowd intelligence in a single discipline derived from parametric design [15]. In a single discipline, designers usually have the same discipline background and conceptual terminology, so it is easy to provide designers with a standard design knowledge architecture with consistent structure and semantics by integrating design parameters and design rules in a specific discipline.

Gent et al. integrated design parameters including area, chord length, wingspan, and design rules such as aspect ratio and tip-to-root ratio to form a wing mechanism knowledge architecture, which helped group members complete rapid design optimization of airfoil shape [16]. Zhang et al. constructed a knowledge architecture composed of parameters such as product structure, spatial location, assembly features, and assembly order to realize the final assembly sequence design of the product [17]. Guan et al. constructed a hull knowledge structure composed of hull engineering structural parameters and principles to support the rapid optimization design of the three-dimensional structure of the hull [18].

Generally, the single-discipline organization of crowd intelligence knowledge provides designers with a standard knowledge architecture with consistent structure and semantics. However, due to the different knowledge structures of various disciplines, it remains a challenge to organize and integrate knowledge from different disciplines.

3.2 Multi-discipline Knowledge Organization

Organization of multi-discipline crowd intelligence knowledge involves more complex knowledge than a single discipline [19]. Due to the variations in technical backgrounds, professional abilities, and knowledge structures among group members, it is a challenging task to achieve a unified representation, syntax, semantics, and granularity for designers through direct integration. Therefore, the multi-discipline design knowledge is usually organized through multi-discipline integration, which builds a unified design knowledge base by integrating existing knowledge models of different disciplines to help designers acquire knowledge in other disciplines.

Topp et al. integrated knowledge models from multiple disciplines, such as robotics, manufacturing process, design, and product component knowledge, and developed a multi-discipline knowledge base that can be used for the design of robotic manufacturing systems to help designers optimize system components and architecture for particular manufacturing tasks [20]. Liu et al. used ontology to unify the flow knowledge model describing hardware and the control knowledge model describing software and realized knowledge organization of hardware and software in the design of mechatronic products [21]. Qiao et al. integrated the existing product knowledge model and assembly knowledge model and designed the optimal assembly process for the products [22]. The team of authors realized the organization of the workpiece knowledge, machining process knowledge, robot knowledge, and programming knowledge to help programmers generate robot manufacturing programs that conform to the process flow of the workpiece automatically [23].

Most of the existing multi-discipline knowledge organization methods are based on the current knowledge models in various disciplines. The organization of multi-discipline knowledge is realized at the structure, semantics, and granularity level by semantic alignment and matching structured or semi-structured knowledge models. However, the organization of crowd intelligence knowledge for multimodal, multi-temporal, and multi-spatial design is rarely mentioned.

As shown in Table 1, the contributions and the limitations of crowd intelligence knowledge acquisition methods are concluded.

Table 1. Contributions and limitations for the organization of crowd intelligence knowledge.

Classification	Contributions	Limitations
Single-discipline methods [16–18]	Designers in a specific discipline are provided with a shared knowledge architecture with consistent structure and semantics	Knowledge from different disciplines is hard to organize and integrate
Multi-discipline methods [20–23]	Knowledge from different disciplines is integrated at the level of structure, semantics, and granularity through semantic alignment and matching of knowledge models	Multimodal, multi-temporal, and multi-spatial crowd intelligence knowledge organization is rarely mentioned

4 Knowledge Acquisition of Crowd Intelligence

It is not enough to simply organize the complex knowledge from different disciplines to make the most of crowd intelligence knowledge. Designers must obtain specific design knowledge effectively and accurately from the multimodal, multi-temporal, and multi-spatial knowledge architecture according to particular design scenarios and requirements. Two widely used methods for acquiring crowd intelligence knowledge are knowledge query and knowledge push.

4.1 Knowledge Query

Previously, knowledge queries were realized through expert consultation and data traverse. As the designers' reliance on knowledge systems increases, most knowledge query methods are based on knowledge retrieval recently [24]. Supported by technologies such as the semantic web [25] and knowledge graph [26], most disciplines have established knowledge systems for group members to query knowledge, as shown in Fig. 1.

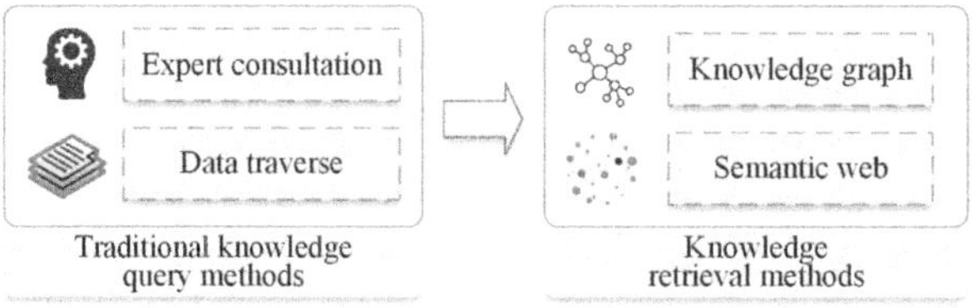

Fig. 1. Traditional and current knowledge query approaches.

Ladj et al. formalized the business rules, such as norms, standards, and constraints that must be followed in product design, and formed a semantically unified design knowledge for designers to query [27]. Camarillo et al. integrated knowledge of different disciplines in the product lifecycle, such as products, processes, and resources, and helped designers find similar cases and solutions to problems encountered in current design based on case-reasoning [28]. Chen et al. applied knowledge in the quality management discipline to product design and summarized 46 sorts of quality management knowledge from 8 design stages to test possible design quality problems [1].

It can be concluded that current knowledge queries are mainly realized through computer-aided methods. However, these approaches have some disadvantages. On the one hand, these methods are time-consuming and involve too much retrieval time. On the other hand, group members may not clearly describe the required design knowledge and may even be unsure whether they need knowledge, making it challenging to complete the query.

4.2 Knowledge Push

Knowledge push transforms the passive "people looking for knowledge" query mode into an active "knowledge looking for people" mode. Knowledge push is developed based on information recommendations [29]. It aims to provide group members with the required design knowledge in a timely and accurate manner by collecting external information such as behavior, design context, and design targets of members.

As illustrated in Fig. 2, the principle of knowledge push can be summarized as follows: The first step is to traverse the items stored in the knowledge system. The second step is to compare the content or semantics of stored items with external information such as the situation of the designer. Finally, the results are filtered and sorted, and timely presented to the knowledge that designers may require.

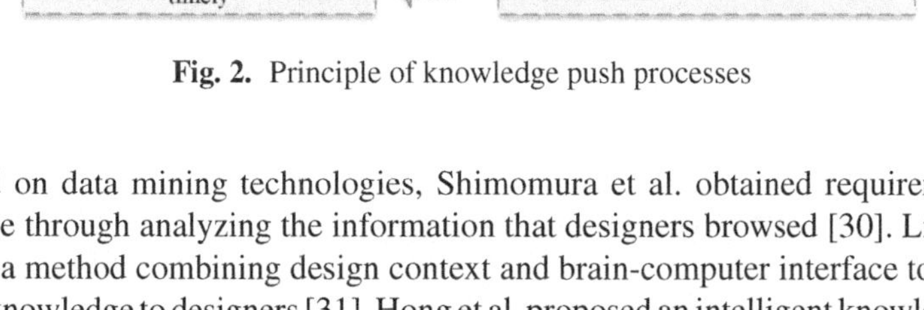

Fig. 2. Principle of knowledge push processes

Based on data mining technologies, Shimomura et al. obtained requirements for knowledge through analyzing the information that designers browsed [30]. Liang et al. proposed a method combining design context and brain-computer interface to push the required knowledge to designers [31]. Hong et al. proposed an intelligent knowledge push method based on feature transfer and provided designers with the necessary knowledge using a similarity matching algorithm to match historical design knowledge [32]. Wang et al. used situational awareness technology to perceive required knowledge and push the knowledge to designers by calculating the correlation between target knowledge and existing process knowledge [33].

However, for the multimodal, multi-temporal, and multi-spatial knowledge generated by large-scale group members, the existing knowledge storage and push methods still have the following shortcomings to be solved, as explained below:

1. The complex product design involving large-scale group participation has massive and complex design knowledge, which consumes much computing time for the system to traverse and match the design knowledge, affecting the real-time knowledge acquisition performance.
2. The results may represent low acquisition efficiency and inaccuracy for the push of knowledge stored in various modalities such as unstructured data, text, pictures, and even natural language. The existing knowledge push methods may have difficulty in solving these problems.
3. The multi-temporal and multi-spatial characteristics of large-scale crowd intelligence knowledge may lead to continuous updating of knowledge stored in the knowledge system, yet the existing push methods cannot guarantee real-time knowledge push. As a result, it may cause inappropriate knowledge or time to push knowledge, which interferes with design activities.

As shown in Table 2, the contributions and the limitations of crowd intelligence knowledge acquisition methods are concluded.

Table 2. Contributions and limitations for the acquisition of crowd intelligence knowledge.

Classification	Contributions	Limitations
Knowledge query methods [1, 27, 28]	Knowledge retrieval is completed and supported by computer-aided methods such as the semantic web and knowledge graph	The knowledge retrieval process is time-consuming, and the required design knowledge may not be clearly described
Knowledge push methods [30, 31, 33]	Transform the passive "people looking for knowledge" query mode into an active "knowledge looking for people" mode	Multi-temporal and multi-spatial characteristics of knowledge affect the efficiency and accuracy of knowledge acquisition

5 Summary and Future Opportunities

In conclusion, crowd intelligence knowledge plays a more important role in modern product design, covering knowledge organization and acquisition throughout product lifestyle. However, although lots of related research has been carried out, several problems remain to be solved for the complex design involving large-scale group participation. The research of crowd intelligence knowledge in product design needs to step further.

5.1 Limits of Current Product Design Methods

Nowadays, crowd intelligence in complex product design represents multimodal, multi-temporal, and multi-spatial characteristics. Accordingly, there are three aspects to be considered for complex product design:

1. Most current multi-discipline knowledge organization methods are based on existing knowledge models in various disciplines, which integrate multi-discipline knowledge at the structure, semantics, and granularity level by semantic alignment and matching of structured or semi-structured knowledge models. However, the organization of multimodal, multi-temporal, and multi-spatial crowd intelligence knowledge in complex design with large-scale group participation is rarely considered.
2. The existing knowledge storage methods based on databases, the semantic web, or knowledge graphs are mainly oriented to structured or semi-structured design knowledge. They cannot provide practical storage for knowledge existing in natural language, image, or video modalities. In addition, the existing knowledge storage methods are only suitable for processing static knowledge, ignoring the temporal and spatial information contained in group design, resulting in disability to represent the temporal evolution of crowd intelligence knowledge.
3. The existing knowledge acquisition methods are mainly based on keyword search, semantic matching, or knowledge graph retrieval, which may bring low query efficiency and inaccurate query results when faced with unstructured design knowledge. On the one hand, it may be unable to query when knowledge is expressed in natural language, image, or video modalities. On the other hand, since the complex product design with large-scale group participation usually composes massive design knowledge, it consumes plenty of computing time for the system to traverse and match, seriously affecting real-time performance.

5.2 Future Perspectives

The human brain is considered the most effective cognitive device, and many researchers have focused on utilizing the cognitive mechanism of the brain for complex product design [34]. As shown in Fig. 3, the brain can continuously receive and integrate external multimodal, multi-temporal, and multi-spatial stimulus information and realize fast and accurate memory backtracking based on the external environment, thanks to its unique perception-backtracking mechanism.

During the situational perception process, the stimulus information of various modalities from the external environment is pushed to the sensory organs and then integrated into different brain regions. The information focused by the brain is filtered, and the useless or redundant information is removed. Then, the rest of the information is updated into short-term memory. The short-term memory that repeats frequently is solidified into prior knowledge and stored in the hippocampus of the medial temporal lobe in the form of long-term memory.

During the knowledge backtracking process, the information in the short-term memory is retrospectively compared with the prior knowledge to infer the external information to complete the active decision.

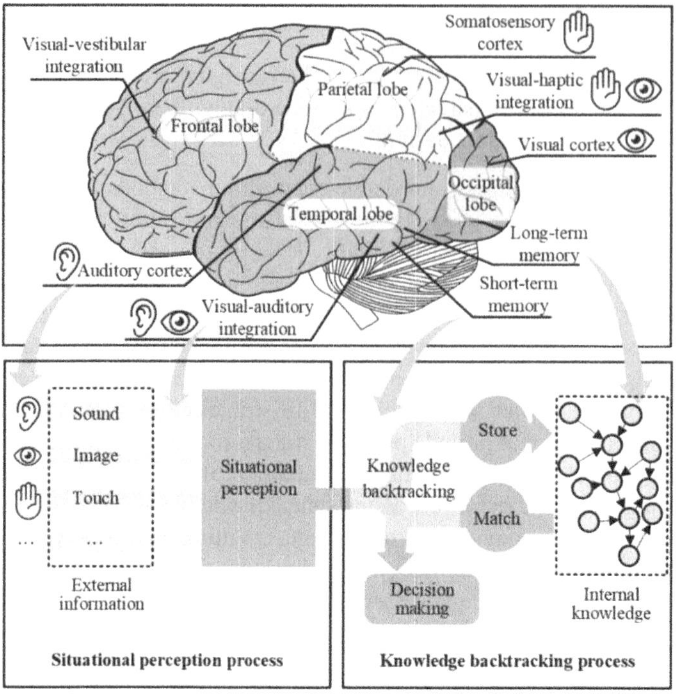

Fig. 3. Biological perception-backtracking cognitive mechanism.

It can be concluded that the cognitive mechanism of the brain can provide new research ideas for developing KBE product design. Many researchers have done research based on the cognitive mechanism of the brain. Zhao et al. proposed a question-answering system that mimics cognitive processes, which can provide quick feedback on complex query questions and give detailed explanations for the rationality of the answers [35]. Qiao et al. pointed out that the imitated cognitive mechanism has great potential to enhance the performance of robots in complex tasks, such as precise control, fast and flexible response, and deep human-robot collaboration [36]. Therefore, it is meaningful to establish an intelligent architecture of human-like cognitive mechanisms in the future to improve the robustness, cognitive ability, and active learning ability of KBE product design.

Acknowledgements. This work was supported by the National Natural Science Foundation of China (Grant No. 72471193) and the Aeronautical Science Foundation of China (Grant No. 2024Z072053001) and the National High-end.

References

1. Yang, W., Fu, C., Yan, X., Chen, Z.: A knowledge-based system for quality analysis in model-based design. J. Intell. Manuf. **31**, 1579–1606 (2020). https://doi.org/10.1007/s10845-020-01535-8

2. Knowledge management: creating context for a connected world, 2020. https://www2.deloitte.com/us/en/pages/human-capital/articles/knowledge-management
3. The future of minds and machines: how artificial intelligence can enhance collective intelligence, 2020. https://www.nesta.org.uk/report/future-minds-and-machines
4. Woolley, A.W., Chabris, C.F., Pentland, A., Hashmi, N., Malone, T.W.: Evidence for a collective intelligence factor in the performance of human groups. Science **330**(6004), 686–688 (2010). https://doi.org/10.1126/science.1193147
5. Achieving competitive advantage through knowledge-based engineering : a best practice guide, 1999. https://search.worldcat.org/zh-cn/title/Achieving-competitive-advantage-through-knowledge-based-engineering-:-a-best-practice-guide/oclc/51374894
6. Aminpour, P., Gray, S.A., Jetter, A.J., Introne, J.E., Singer, A., Arlinghaus, R.: Wisdom of stakeholder crowds in complex social-ecological systems. Nat. Sustaina. **3**(3), 191–199 (2020). https://doi.org/ https://doi.org/10.1038/s41893-019-0467-z
7. Mcharek, M., Azib, T., Hammadi, M., Choley, J.Y., Larouci, C.: Knowledge sharing for mechatronic systems design and optimization. IFAC-PapersOnLine **51**(11), 1365–1370 (2018). https://doi.org/10.1016/j.ifacol.2018.08.338
8. Kerzner, H.: Project Management: A Systems Approach to Planning, Scheduling, and Controlling, 13th edn. Wiley, New York (2022)
9. Bahill A.T., Madni A.M.: Discovering system requirements. In: Tradeoff Decisions in System Design, pp. 373–457. Springer, Cham (2017). https://doi.org/10.1007/978-3-319-43712-5_4
10. Ciampa P.D., Moerland E., Seider D., Baalbergen E., Lombardi R., D'Ippolito R.: A collaborative architecture supporting AGILE design of complex aeronautics products. In: 18th AIAA/ISSMO Multidisciplinary Analysis and Optimization Conference. American Institute of Aeronautics and Astronautics, Reston, Virginia (2017). https://doi.org/10.2514/6.2017-4138
11. Luo S.: New thought of crowd intelligence design. J. Mach. Des. **37**(3), 121–127 (2020). https://doi.org/10.13841/j.cnki.jxsj.2020.03.020
12. Yang, J., Wang, C., Jiang, B., Song, H., Meng, Q.: Visual perception enabled industry intelligence: state of the art, challenges and prospects. IEEE Trans. Ind. Inf. **17**(3), 2204–2219 (2021). https://doi.org/10.1109/TII.2020.2998818
13. Yang, G., Dong, Z., Renke, K., Yan, B., Guo, D.: Research progress of helical milling technology. Chin. J. Aeronaut. **41**(7), 12–26 (2020). https://doi.org/10.7527/S1000-6893.2019.23311
14. Ley, T.: Knowledge structures for integrating working and learning: a reflection on a decade of learning technology research for workplace learning. Br. J. Edu. Technol. **51**(2), 331–346 (2020). https://doi.org/10.1111/bjet.12835
15. Dong, Y.: Research on the modeling method of die casting channel based on semantics. J. Mech. Eng. **54**(9), 224–232 (2018). https://doi.org/10.3901/JME.2018.09.224
16. Gent, I., Aigner, B., Beijer, B., Jepsen, J., La Rocca, G.: Knowledge architecture supporting the next generation of MDO in the AGILE paradigm. Prog. Aerosp. Sci. **119**, 100642 (2020). https://doi.org/10.1016/j.paerosci.2020.100642
17. Peng, G., Wang, H., Zhang, H.: Knowledge-based intelligent assembly of complex products in a cloud CPS-based system. In: 2019 IEEE 23rd International Conference on Computer Supported Cooperative Work in Design (CSCWD), pp. 135–139. IEEE, Porto, Portugal (2019). https://doi.org/10.1109/CSCWD.2019.8791877
18. Guan, G., Lin, Y.: Knowledge-based quick design and optimization for hull structure. J.Ship Mech. **21**(4), 472–483 (2017). https://doi.org/10.3969/j.issn.1007-7294.2017.04.012
19. Li, W., et al.: Crowd intelligence in AI 2.0 era. Front. Inf. Technol. Electron. Eng. **18**, 15–43 (2017). https://doi.org/10.1631/FITEE.1601859

20. Topp, E.A., Stenmark, M., Ganslandt, A., Svensson, A., Haage, M., Malec, J.: Ontology-based knowledge representation for increased skill reusability in industrial robots. In: 2018 IEEE/RSJ International Conference on Intelligent Robots and Systems (IROS), pp. 5672–5678. IEEE, Madrid, Spain (2018). https://doi.org/10.1109/IROS.2018.8593566
21. Cao, Y., Liu, Y., Ye, X., Zhao, J.: An automated approach for execution sequence-driven software and physical co-design of mechatronic systems based on hybrid functional ontology. Comput. Aided Des. **131**, 102942 (2021). https://doi.org/10.1016/j.cad.2020.102942
22. Qiao, L., Qie, Y., Zhu, Z., Zhu, Y., Zaman, U.K., Anwer, N.: An ontology-based modelling and reasoning framework for assembly sequence planning. Int. J. Adv. Manuf. Technol. 94, 4187–4197 (2018). https://link.springer.com/arti-cle/10.1007%2Fs00170-017-1077-4
23. Zheng, C., et al.: Knowledge-based program generation approach for robotic manufacturing systems. Robot. Comput.- Integr. Manuf. **73**, 102242 (2022). https://doi.org/10.1016/j.rcim.2021.102242
24. Osman, M.A., Noah, S.A.M., Saad, S.: Ontology-based knowledge management tools for knowledge sharing in organization-a review. IEEE Access **10**, 43267–43283 (2022). https://doi.org/10.1109/ACCESS.2022.3163758
25. Zangeneh, P., McCabe, B.: Ontology-based knowledge representation for industrial megaprojects analytics using linked data and the semantic web. Adv. Eng. Inform. **46**, 101164 (2020). https://doi.org/10.1016/j.aei.2020.101164
26. Bonatti, P.A., Decker, S., Polleres, A., Presutti, V.: Knowledge graphs: new directions for knowledge representation on the semantic web. Dagstuhl Rep. **8**(9), 29–111 (2019). https://doi.org/10.4230/DagRep.8.9.29
27. Ladj, A., Wang, Z., Meski, O., Belkadi, F., Ritou, M., Da Cunha, C.: A knowledge-based digital shadow for machining industry in a digital twin perspective. J. Manuf. Syst. **58**, 168–179 (2021). https://doi.org/10.1016/j.jmsy.2020.07.018
28. Camarillo, A., Ríos, J., Althoff, K.-D.: Knowledge-based multi-agent system for manufacturing problem solving process in production plants. J. Manuf. Syst. **47**, 115–127 (2018). https://doi.org/10.1016/j.jmsy.2018.04.002
29. Zhang, K., Zhao, W., Wang, J., Chen, L., Guo, X.: Knowledge push technology based on quality function knowledge deployment. Proc. Inst. Mech. Eng. Part C: J. Mech. Eng. Sci. **233**(4), 1119–1138 (2019). https://doi.org/10.1177/0954406218768843
30. Shimomura, Y., Nemoto, Y., Ishii, T., Nakamura, T.: A method for identifying customer orientations and requirements for product-service systems design. Int. J. Prod. Res. **56**(7), 2585–2595 (2018). https://doi.org/10.1080/00207543.2017.1384581
31. Ye, L., Lemiao, Q., Xiaojian, L., Shuyou, Z., Jianrong, T.: Technology for product design knowledge need awareness based on design context and brain-computer interface. J. Mech. Eng. **56**(7), 151–163 (2020). https://doi.org/10.3901/jme.2020.07.151
32. Hong, Y., Li, W., Li, C., Xiang, H., Ling, S.: Adaptive knowledge push method of product intelligent design based on feature transfer. Adv. Eng. Inform. **59**, 102260 (2024). https://doi.org/10.1016/j.aei.2023.102260
33. Wang, Y., Gao, Q.: Design and research of process knowledge push scheme in CAPP system. Modul. Mach. Tool Autom. Manuf. Tech. (5), 134–137 (2019). https://doi.org/10.13462/j.cnki.mmtamt.2019.05.033
34. Qiao, H., Wu, Y.X., Zhong, S.L., Yin, P.J., Chen, J.H.: Brain-inspired intelligent robotics: theoretical analysis and systematic application. Mach. Intell. Res. **20**(1), 1–18 (2023). https://doi.org/10.1007/s11633-022-1390-8
35. Zhao, X., Chen, H., Xing, Z., Miao, C.: Brain-inspired search engine assistant based on knowledge graph. IEEE Trans. Neural Netw. Learn. Syst. **34**(8), 4386–4400 (2023). https://doi.org/10.1109/TNNLS.2021.3113026

36. Qiao, H., Chen, J., Huang, X.: A survey of brain-inspired intelligent robots: integration of vision, decision, motion control, and musculoskeletal systems. IEEE Trans. Cybern. **52**(10), 11267–11280 (2022). https://doi.org/10.1109/TCYB.2021.3071312

Advancing Industrial Digital Twins: Towards an Open Platform Aligned with Standards

Abdelhadi Belfadel(✉), Stephen Creff, and Amira Ben Hamida

IRT SystemX, Palaiseau, France
{abdelhadi.belfadel,stephen.creff,
amira.benhamida}@irt-systemx.fr

Abstract. Digital Twins are a major driver in the digital transformation of industry. It is gaining in adoption in industry and interest in research and standardization. One of the characteristics of digital twins is the coupling and synchronization between the asset and its virtual representation, raising the issue of interoperability. To date, there is no unique reference architecture for digital twins, nor any implementation covering the full range of capabilities needed to implement digital twins for industrial systems, which may in practice pose a barrier to their adoption and dissemination. Our approach is to move towards an open platform aligned with standards to promote interoperability. The contributions of the paper are twofold: i) a proposal for a high-level architecture of a standards-compliant digital twin platform; ii) a proposal for an open-source implementation on the restricted perimeter of asset/digital twin coupling. This implementation is aligned with the IEC 63278 standard and the Asset Administration Shell vision of Industry 4.0. The article presents 19 criteria applied to the selection of existing open-source implementations, as well as the integration of these into a technical architecture and its implementation.

Keywords: Digital Twin · Industry 4.0 · ISO 23247 · IEC 63278 · Asset Administration Shell · Interoperability

1 Introduction

Digital Twins (DTs) are a major driver in the digital transformation of industry [42]. Gartner identified DTs as one of the "Top 10 Strategic Technology Trends" for the period 2017–2019, e.g. [2]. Besides, MarketsAndMarkets [32] estimates that: "The global digital twin market size in terms of revenue was estimated to be worth USD 10.1 billion in 2023 and is poised to reach USD 110.1 billion by 2028, growing at a CAGR[1] of 61.3% from 2023 to 2028.", while Gartner [11] states that: "The digital twin market will cross the chasm in 2026 to reach $183 billion in revenue by 2031". DTs are therefore gaining in adoption in industry, and have lately been a subject of many interests in different

[1] Compound Annual Growth Rate.

Published by Springer Nature Switzerland AG 2025
P. Sureephong et al. (Eds.): PLM 2024, IFIP AICT 741, pp. 110–124, 2025.
https://doi.org/10.1007/978-3-031-93323-3_9

research communities [5, 8, 38, 40], and a matter of concern in the field of standardization, e.g. in standardization bodies such as ISO/IEC JTC1/SC41, ISO TC184/SC4, IEC TC65, or consortia such as the Digital Twin Consortium[2] (DTC), the Industry IoT Consortium[3], or the Industrial Digital Twin Association[4] (IDTA).

Grieves' original description defines a digital twin as a virtual representation of a physical product containing information about the product's characteristics, and has its roots in the field of Product Lifecycle Management (PLM) [14]. From ISO 23247-1 [20], we keep the following definition: a digital twin is a "fit for purpose digital representation [...] of an [...] element with synchronization between the element and its digital representation". We consider the element in a broad sense, i.e. a physical asset, a process, or a system. DTs are complex systems, coupled (interfaced and synchronized) with the physical world. They offer the ability to model real-world entities, processes and systems, and simulate their behavior, while being connected to them. DTs are not objectives in themselves, but means of increasing the operational efficiency of industrial systems. They feature enhanced capabilities for understanding, analyzing, predicting and optimizing the operation and control of the of the real entity [20]. Digital twins enable proactive decision-making and interventions for scenarios like predictive maintenance and logistics optimization. However, they consequently pose challenges such as cyber threats and interoperability issues.

Digital Twin systems are intrinsically complex in their nature and the lack of guidelines makes difficult for corporation to adopt them [1]. To date, there is no unique reference architecture for digital twins, nor any implementation covering the full range of capabilities needed to implement digital twins for industrial systems, which may in practice pose a barrier to their adoption and dissemination. To name few, architectures are defined at different levels: functional and logical reference architecture in ISO 23247-2 [21], or as a reference architecture framework in the upcoming ISO/IEC 30188 ED1 [19], defined by a context diagram and a functional view in ISO/IEC 30173 [23], or at the capacity level and service level [12, 33] by the DTC, or in various articles, as reference architecture e.g. [35, 41], or functional one [5].

Our approach is to move towards an open platform aligned with standards to promote interoperability, defined as: "the degree to which two or more systems, products or components can exchange information and use the information that has been exchanged" [22]. Indeed, the lack of a universal standard poses interoperability challenges, and at the same time we are witnessing the emergence of several proprietary and heterogeneous models, which prevents from the efficient adoption of DTs. Interoperability has to be addressed at the architecture level, and at a data and information one. In that context, the Asset Administration Shell (AAS, specified in IEC 63278 [18]) proposes a standardized electronic representation of industrial assets to enable interoperability.

The contributions of the paper are twofold: i) a proposal for a high-level architecture of a standards-compliant digital twin platform; ii) a proposal for an open-source implementation on the restricted perimeter of asset/DT coupling. Indeed, this implementation is aligned with the IEC 63278 [18] standard that defines the AAS digital representation

[2] https://www.digitaltwinconsortium.org/.
[3] https://www.iiconsortium.org/.
[4] https://industrialdigitaltwin.org/.

coming from Industry 4.0. In addition, we present the criteria applied to the selection of existing open-source implementations, as well as the integration of these into a technical architecture and its implementation.

The remainder of the paper is organized as follows. Section 2 introduces the required capabilities and requirements for a DT. Then, in Sect. 3, the high-level architecture of a standards-compliant digital twin platform is introduced. Section 4 describes the DT Data Model Building System, considering the criteria to select existing solutions, the evaluation and integration of the final implementation. Later, Sect. 5 presents some related work. Finally, the last Section draws our conclusion and narrows down possible future works.

2 Digital Twin Capabilities and Requirements

On the one side, Digital twin may have the capabilities of connection, integration, analysis, simulation, visualization, optimization, etc. [23]. These are best described by the DTC. Indeed, as per the DTC, the *Capabilities Periodic Table* (CPT) [12, 33] serves as an architecture- and technology-neutral framework for defining requirements. It targets organizations involved in designing, developing, deploying, and operating digital twins based on capability requirements aligned with specific use cases, as opposed to the features of technology solutions. Following a periodic table approach, the CPT organizes capabilities into clusters based on common characteristics. This structure makes it easily comprehensible in both boardroom discussions for securing funding for digital twin projects and on the shop floor when gathering requirements for digital twin applications. The CPT offers visual guidance to facilitate collaboration, brainstorming, and the explicit articulation of capability requirements.

In this context, capability refers to the capacity to perform specific actions or achieve desired outcomes. The CPT specifically focuses on technology-based capabilities while remaining agnostic to particular technology or product solutions. The capabilities of digital twins are broken into six key high-level categories: *CPT1: data services, CPT2: integration, CPT3: intelligence, CPT4: user experience (UX), CPT5: management and CPT6: trustworthiness*. Each of these categories further expands into 62 discrete top-level capabilities.

On the other side, regarding requirements enabling DT platform, several contributions such as in [10] offers an initial synthesis of DT requirements, identified through a literature review and industry interviews. The highlighted concerns include real-time data, integration, fidelity, interaction, communication, convergence, automation, updates, autonomy, connectivity, data-related aspects, efficiency, expansibility, and scalability. This integration results in a comprehensive set of functional and non-functional requirements as compiled by [34]. These requirements are related to interoperability of the DT, with the assets, or other services.

The contribution made by [25] introduced a set of requirements that can also be relevant to our broader context, facilitating the safe and harmonious operation of our digital twin platform alongside its physical counterpart.

Considering standardization bodies, in 2021, the ISO launched efforts with the release of the four parts ISO 23247 series for DTs in manufacturing. Part 1 [20] is related to requirements, and is concerned with some functional requirements (Communication, Data acquisition, Data analysis, Identification, Management, Product life-cycle, Simulation, Synchronization, Viewpoint, Hierarchical modelling of digital twin for manufacturing), and some "-ilities" (Accuracy, Data integrity, Extensibility, Granularity, Security).

These requirements are considered when building the architecture described in the following section.

3 High-Level Digital Twin Platform Functional Architecture

ISO 23247 Part 2 [21] of the series defines a reference architecture for DTs in the manufacturing domain, that includes (i) an entity-based reference model, and (ii) a functional view of the entity-based reference model with specified functional entities. The entity-based reference model uses a segmented layered pattern with sidecar [3], each of which has its own specified process logic:

The Device Communication entity (23247-DCE): in charge of for collecting data from the Observable Manufacturing Elements (OMEs) including personnel, equipment, materials, manufacturing processes, facilities, environment, products, and supporting documents and controlling the OMEs;

The Digital Twin entity (23247-DTE): responsible for modelling the data.

collected from the Device Communication entity and providing functionalities such as realization, management, synchronization, and simulation. The digital twin entity digitally represents OMEs;

The User entity (23247-UE): that host the application of the framework;

The Cross-System entity (23247-CSE): that resides across all the other entities to provide common functionalities such as data translation, data assurance, and security support.

Following a component architecture pattern, we define a high-level functional architecture aligned with the ISO 23247-2. The proposed high-level architecture, cf. Figure 1, offers a comprehensive overview of the technology foundations, ensuring a thorough understanding of the system's structure. It is important to note that interactions between these components will occur either through Application Programming Interfaces (APIs) or direct calls within each component. This architectural representation has been deliberately kept at a high level of abstraction to accommodate page limits and maintain consistency across the paper.

- **Platform UI**: This user-friendly dashboard provides customizable interfaces tailored to each platform actor, facilitating dynamic what-if scenarios runs and result exploration in a workflow-driven manner. Users can create DT data models, what-if scenarios with specific metadata or Key Performance Indicators (KPIs), import or design data models, and define threshold parameters. The interface orchestrates simulation scenarios as a pipeline of models (simulation models, AI models, data transformers, etc.), with KPI metrics displayed for both the physical and digital worlds. This component is mapped to *23247-UE* and *CPT4*.

- **Analytics**: A dashboarding component that collects from the data and general storage of the platform, monitors, and displays complex data via interactive and customizable dashboards and reports. This component is mapped to *23247-UE* and *CPT3*.
- **APIs Interoperability for IT**: This component plays a crucial role in seamlessly integrating the digital twin platform with external IT systems. By exposing its APIs, it enables external DT applications to effortlessly consume platform features, facilitating enhanced interoperability and streamlined communication between systems. Additionally, it oversees essential security measures and administrative tasks such as authentication, metric collection, input validation, and response transformation, ensuring the smooth and secure operation of the integrated ecosystem. This component is part of *23247-UE* and *CPT2*.
- **DT Data Model Manager (DTDMM) & Visualization**: This component manages digital twin data models, updating them in the database based on synchronization from physical objects or decisions from the decision system. It also offers real-time visualization of the digital twin replica. This component is part of *23247-DCE* and *CPT1*.
- **Data Integration**: This component synchronizes data from external Information Systems and physical assets, forwarding updates to the Data Model Manager for persistence in the appropriate DT Data Model instance. This component is part of *23247-DCE* and *CPT2*.
- **Security Support**: This vital component ensures strict access control to safeguard an organization's data and resources. It oversees authentication processes, user management, role assignments, service access permissions, and DT data model consumption rights, guaranteeing that only authorized individuals can access sensitive information and perform designated actions. This component is part of *23247-CSE* and *CPT6*.
- **Application Layer**: This component empowers users to design workflow scenarios using registered AI or simulation models and facilitates the deployment and initiation of simulation environments. This component is part of *23247-DTE* and *CPT3*.
- **Modeling & Simulation**: This component integrates model registration, workflow execution, and decision support functionalities. It enables the design and execution of workflow scenarios, aids decision-making based on simulation outcomes and predefined KPIs, and seamlessly executes workflows in external environments. This component is part of *23247-DTE* and *CPT3*.
- **Data and General Storage**: This component handles diverse data storage needs within the platform, offering multiple APIs for access to various storage types tailored to each component's specific requirements and constraints. This component is part of *23247-DTE* and *CPT1*.

4 DT Data Model Building System

The growing necessity for the adoption and seamless creation, processing, and integration of engineering digital twins emphasizes the importance of adhering to standards. Considering the DT Data Model Building System, the latest standard published by IEC 63278-1:2023 [18] specifies the configuration of a standardized digital representation of an asset, referred to as the AAS. The AAS ensures consistent access to information

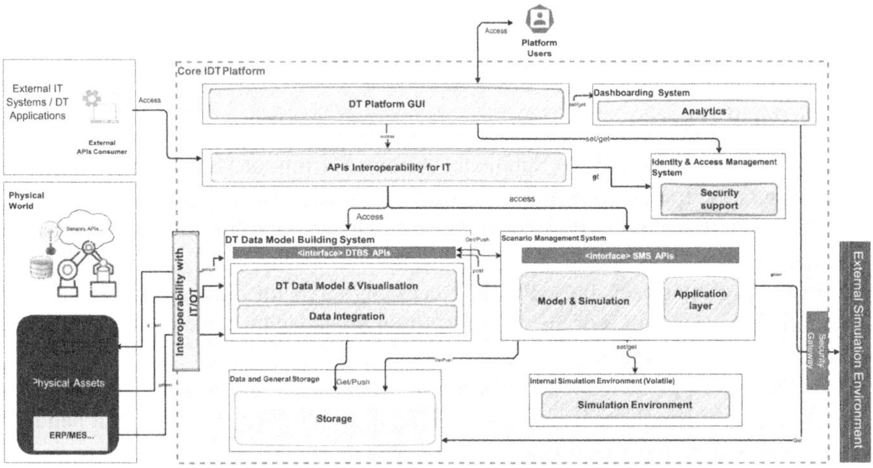

Fig. 1. DT Platform High-level Functional Decomposition Architecture Diagram

and services. Its primary objective is to facilitate the secure and trustworthy exchange of information between two or more software applications, allowing them to mutually utilize the exchanged information.

4.1 Overview of the Functional Analysis

The functional analysis per each component is made from three perspectives. *Behavior and Functionality* detailing a story map [36] outlining features and associated user stories, *Interaction Descriptions* which outline interactions between components and users, and *UI mock-ups and sequence diagrams* providing insight into component interactions with users or other platform components as necessary.

In terms of the behavior and functionality, story maps describe the functionality on different levels of aggregation/abstraction. User stories have been defined for the two components of the *DT Data Model Building System*: 1) For the *DT Data Model Manager* (DTDMM): create, update, delete, list, import, and export data model; 2) For the *Data Integration Manager*: (From physical asset to digital asset) receive data updates from the IT/OT, (From physical asset to digital asset) Send data to DTDMM, (From digital asset to physical asset) Receive DM updates from DTDMM, (From digital asset to physical asset)Send update request to the IT/OT; and Data transformation from DTDMM to send to the IT/OT, and Data transformation to send to DTDMM.

4.2 Criteria for AAS Open-Source Implementations Selection

Several prior contributions, such as those proposed by [24] and [26], presented an overview of existing AAS implementations with the goal of identifying potential candidates for implementing our targeted AAS-based DT platform. However, new contributions have emerged in GitHub repositories, and updated versions have been released since then. In order to conduct a thorough review and improvement of the identified open-source solutions, aiming to select those that comprehensively cover the functional aspects of the AAS specification, we define two types of criteria functional and technical ones, listed below.

Functional criteria:

- **F1:** Creating/Generating a Submodel Instance from an Established Submodel Template/Type (publicly available or from a manufacturer's specification).
- **F2:** Concept description: It refers to the ability to describe (or link to) the semantics of a property
- **F3:** Offering Interfaces and APIs for Accessing AAS Information: Functionalities, operations, and methods of the Administration Shell can be made accessible through an API.
- **F4:** Registry implementation: Infrastructure that hosts and interconnects multiple AASs together (implement registry services, discovery services, endpoint handling, and other related functionalities).
- **F5:** Event Management (In/Out): Events are a very versatile mechanism of the AAS. The different types are: Forward events; Reverse events; Tracking events and Value push events across clouds.
- **F6:** AAS Security Management: It involves establishing access control mechanisms and defining access permission rules that specify the permissions granted to authenticated subjects for accessing specific objects.
- **F7:** AASX file formats: Management of file formats to exchange the structure of the AAS (incl. Associated values and persistence management).
- **F8:** Filtering Import/Export Information.
- **F9:** Facilitating Information Sharing Across Various Systems within the scope of the entire RAMI4.0. Various data formats, such as OPC-UA (IEC62541), AutomationML (IEC 62714), XML (ISO 8879) / JSON (ISO/IEC 21778), and Resource Description Framework, are utilized or recommended for use in different product life-cycle phases.
- **F10:** Referencing in AAS: Within the context of AAS, two distinct types of references are identified: *Global references* (unique identifiers, possibly concatenated from multiple sources) and *Model references* (ordered keys, enabling precise element access via concatenated paths).

Technical criteria:

- **T1:** Modeling language: Technical approach to systems engineering that focuses on the creation and operation of the AAS model. Several paradigms might be used such as Object-Oriented Programming (OOP), Model-based systems engineering (MBSE), Business Process Modeling Notation (BPMN), Service-Oriented Architecture (SOA) and others.
- **T2:** Communication protocol: For example OPC-UA, REST, MQTT (ISO/IEC 20922), MTConnect (ANSI/MTC1.4), CoAP (RFC 7252), SOAP (W3C), AMQP (ISO/IEC 19464), etc.
- **T3:** ASS Registry: the AASX Registry stores registry information from AASX Servers and other AAS servers. It allows to find AAS endpoints by AssetIDs.
- **T4:** Data storage: the distributed AAS approach allows for utilizing AAS and Submodels as decentralized data storage and a unified interface to data sources. This

means static data, databases, and tools can be outfitted with a Submodel interface to offer different types of data as AAS Submodels.
- **T5:** Type of solution: Can be a Software Development Kit (SDK), a Docker container, a JAR file, or a library.
- **T6:** Containerization: simplifies software deployment by packaging an application's code and dependencies into a single (OS independent) container.
- **T7:** License and Copyright type.
- **T8:** Documentation: Comprehensive, appropriate, well-structured user documentation.
- **T9:** User interface: Does the application offer a user interface?

4.3 Evaluation of Existing Solutions

The assessment of most criteria has been conducted via documentation review, application deployment, and rigorous testing, or through a comprehensive analysis of the codebase, leading to Table 1. Based on our assessment, the optimal solutions for each category, enabling the development of a cloud-based DT platform and adhering to the specified criteria, standards and constraints, are as follows:

- **Back-end solution**: Basyx Java Server SDK V2[5], implementing both Part 1 and Part 2 specifications of AAS V3.
- **Front-end solution**: AAS Web Client (Faunhover AAS Graphical User Interface GUI).[6]
- **Data integration solution**: Basyx Data Bridge component.[7]

The Basyx Java Server SDK V2 back-end presents numerous advantages. It is constructed on the AAS4J data model library, which is accessible in the Eclipse project repository. This library adheres to the latest AAS V3 specification release and complies with IEC 63278 [18]. This solution seamlessly integrates multiple independent components, encompassing an AAS Repository server, an AAS Registry server, a Submodel Repository server, and a Submodel Registry server.

Some noteworthy features include the ability to deploy multiple server instances for scalability, with each instance exposing its own APIs and Swagger interface. It also simplifies model storage in a MongoDB database, utilizing memory storage by default, and supports event communication through Kafka or MQTT protocols. However, it is essential to highlight a potential limitation: the incorporation of the Basyx data bridge component is mandatory for data integration.

4.4 Platform Implementation and Integration

In addition to the major integration effort of different components, it is important to mention some functional improvements done comparing to the state of the art such as: an automatic registration of an AAS model to the registry, and the possibility to import

[5] https://github.com/eclipse-basyx/basyx-java-server-sdk.
[6] https://github.com/eclipse-basyx/basyx-applications/tree/main/aas-gui.
[7] https://github.com/eclipse-basyx/basyx-databridge.

Table 1. Tools analysis and evaluations

Solution	F1	F2	F3	F4	F5	F6	F7	F8	F9	F10	T1	T2	T3	T4	T5	T6	T7	T8	T9
BaSyx SDK V1 (aas spec. v2.0)	X		X	X	(X)	X	X			X	OOP (Java), BPMN engine (camunda), SOA	OPC-UA, HTTP REST, MQTT, Basyx native TCP/UDP, JSON serialization supports ActiveMQ or Kafka	X	InMemory	SDK, Compliant with Details of the AAS Part 1 V2.0.1	Maven, JAR	Eclipse Public License - v 2.0	X	
BaSyx SDK V2 (aas spec. v3.0)	X	X	X	X	(X)	X	X		(X) interface with DataBridge Component	X	OOP (Java)	HTTP(s), MQTT, Kafka	X	InMemory (by default), MongoDB,	SDK	Maven Jar, Docker	MIT Licence	X	
AAS Server	X	X	X	X	(X)	X	X		X	X	OOP (Java)	HTTP, MQTT		InMemory (by default), MongoDB	Backend Web App	Docker	MIT X Licence		
AAS DataBridge Component					X			X	X		OOP (Java), SOA	OPC UA, HTTP REST, MQTT, ActiveMQ, Kafka		In-Memory; File-based; Will support database storage in version 1.0.0	Backend Web App	Docker	Apache License V2.0	X	
AAS Web Client (AAS GUI)	X						X					HTTP REST	X		Web App GUI	Docker	MIT Licence	X	X
FAAAST-Service	X	X	X		(X)	X	X	X	(X) OPC UA		OOP (java)	HTTP(S), OPC UA, MQTT		In-Memory; File-based; will support database storage in version 1.0.0	Web App (Back-end)	Docker	Apache License Version 2.0	X	
PyI40AAS	X						X			X	OOP (python)			In-memory; CouchDB;	SDK		MIT License	X	
Aasx-server	X	X	X		X	X	X	X	X (HTTP(S) OPC UA MQTT, RDF, XML, XMI)	X		HTTP(S); OPC UA	X	File-based	Web App (Back-end)	Docker	Apache License 2.0	X	X
Aasx-package-explorer	X	X	X		X	X	X	X	X (HTTPS, OPC UA, MQTT, File with different types (json, xml, aml, aasx..)	X	JSON; XML; AASX files; AutomationML (AML);	HTTP(S); OPC UA; MQTT		File-based	Windows Software	Binaries package for 2.0 Windows 10	Apache License	X	X
NOVAAS	X				X	X	X	X	X HTTP; MQTT; TCP socket connections		Low-code development platform (LCDP)	HTTP; MQTT; OPCUA; TCP		file-based; In-Memory; SQLite	Web App	Docker	European Union Public License 1.2	(X)	X

AAS file in different format for DT model instantiation in our platform. Derived from the solutions chosen in the analysis stage, Fig. 2 illustrates the technical architecture for this initial platform release[8].

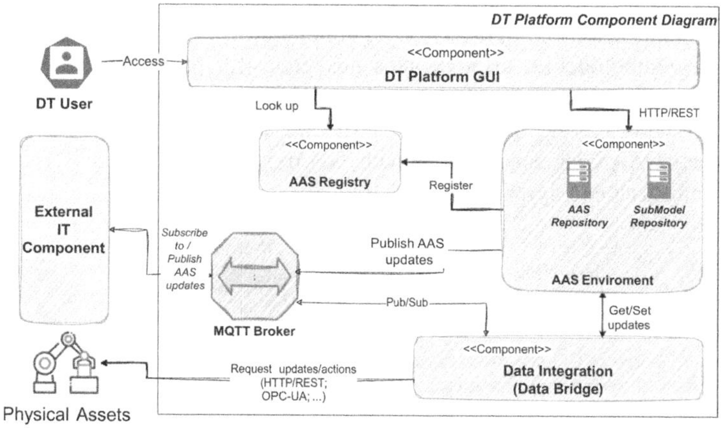

Fig. 2. Industrial DT Integrated Platform Component Diagram

The AAS server component is used to host several AAS and Submodels. It provides access to reactive AAS and Submodel instances through an AAS API. In this version of the platform, Submodel instances can be serialized separately and can be hosted on separate AAS server applications, which can be also accessed by the AAS API (a deployment of a Submodel registry is needed in this case). Each AAS Server Application and each Submodel may have an endpoint for the API access. Such an endpoint defines an address to access the AAS Server Application by the AAS API specified protocols and operations. In addition, a registry specification is developed to register endpoints and descriptions for AAS Server Applications. Such AAS Registry contains Endpoints and IDs of AAS instance and IDs of Submodel instance from a related AAS server application. These entries can be found by requesting the registry.

The Data Integration component, facilitated by the Data Bridge SDK, supports seamless integration with various protocols including OPC UA, HTTP REST, MQTT, ActiveMQ, and Kafka. It not only enables data transmission but also empowers transformation across these protocols. This component embraces two integration patterns: On-demand delegation, where data is queried, transformed, and returned upon request by the AAS server (pull); and cyclic integration, where data is periodically queried, transformed, and integrated into the AAS or triggered by events (push).

Interactions within the system outline the sequence of operations executed by client applications to accomplish specific goals in various use cases. In upcoming iterations of this platform, users will gain the ability to craft workflows tailored to what-if scenarios, thereby augmenting functionality and expanding their capabilities.

[8] The source code will be available at this public repository https://github.com/IRT-SystemX/jni-platform.

During platform UI enhancements, a new feature has been introduced to facilitate the importation of AAS files. Additionally, several updated has been added to the reused AAS GUI to comply to the latest V3 AAS specification and dynamically retrieve AAS from discovered AAS Repository endpoints and gathers information about their Submodel references. Leveraging these references, along with a set Submodel registry and distributed architecture deployment, the UI communicates with the Submodel Registry for respective Submodel Descriptors and sends requests to the Submodel Registry, then to the Submodel repository interface based on defined endpoints.

Figure 3 showcases the integrated GUI providing access to platform features. Furthermore, a reactive AAS data model, along with its corresponding submodels, is instantiated for a real industrial energy-oriented use case.

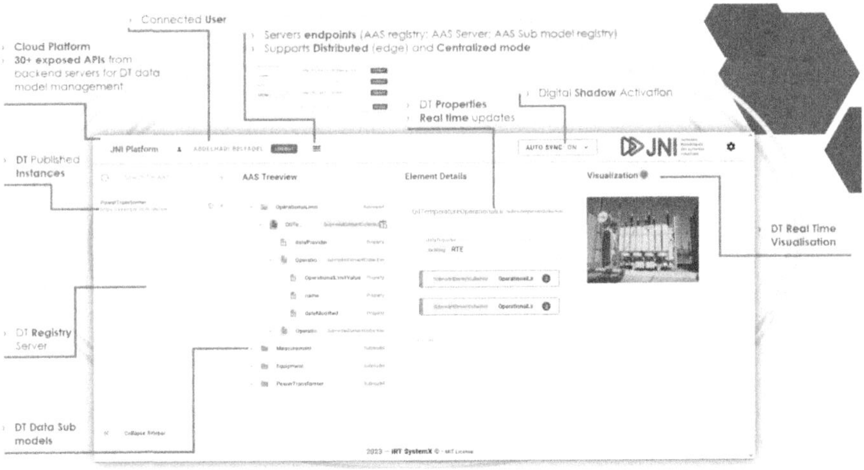

Fig. 3. Industrial DT Integrated Platform GUI and an AAS instance displayed

5 Related Works

In Industry 4.0, smart manufacturing techniques like dynamic modeling and simulation, coupled with Digital Twin (DT) technology, are enhancing production capabilities by mirroring physical systems virtually, facilitating real-time communication between cyber and physical spaces for improved manufacturing efficiency [9, 17, 29, 39]. It fosters continuous improvement towards human wellbeing and overall quality of life [9]. Additionally, DT unlocks new possibilities for Industry 4.0, making it a strategic priority for leading organizations. Leveraging DT simulations and optimized decision-making, enterprises develop self-aware smart products, gaining competitive advantages [13, 37]. With benefits spanning product design to maintenance [31], DT also plays a pivotal role in developing smart machine tools [43] and has potential as a cloud service, shaping future technologies.

In this regard, authors in [28] outline the technology stack for DT establishment. They start with the Service layer, detailing services for shop-floor assets and modeling. Next, the Data Computation layer covers sensor data querying, aggregation, and processing. Data Representation, storage tools, and machine learning methodologies are summarized. The Data Management and Connectivity layer examines tools for data acquisition and transmission.

Regarding DT perspectives in PLM stages, [28] analyzes them using a generic competitive process framework proposed by [7], outlining DT's implementation structure and process. This framework allows companies to focus on relevant PLM aspects for product enhancement. The contributions identified involve engineering PLM phases and highlight various industry applications. For instance, the Cyber Guided Vehicle in M2M logic project [4] utilizes a Discrete Event Simulation (DES) approach with sensors primarily focused on proximity detection. Similarly, the Framework for CPPS-DT implementation [16] also employs DES and utilizes communication protocols like MQTT, OPC, and WebSocket. Conversely, the Multi-dimensional DT for prestressed steel project [30] relies on simulation tools such as ABAQUS and ANSYS, employing serial communications with sensors measuring cable force and displacement. Lastly, the Digital Twins on High Voltage Transmission Line Live Working Scenes project [15] uses a Static Electrical Field Finite Element (FE) approach and incorporates sensors such as accelerometers, gyroscopes, and magnetic field sensors.

Finally, in the domain of cybersecurity, [6] assesses the current threat landscape concerning industrial control systems communicating with the AAS, as well as IT systems hosting the AAS. The authors evaluate the relevance of these threats to the AAS and delve into detailed analyses of the most pertinent threats, specifically Basic Web Application Attacks and Malware Infections. Comparisons of recommended countermeasures for these threats with those outlined in BSI [6] and IEC 62443-4-2 [27] reveal gaps and missing countermeasures, highlighting areas for further research to bolster the overall security of the AAS.

6 Conclusion and Future Work

We presented in this contribution i) a high-level architecture of a standards- compliant digital twin platform (ISO 23247), and ii) an industrial DT platform that is designed in alignment with IEC 63278. With a primary goal to facilitate the design of industrial digital twin systems for various use cases. The produced open platform integrates and enhances existing open-source tools. This strategic approach aims to accelerate the implementation of digital twins, fostering their widespread adoption within the industry while ensuring compliance with industry standards and cybersecurity protocols.

The ongoing development efforts prioritize several critical aspects, including the adoption of a risk-based analysis approach, which facilitates a cyber-by- design approach to support the life-cycle of Digital Twins and ensure cybersecurity assurance. In addition, key components such as Identity & Access Management, Scenario Management System for what-if scenarios and Decision System are under conception and development for future releases. These components aim to enhance security, streamline workflow execution, assist decision-making processes, and optimize resource utilization within the integrated industrial Digital Twins platform. Their implementation represents significant strides towards achieving greater efficiency, effectiveness, and adaptability in industrial operations.

Acknowledgements. This work has been supported by the French government under the "France 2030" program, as part of the SystemX Technological Research Institute.

References

1. Ahmad, B., Harrison, R., Vera, D.: A connective framework to support the lifecycle of cyber-physical production systems. Proc. IEEE **109** (2021). https://doi.org/10.1109/JPROC.2020.3046525
2. Analyst(s): Marc Halpern, Simon Jacobson, R.F.: Hype cycle for it evolution in manufacturing, id: G00340295. Tech. rep., Gartner (2018)
3. Bass, L.J., Clements, P.C., Kazman, R.: Software Architecture in Practice. SEI Series in Software Engineering, Addison-Wesley-Longman, Boston (1999)
4. Bottani, E., Cammardella, A., Murino, T., Vespoli, S., et al.: From the cyberphysical system to the digital twin: the process development for behaviour modelling of a cyber guided vehicle in m2m logic. XXII Summer School Francesco Turco Industrial Systems Engineering, pp. 1–7 (2017)
5. Boyes, H., Watson, T.: Digital twins: an analysis framework and open issues. Comput. Ind. **143**, 103763 (2022). https://doi.org/10.1016/j.compind.2022.103763
6. Broring, A., Ehrlich, M., Trsek, H., Wisniewski, L.: Secure usage of asset administration shells: an overview and analysis of best practices (2021). https://opendata.unihalle.de//handle/1981185920/41505
7. Casadesus-Masanell, R., Ricart, J.E.: From strategy to business models and onto tactics. Long Range Plan. **43**(2–3), 195–215 (2010)
8. Dalibor, M., et al.: A cross-domain systematic mapping study on software engineering for digital twins. J. Syst. Softw. **193**, 111361 (2022). https://doi.org/10.1016/j.jss.2022.111361
9. Davis, J., Edgar, T., Porter, J., Bernaden, J., Sarli, M.: Smart manufacturing, manufacturing intelligence and demand-dynamic performance. Comput. Chem. Eng. **47**, 145–156 (2012)
10. Durão, L.F.C., Haag, S., Anderl, R., Schu¨tzer, K., Zancul, E.: Digital twin requirements in the context of industry 4.0. In: Product Lifecycle Management to Support Industry 4.0: 15th IFIP WG 5.1 International Conference, PLM 2018, Turin, Italy, 2–4 July 2018, Proceedings 15, pp. 204–214. Springer (2018)
11. Gartner: Emerging technologies: Revenue opportunity projection of digital twins (2022). https://www.gartner.com/en/documents/4011590. Accessed 26 Feb 2024
12. Geater, J., Hirsch, F., Richter, D., Robkin, M., Zahavi, R.: Iot security maturity model digital twin profile an industry iot consortium and digital twin consortium whitepaper. Tech. rep., DTC (June 2022)
13. Ghobakhloo, M.: The future of manufacturing industry: a strategic roadmap to-ward industry 4.0. J. Manuf. Technol. Manag. **29**(6), 910–936 (2018)
14. Grieves, M.: Digital twin: manufacturing excellence through virtual factory replication. White Paper **1**(2014), 1–7 (2014)
15. Guo, J., Wu, Q., Guo, K., Xiong, S., Feng, W., Xue, J.: Study on the construction and application of digital twins on high voltage transmission line live working scenes. IEEE Access **9**, 111587–111594 (2021)

16. Harrison, R., Vera, D.A., Ahmad, B.: A connective framework to support the lifecycle of cyber–physical production systems. Proc. IEEE **109**(4), 568–581 (2021)
17. I-Scoop: Digital twin technology and simulation: Benefits, usage and predictions 2018. pp. 1–6 (November 11 2017)
18. IEC: Asset administration shell for industrial applications - part 1: Asset administration shell structure IEC: Iso/iec 30188 ed1 digital twin - reference architecture (2024). Accessed 26 Feb 2024
19. ISO 23247–1:2021 - Automation systems and integration Digital twin framework for manufacturing - Part 1: Overview and general principles. Standard, International Organization for Standardization, Geneva, CH (October 2021)
20. ISO 23247–2:2022 - Automation systems and integration Digital twin framework for manufacturing — Part 2: Reference architecture. Standard, International Organization for Standardization, Geneva, CH (October 2021)
21. ISO/IEC 25010:2023 - Systems and software Quality Requirements and Evaluation (SQuaRE) Product quality model. Standard, ISO/IEC, Geneva, CH (November 2023)
22. ISO/IEC 30173:2023 - Digital twin – Concepts and terminology. Standard, ISO/IEC, Geneva, CH (November 2023)
23. Jacoby, M., et al.: Open-source implementations of the reactive asset administration shell: a survey. Sensors **23**(11), 5229 (2023)
24. Kasper, L., Birkelbach, F., Schwarzmayr, P., Steindl, G., Ramsauer, D., Hofmann, R.: Toward a practical digital twin platform tailored to the requirements of industrial energy systems. Appl. Sci. **12**(14), 6981 (2022)
25. Kilian, L.: Implementing an Asset Administration Shell based on Open-Source Frameworks. Bachelor's thesis (2023)
26. Kronfuss, E.: Industrial cyber security standard-iec 62443. Tech. rep. (2018)
27. Lim, K.Y.H., Zheng, P., Chen, C.H.: A state-of-the-art survey of digital twin: techniques, engineering product lifecycle management and business innovation perspectives. J. Intell. Manuf. **31**, 1313–1337 (2020)
28. Liu, J., et al.: Dynamic evaluation method of machining process planning based on digital twin. IEEE Access **7**, 19312–19323 (2019)
29. Liu, Z., Bai, W., Du, X., Zhang, A., Xing, Z., Jiang, A.: Digital twin-based safety evaluation of prestressed steel structure. Adv. Civ. Eng. 1–10 (2020)
30. Mabkhot, M.M., Al-Ahmari, A.M., Salah, B., Alkhalefah, H.: Requirements of the smart factory system: a survey and perspective. Machines **6**(2), 23 (2018)
31. MarketsAndMarkets: Digital twin market size, share, statistics and industry growth analysis report by application [...], industry [...], enterprise and geography - global growth driver and industry forecast to 2028 (2022). https://www.marketsandmarkets.com/Market-Reports/digital-twin-market-225269522.html. Accessed 26 Feb 2024
32. McKee, D.: Platform stack architectural framework: An introductory guide - a digital twin consortium whitepaper. Tech. rep., DTC (7 2023)
33. Moyne, J., et al.: A requirements driven digital twin framework: specification and opportunities. IEEE Access **8**, 107781–107801 (2020)
34. Newrzella, S.R., Franklin, D.W., Haider, S.: Three-dimension digital twin reference architecture model for functionality, dependability, and life cycle development across industries. IEEE Access **10**, 95390–95410 (2022). https://doi.org/10.1109/ACCESS.2022.3202941
35. O'hEocha, C., Conboy, K.: The role of the user story agile practice in innovation. In: International Conference on Lean Enterprise Software and Systems, pp. 20–30. Springer (2010)
36. Posada, J., et al.: Visual computing as a key enabling technology for industrie 4.0 and industrial internet. IEEE Comput. Graph. Appl. **35**(2), 26–40 (2015)

37. Qi, Q., et al.: Enabling technologies and tools for digital twin. J. Manuf. Syst. **58**, 3–21 (2021). https://doi.org/10.1016/j.jmsy.2019.10.001
38. Schneider, G.F., Wicaksono, H., Ovtcharova, J.: Virtual engineering of cyber-physical automation systems: the case of control logic. Adv. Eng. Inform. **39**, 127–143 (2019)
39. Segovia, M., Garcia-Alfaro, J.: Design, modeling and implementation of digital twins. Sensors **22**(14) (2022). https://doi.org/10.3390/s22145396
40. Tao, F., et al.: Maketwin: a reference architecture for digital twin software platform. Chin. J. Aeronaut. **37**(1), 1–18 (2024). https://doi.org/10.1016/j.cja.2023.05.002
41. Tao, F., Zhang, H., Liu, A., Nee, A.Y.C.: Digital twin in industry: state-of-the-art. IEEE Trans. Ind. Inf. **15**(4), 2405–2415 (2019). https://doi.org/10.1109/TII.2018.2873186
42. Xu, X.: Machine tool 4.0 for the new era of manufacturing. Int. J. Adv. Manuf. Technol. **92**, 1893–1900 (2017)

Overview of Digital Thread from a Traceability Perspective: From Engineering, Manufacturing to Operations

Saina Sadeghzadeh[1(✉)], Brendan P. Sullivan[2], Dominik Knoll[1], Monica Rossi[2], and Clement Fortin[1]

[1] Skolkovo Institute of Science and Technology, Moscow, Russia
Saina.Sadeghzadeh@skoltech.ru

[2] Department of Management, Economics and Industrial Engineering, Politecnico di Milano, Milano, Italy

Abstract. In developing large or complex systems, it is critical to maintain relationships among data across all the different phases of the product lifecycle. The concept of a digital thread connecting product related information is very promising and enabled by Industry 4.0 and Industry 5.0 initiatives, but faces the difficulty of heterogeneous data formats, storage and tools. Traceability is the ability to navigate between artifacts of the digital thread along the product/system evolution, from early engineering design to manufacturing, operations and end of life. In this context, through a conceptual study we identified a particular challenge during operations due to limited linkage of operation manuals to engineering data. While there already exist approaches for traceability in manufacturing, this work identifies the need for improved traceability in industrial operations. Similarly, at the transition from operations to end-of-life, the traceability is analyzed, and an approach is presented and discussed. Our study shows that the digital thread is not a single continuous line though the lifecycle, because industrial operations are given limited access to information of engineering and almost none from manufacturing. These limitations on traceability are mainly caused by organizational boundaries.

Keywords: Traceability · Digital Thread · PLM · Industrial Operations · System Engineering

1 Introduction

Advances in Industry 4.0, and nowadays Industry 5.0, have brought about a transformative period in manufacturing, distinguished by the development and production of highly customized products, the adoption of flexible production systems, and the integration of advanced technologies into cyber-physical systems. Central to this transformation is the concept of the Digital Thread (DTH), a data-driven architecture that covers the entirety of a product's lifecycle, from its inception through engineering, manufacturing,

Published by Springer Nature Switzerland AG 2025
P. Sureephong et al. (Eds.): PLM 2024, IFIP AICT 741, pp. 125–139, 2025.
https://doi.org/10.1007/978-3-031-93323-3_10

operations, and maintenance [1–3]. The DTH stands as the cornerstone for achieving unparalleled levels of consistency, traceability, and transparency, thereby addressing the complexities inherent in modern product development. Traceability, fundamentally pertains to the capacity to record and analyze all aspects of a system throughout its lifecycle, ranging from the sourcing of raw materials to the stages of production, consumption, and operation. Traceability is defined by [4], as the ability to link and investigate the relationships among data stored across various artifacts. Moreover, the International Organization for Standardization (ISO 9001) describes traceability as the capacity to trace an object's history, application, or location by elucidating the origins of materials and parts, processing history, and distribution [5]. Through vertical and horizontal processes, traceability traverses the lifecycle of a system. Enabling the tracking and monitoring of requirements from conception, through implementation within the design phase, through to deployment and ultimately retirement.

In the context of Industry 4.0 the increase of automated production process and cyber-physical production systems in manufacturing operation, asset traceability problems have become more relevant in different stages of supply chain [6]. The available data for improved decision making and traceability will support greater control on product and process quality due to the identification, transform/routing documentation and tracing of product attributes [7]. The integration of the DTH in industrial engineering, manufacturing and operations represents a paradigm shift in managing the complexities and demands of contemporary production systems. This paper aims to conduct an indepth analysis of the DTH, highlighting its crucial role in enhancing traceability within the Industry 4.0 framework. Based on the findings from literature, this research seeks to elucidate how the DTH enables detailed traceability of products and assets, thus supporting operational goals of organizations faced with the challenges of designing and managing highly customized and flexible production systems.

1.1 Paper Objective and Problem Definition

The findings of this research are intended to inform practitioners and researchers that all phases of the product system lifecycle can be linked to maintain a harmonized data stream using multiple sources. The DTH acts as a pivotal element in ensuring comprehensive forward, backward and bi-directional traceability, to increase product quality and support decision making. In particular, this research work wants to show that, industrial operations could benefit from stronger integration into the DTH complete traceability during product development cycle. Our main finding is that Engineering, Manufacturing and Industrial operations traceability in practice must be adjusted. The paper is organized using the following sections: Sect. 2 presents the current state of research related to DTH and traceability, Sect. 3 presents how traceability and DTH engage throughout the lifecycle, Sect. 4 describes the process of implementing traceability in engineering, manufacturing and operation processes, and Sect. 5 concludes with a summary of the findings and recommendations for future research.

2 Digital Thread and Traceability

The literature review explores the interplay between traceability and the DTH as documented in recent scholarly works. The search was executed using both Google Scholar and Scopus, employing the key terms "Traceability" and "Digital Thread" with a focus on literature published after 2001, ensuring contemporaneity and relevance to current industry standards and challenges. From this targeted search the publications were reviewed first to ensure relevance to traceability, specifically in relation to product, process of requirements traceability. Second, publications were considered if it presented digital engineering technologies related to product lifecycle management (PLM). This dual-lens criterion coupled with a snowball search strategy ensured that the selected literature not only conceptualizes traceability within the theoretical domain of the DTH but also traverses the practical landscape where such theoretical constructs can be operationalized.

2.1 Digital Thread

The DTH is a contemporary paradigm within systems engineering that encapsulates a digital architecture designed to streamline the flow of data across the entire product lifecycle phases Begin-of-Life (BoL), Mid-of-Life (MoL), End-of-Life (EoL) [1–3]. It has its roots in the aerospace industry, intended to digitize and manage the continuum from 3D CAD system component design through to manufacturing, assembly, and delivery. The DTH extends these principles to encompass the entire lifecycle, including research, design, manufacturing, operations, sustainment, and training [3]. Whereby enabling an integrated flow of data, creating a unified and coherent data stream that not only charts the trajectory of a product's development but also ensures that each phase of its lifecycle is interlinked and accessible. Nigischer et al. [8] proposed, *"with the new requirements of a volatile future market and to exploit benefits of flexible digital production environments, it is crucial to adapt integration concepts of Industry 4.0, e.g. enhance information exchangeability by improving interconnectivity between software solutions, utilize all sources of data from a products production and use phase, and maintaining an up-to-date digital twin of the product and production assets"*.

The value of the DTH within the evolving Industry 4.0 landscape is in its ability to connect disparate data across different stages of the product lifecycle, ensuring accessibility, to accurate and real-time information [1]. By integrating data from the processes, products and insights from engineering processes, DTH can allow for real-time evaluations of product quality, project costs, schedules, and performance metrics. This integration enhances process traceability, fosters collaborative development, and facilitates real-time collaboration among distributed teams/organizations [1, 3, 9]. However, despite the clear value, DTH implementation is a complicated task requiring the aggregating of disparate data types across the lifecycle phases of the product and processes used in its creation. This requires sophisticated interoperability solutions and the development of frameworks that ensure data homogeneity and security across the product innovation chain, the enterprise value chain, and the field and services chain [1]. Requiring the use of advanced modeling and simulation tools, as well as commercial software platforms capable of managing centralized data storage. Moreover, the need to balance the availability of information with the safeguarding of intellectual property rights imposes additional layers of complexity, requiring tiered access to sensitive data. Therefore, the DTH not only demands advanced technological infrastructure but also a strategic approach to data governance. As a digital conduit for lifecycle data, the DTH facilitates multifaceted support for product and process traceability bringing together data from different phases of the lifecycle by enabling feedback loops, iterative processes. This is further enhanced by big data analytics, which is capable of supporting the operational goals of organizations, particularly in the context of customized and flexible systems where detailed traceability of products and assets is crucial.

2.2 Traceability

Traceability in the context of systems engineering and the DTH is a method of tracking and documenting the lifecycle of a product from beginning, to middle and finally the end-of-life. The ability to track and monitor all aspects of an artifact leveraging a comprehensive record of every change, decision, and processes that has impacted the product. This record enabled through the DTH provides a chronological trail that allows stakeholders to follow the evolution of a product lifecycle. This ability to track and trace can help the organization in the event of issues or defects to identify the point of failure within the product's history, facilitating quicker resolution. According to the [10], the challenge for traceability is that each element needs to be acquired, represented and stored and then subsequently retrieved as a trace to enable software and systems engineering activities and tasks. Without particular attention, track is loss of artifacts in operation, manufacturing, maintenance records and actions also in the event of an accident. Traceability helps to improve the retrieval of information and to prevent a chain of errors. Based on the scope and capabilities of the digital thread it is possible to view traceability through three distinct perspectives: requirements traceability, product traceability, process traceability. While these three forms of traceability focus on different aspects of the lifecycle, they are interrelated and collectively form a comprehensive approach that is crucial for ensuring the integrity and success of the entire product development process.

As illustrated in Table 1 the foundational concepts of requirements, product, and process traceability are distinct yet interconnected components within the digital thread framework. These specific types facilitate a holistic view of the product lifecycle, ensuring that each step from inception to final delivery is traceable and aligned with the original objectives, thus allowing for comprehensive oversight and precise navigation through the product's evolution.

Literature distinguishes three types of traceability links: forward, backward and bidirectional traceability. The referential work on traceability by Gotel et al. [10], describes forward traceability, as the potential to follow trace links in the direction of the development path. The backward traceability is the potential to follow trace links to antecedent steps on the development path. Consequently, the bidirectional traceability is given, if there is potential to follow trace links up and down the development path.

Table 1. Interconnected components of traceability.

	Requirements [11]	Product	Process
Definition/Focus	The ability to follow the life of a requirement, in both a forward and backward direction	Tracking the origin, production, and distribution history of a product	Monitoring the sequence and history of operations performed during the manufacturing process
Associated Tools	Requirements management tools, change management systems	Product lifecycle management (PLM) tools, serialization and barcoding systems	Manufacturing execution systems (MES), process management software
DTH Relationship	Links requirements with all lifecycle stages and associated data	Ensures each physical product is matched with its digital twin and lifecycle data	Tracks and validates each step of the manufacturing within the digital thread framework
Benefits	Facilitates changes, verifies final product compliance	Quick response to recalls/issues, regulatory compliance and quality	Optimizes manufacturing efficiency and quality control
Challenges	Maintaining accuracy and consistency of requirement changes over time	Complex when dealing with many components from diverse suppliers	Adapting to changes in processes and technology adoption

3 Traceability and Digital Thread Across the Lifecycle

The virtual representation of a product lifecycle, connecting all relevant information from its design, and development to manufacturing, operations and maintenance is referred to as a Digital Thread. The Digital Thread will be determined as a digital tool to integrate all the data and history of the product from engineering to manufacturing and service. Data from manufacturing process, bills of materials, geometry, supply chain, design, service management and other elements are involved as a part of digital thread information. According to the [12], "Linking heterogeneous lifecycle data is critical to maintain associativity and traceability in a digital thread". By implementing a Digital Thread, industries and companies have more capabilities to improve their future performance, reduce costs and enhance product quality. Traceability would facilitate tracking flow of digital information and data during the whole lifespan of the product. The key aspect of traceability in digital thread is to integrate all data to provide consistent information from various disciplines, involved in product lifecycle. Maintaining the traceability throughout the lifecycle will support better synchronization between final product and initial requirements. Hence, traceability is required as a method to integrate various data from different sources and domains.

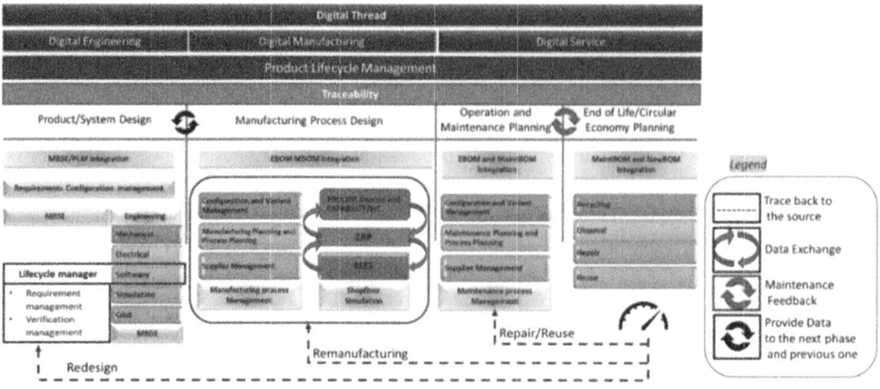

Fig. 1. Digital thread and traceability between lifecycle phases

In this research, we discuss the concept of Digital Thread over the entire lifecycle of the product, including Digital Engineering, Digital Manufacturing, and Digital Service, as shown in Fig. 1. The Digital Thread, consists of a virtual representation of the product, encompassing all relevant data from its design phase and development to manufacturing, operations and maintenance phase till the end of life of the product. The activity of maintaining product related digital artifacts is commonly referred to as Product Lifecycle Management (PLM), and PLM systems are instruments to support this activity. However, the promise of many PLM system vendors is that one single system can cover all lifecycle. In practice, a single all-encompassing PLM system is often hard to realize due to reasons such as company legacy, which is extremely hard to migrate into a new system, or specific needs not covered by a new system of choice, or organizational boundaries. The latter is specifically true for products which are sold, and hence the operations and service is done by an organization that differs from the manufacturer. Kirova et al. [20], suggested that traceability has the potential to yield substantial benefits when performed properly, therefore, the strategy for implementing traceability must be carefully recognized by examining and correlating the set of business, project, product and technology.

3.1 Product/System Design

The initial stage of product development, as indicated in Fig. 1, takes a design concept of product/system and evolves it into detailed design, components specifications, materials and function to effectively meet all the requirements and users needs among various disciplines of the lifecycle. Model Based Systems Engineering (MBSE) utilizes comprehensive modeling languages such as SysML to represent structure, behavior, architecture, requirements and other aspects of product system that feed into the digital thread. The abstract models of MBSE will assure consistency all over the lifecycle. For effective change management control and consistent traceability path, requirements, and configuration management in the context of design engineering is essential. It is therefore crucial to identify requirements at the early design stage so that they can be tracked and undergo change control and configuration process. The requirement for design engineering is composed of Mechanical, Electrical, Software, Simulation and Cost defined as main parts of system requirements. Therefore, to ensure consistent requirement management, an integral part of engineering software is the lifecycle manager, handling verification and requirement management, during design phase. The lifecycle manager effectively controls all system and software development processes to robust comprehensive lifecycle traceability. As one of the design engineering needs, the cost essentially comprises assessing the expenses associated with developing, producing, operating and maintaining of the product. It is also directly related to the supply chain by providing budget for parts distribution, delivery, orders and storage of the product. Model based digital engineering (MBDE) encompasses a broader view of the concept of model-based system engineering by leveraging digital models. For example, simulating behavior plays an important role in detailed design, and the ability to link the results of simulation to certain model components aids understanding system functioning in various scenarios. Hence, effective Product System design necessitates crossdisciplinary collaboration through engineering and design components to provide coherence data to manufacturing process design. A major artifact of the engineering process is a product physical definition in the form of the Engineering Bill of Material (EBOM) containing the product data structure and features.

3.2 Manufacturing Process Design

Manufacturing process selection often involves considering complex incorporation between characteristics of design, materials and the process [13]. As the next stage, to establish traceability fed into DTH, the data from requirement configuration management, through design requirement engineering to product system design will be transmitted to the manufacturing process, which in turn represents a forward traceability direction. The integration between Engineering and manufacturing is facilitated via the Engineering Bill of Material (EBOM) and Manufacturing Bill of Material (MBOM). EBOM represents the hierarchal structural breakdown of a product which involves detailed specification such as components, materials, part numbers, names, revision, etc. Since the entire product must be manufactured and assembled along with much more information such as tooling, stocks and manufacturing processes, thus MBOM represents all the components of EBOM, but also illustrates the production process sequences and assembly steps.

To ensure that the manufacturing process is aligned with designed product, configuration and variant management is an essential factor throughout the whole manufacturing process. Any challenges and issues throughout the production process are fed back to the product system design for modification and improvement of part, which ultimately represents backward traceability direction. Indeed, the feedback from manufacturing process design will help to improve digital engineering and product reliability. Manufacturing process management (MPM) has been demonstrated in huge industrial as a collaborative platform to transfer all the data from a virtual 3D model to manufacturing process [14, 15]. It particularly begins with product development strategy planning and ends when the production ramp-up process is accomplished [15]. Thus, the MPM maintains the product reliable, ensures configuration management and removes numerous inconsistencies [15]. To achieve seamless manufacturing and operational efficiency, effective supply management coordinates the flow of product and services including product authentication and specification, product delivery and service management, and an efficient communication with stakeholders and maintenance. To ensure high quality and reliability in the manufacturing process bi-directional traceability plays a crucial role in providing forward and backward traceability.

Shopfloor Simulation. Shopfloor simulation has such capabilities to achieve optimal production schedule and the allocation of resources to effectively optimize throughput and reduce lead-time in manufacturing process. As shown in Fig. 1, the shop floor simulation is connected with enterprise resource planning (ERP) and manufacturing. The ERP consolidates and monitors business processes such as orders and shipments, finance and supply chain, to robust planning for scheduling materials. Thus, ensuring that materials are available for production as needed. The manufacturing Execution system (MES) is a software system focused on the monitoring and orchestrating manufacturing data and operation on the shop floor from raw materials to final products. There is a gap between the data in ERP and MES as they cover different data. MES particularly receives all the information from after it has been scheduled to the ERP and integrates it with manufacturing data. Hence to carryout traceability, once all the sequences are done on the shopfloor, and scheduled on ERP, then it will provide all the relevant data which is directly related to MES contains production planning and scheduling, materials orders and resource availability. This integration will support proper traceability by detailed scheduling of data.

Process Control and Capability/IoT. In manufacturing process monitoring and managing variables are a part of process control, while process capability conducts assessment of the manufacturing process. This could be considered to evaluate and control variations, trace material movement, data exchange and improve overall quality at the shopfloor. Moreover, this process uses IoT technology to optimize manufacturing processes by providing smart devices and tools, which provide real-time data through various sensors for manufacturers and maintenance requirements. For instance, by using sensors on the machines, multiple parameters can be measured in real time, including vibration, temperature, tolerance, noise, etc. In case that vibrations above the minimum level are detected, corrective actions will be taken based on the identification of root cause analysis in comparison to normal operating conditions. Thus, IoT in manufacturing can enhance manufacturing process and facilitate operation by reducing downtime cost and extend lifespan of equipment [16].

3.3 Operation and Maintenance Planning

Operation and maintenance planning includes configuration and variant management, maintenance planning and process planning, supplier management and maintenance process management (see Fig. 1). The configuration and variant management will provide consistent information after receiving technical data from engineering and MaintBOM. The highest level of MaintBOM represents the entire part with the number of assemblies. The lower levels illustrate the technical instructions and specifications of each assembly to provide precise technical information of all components for maintenance, as shown in Fig. 2. For example, it shows Assembly 1 made up of individual subassemblies and spare parts, along with their technical information which involves part numbers, manufacturing serial numbers, assembling serial number, quantities and other technical specification. After the maintenance process is complete, the individual components of the part will be updated in the MaintBOM.

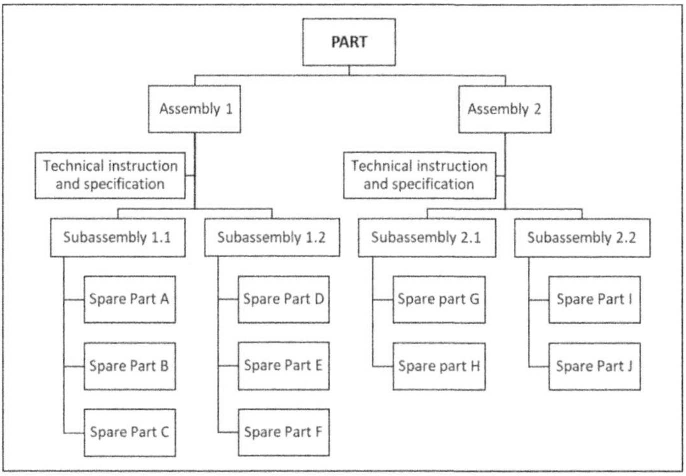

Fig. 2. Example structure of a Maintenance BOM

Subsequently the data will be managed in accordance with maintenance planning process planning. Maintenance planning and process planning will establish and define maintenance work order in response to operation inquiries, to provide any required periodic, un-schedule and on-time maintenance task such as inspection, removal, installation, testing, etc. Hence, once maintenance task has been completed, the supplier management will be notified if any new or replacement spare parts, assemblies or subassemblies shall be ordered. Maintenance process management effectively monitors all the processes in operation and maintenance for efficiency and compliance of each element associated with this stage. Moreover, it is responsible for complying with the requirements and standards of the organization. This process ensures that all maintenance tasks are carried out accurately and on time. As a result, it assures that all the defects that affect the safe operation are rectified.

Maintenance Planning and Process Planning. The maintenance process covers the ability to conduct procedures and address the issues for maintaining the product and improving its serviceability. The structure of the maintenance process and planning is illustrated in Fig. 3.

It starts when operation is directing a task to be performed, commonly referring to a single part, assembly or subassembly. To facilitate seamless maintenance planning, the technical data relevant to the particular part is provided to the maintenance planners. Thus, after maintenance planning begins to create maintenance job task/workorder, the work order will be performed by considering track of the changes which occurred to the

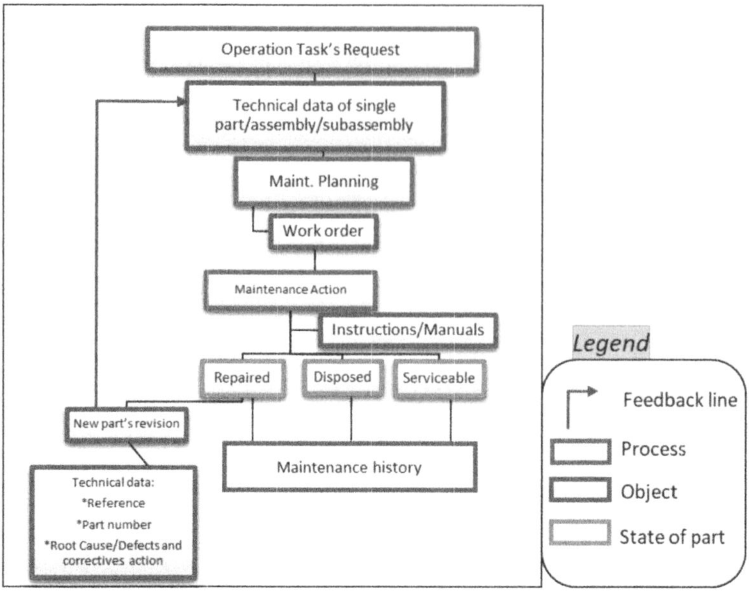

Fig. 3. Maintenance planning and process planning

bill of material. A specific task or activity which conducts multiple stages by maintenance to ensure the proper functioning and reliability of a particular part, is known as maintenance action. The instruction, manuals and MRO (Maintenance repair and overhaul), are the primary maintenance resources that assist in taking proper actions on the part identifying its status as repaired, disposed or serviceable. When a particular part needs to be repaired, it indicates what changes should be made in accordance with the failure or malfunctions to restore it to its normal operating condition. These changes may include trouble shooting, replacement or calibration which leads to extension of its operating life. Therefore, a new revision will be made containing the new part number. This is then use by maintenance planning incorporating all the technical information to guarantee that the required actions are performed in a proper manner and on time. Any modification should be recorded in the whole part's/subassemblies technical data to indicate the precise root cause of the failure or defect, together with references used as manuals/instructions to implement corrective actions. A particular part/subassembly will be determined as a disposed, when it reaches the expiry of the service life, involvement in an incident affecting its serviceability and evidence of defects or malfunctions. Thus, it shall be removed and replaced. This process also should be done in accordance with the engineering design data provided to maintenance. When a part/subassembly is operating normally, it fulfills the required functions for which it was designed. For instance, in the matter of operational life expectancy and serviceability, it is not prone to any defect or malfunctions. Maintenance history shall contain all the records for modifications, repairs, disposal, replacement and the conditions of serviceable and unserviceable. An effective maintenance planning and process planning is essential to ensure reliability and performance, reduce downtime and to achieve seamless operational enhancement.

EBOM and MaintBOM Integration. Above the previous section, we described how operations and, maintenance planning and process planning needs to manage the composition of a product and the connected technical documentation. This information we described as MaintBOM, shown in Fig. 2. To keep the Digital Thread and enable traceability we analyse the flow of information MaintBOM. Previous work proposed an integration of EBOM and MBOM through complementary data structures [14, Fig. 2]. Here in Fig. 4, we adopted this approach through the integration of EBOM and MaintBOM. In practice the information provided to maintenance is technical documentation, manuals and instructions, with reduced complexity. In fact, it will not contain details from the manufacturing, such as production technologies or assembly steps which could be relevant for service and operations. The most relevant for operations is the composition of the final product as it was designed. Hence, we can consider this as a simplified version of the EBOM. The integration between EBOM and MaintBOM, is meant to allow for comprehensive traceability. The connection between BOMs ensures data consistency during the lifecycle of the product. We represent the links between these two entities in Fig. 4. The maintenance plan refers to parts in the MaintBOM, and is based on the maintenance processes specifying the location of defects/malfunctions accurately.

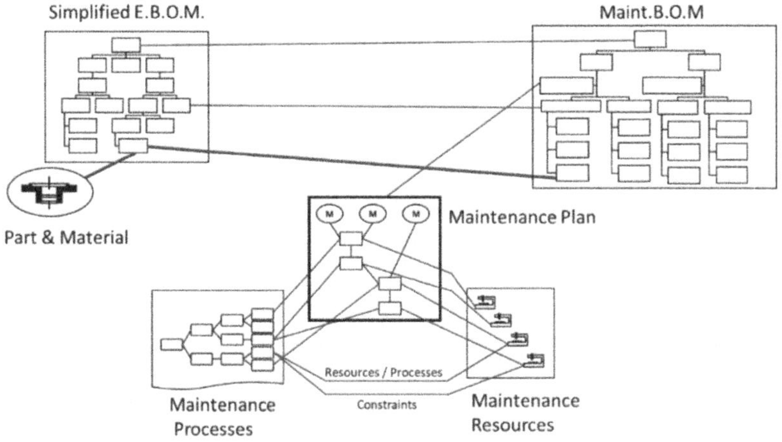

Fig. 4. Traceability data structure for industrial operations

Importantly, the replaced components keep their traceability with the respective items in the EBOM. While the connection within EBOM and MaintBOM is clearly evident (Fig. 4), there are no evident links between MBOM and MaintBOM.

Therefore, the inclusion and analysis of the manufacturing BOM is in this paper omitted, since it will only contribute to the manufacturing phase itself and being the focus the ability to encompass the traceability of the overall lifecycle, we decided to focus on EBOM and MaintBOM as a successful extent to connect beginning and middle of phases. However, the feedback loops shown at the bottom of Fig. 1 encompass all phases, including manufacturing phase.

3.4 End of Life, Circular Economy Planning

We distinguish the following statuses at EOL in recycling, disposal, repair and reuse. Indeed, the decision making of maintenance from previous phase will have an impact of the status of product in EOL phase. In turn, circular economy planning will therefore allow the product to continue using resources and eliminate waste after it has been established to be due for disposal. In fact, the purpose of recycling is to reuse the parts that are deemed waste, into a process that can be rehabilitated as new parts over the period of lifecycle. As shown in Fig. 1, Repair and Reuse can restore a part into a serviceable condition in the normal course of operation after repair. The integration between MaintBOM and New BOM will ensure data synchronization across both engineering and maintenance activities. When maintenance activities and processes are performed on the product, then valuable feedback may be generated leading to changes in product design, specifications, spare parts and components. Hence, the changes will require updates in BOM to indicate the revised product configuration. As a result, a new BOM will be provided with updated revisions to align the latest product revisions with engineering and operation.

In essence, maintenance feedback is used to inform the next phase about product's status and to modify configuration and variant management upon receiving input from EOL phase which in turn assures data conformance and traceability. As an example, Fig. 1, shows the backward traceability from EOL/Circular economy planning to operation and Maintenance planning, and from Manufacturing process design to product system design. By establishing robust backward traceability, any issues and failures can be recognized based on the insight obtained from EOL analysis which can link the product back to the resources for Repair/Reuse, Remanufacturing or Redesigning as antecedent steps. As a result, Backward traceability will give a holistic view of product lifecycle to organizations and stakeholders to enhance quality of product, decision making and efficiency.

4 Implementation of Traceability in Manufacturing and Operation

Establishing traceability between manufacturing and operations presented above aims to realize continuity of the Digital Thread. Like in general, the Digital Thread ideally can span different engineering disciplines and also departments or even organizations. In practice organizational boundaries pose a major challenge to the actual implementation. For any product which is operated by a different company than the manufacturer, the protection of intellectual property strongly limits the will for data sharing. For example, an airplane manufacturer will provide only as much information as strictly necessary to the airline operating the aircraft. It is likely to be part of the contract which kind of maintenance documentation will be provided. In a highly regulated area such as Civil Aviation, this also motivated by certification requirements. For the Digital Thread it means that the product traceability back to manufacturing is almost nonexistent.

In companies which follow a product from design to operation, the challenge still is to establish the traceability links between product related information generated and maintained from different disciplines. Within a single organization it is also possible to reach a high level of consolidation through the use of a powerful PLM system. In case the product related information is maintained in separate systems, there is an industry standard, which allows to establish and maintain links between heterogenous data called Open Services for Lifecycle Collaboration (OSLC) [17]. Since its latest version from 2021, this standard covers domains such as requirements management, change management, configuration management and quality management. Tools implementing this standard allow to define and navigate links between different product related information. Different types of traceability during operation can guide manufacturers and operators of products to make informed of decisions about how to handle incidents such as defects or other problems along the way even in critical situations. The requirements traceability may call for different tracing directions including Forward, Backward and Bidirectional traceability, as illustrated in Fig. 1 with circular arrows. Traceability could enhance the processes related to the operation by gathering, recording and reporting all information about every important activity during production. That important information may be used throughout the operation or to resolve seemingly unrelated challenges that might happen during the operation [5].

Digital Thread can also serve for service management improvement and operational activities. The recognition by manufacturers that they can better meet long term customer requirements by providing a service associated with product which has become a spotlight in manufacturing and service management process [18]. Effective operation and maintenance processes are critically important for high quality product service management. It can have an impact on product lifecycle traceability due to the cost associated with downtime and faulty products [19].

5 Discussion and Conclusion

An end-to-end traceability is crucial to achieve seamless data exchange and to maintain links and relationships over the lifecycle from detailed design to manufacturing, operation and maintenance to end of life and conversely. To analyze the importance of traceability through the Digital Thread, we present a comprehensive diagram to illustrate the implementation of various modes of traceability through relationships, enabling consistency at each transition between phases of the product lifecycle. Furthermore, we described a data structure to support traceability in industrial operation. The novelty of this work consists in the linkage of the simplified EBOM and MaintBOM as integrated models of data which is required to support traceability in industrial operations. In practice, during operation and maintenance, the engineering design and manufacturing process definitions are not passed to operators due to their intellectual property protection. Although the subsequent phase is operation and maintenance planning, it takes information from engineering design, in the form of a simplified EBOM, and uses a MaintBOM providing essential information about spare parts, assembly and subassemblies needed to modify.

The compatibility of the process between EBOM and MaintBOM is pivotal since it is extending the Digital Thread into maintenance phase where it allows to document defects or malfunctions happening during the operations. This comprises the corrective actions taken through maintenance processes by creating new revisions of any components that have undergone changes. Some changes captured in the MaintBOM based on the maintenance activities may lead to potential updates of the next version of the product. In essence, this approach would support the alignment of design requirements with maintenance requirements to increase the efficiency of maintenance, and alleviate the problem of engineers to be less overwhelmed with product recalls and managing revisions. Consequently, will improve service delivery and customer satisfaction. This would be particular importance in the case of Product Service Systems which will be considered in future work. It is also important to note that, there are different needs for traceability during different phases of the lifecycle. Specifically, they are different in operation and product development. In operations as we have shown, traceability is required to the simplified EBOM only, while in product/system design forward and backward traceability are required across all disciplines throughout product development cycle. This approach could be called modulated traceability, which needs to be further investigated. Including concrete case study which can promote this method with practitioners.

Acknowledgments. Aspects of this study were carried out within the MICS (Made in Italy – Circular and Sustainable) Extended Partnership and received funding from the European Union Next-Generation EU (PIANO NAZIONALE DI RIPRESA E RESILIENZA (PNRR) – MISSIONE 4 COMPONENTE 2, INVESTIMENTO 1.3 – DD 1551.11-10-2022, PE00000004).

This manuscript reflects only the authors' views and opinions, neither the European Union nor the European Commission can be considered responsible for them.

References

1. Pang, T.Y., Pelaez Restrepo, J.D., Cheng, C.T., et al.: Developing a digital twin and digital thread framework for an 'industry 4.0' shipyard. Appl. Sci. (Switzerland) **11**, 1–23 (2021)
2. Singh, V., Willcox, K.E.: Engineering design with digital thread. AIAA J. **56**, 4515–4528 (2018)
3. Zhang, Q., Zheng, S., Yu, C., et al.: Digital thread-based modeling of digital twin framework for the aircraft assembly system. J. Manuf. Syst. **65**, 406–420 (2022)
4. Gotel, O., et al.: Traceability Fundamentals. In: Software and Systems Traceability, pp. 3–22. Springer, London (2012)
5. GS1. GS1 Global Traceability Standard GS1's framework for the design of interoperable traceability systems for supply chains (2017)
6. Granillo-Macías, R., Simón-Marmolejo, I., González-Hernández, I.J., Zuno-Silva, J.: Traceability in industry 4.0: A case study in the metal-mechanical sector. Acta Logist. **7**, 95–101 (2020)
7. Schuitemaker, R., Xu, X.: Product traceability in manufacturing: a technical review. Prcedia CIRP 700–705. Elsevier B.V. (2020)
8. Nigischer, C., Hennig, M., Gerhard, D.: A concept to integrate manufacturing execution and product data management systems. In: IFIP Advances in Information and Communication Technology, pp 315–324. Springer (2019)
9. Lu, Y., Xu, X., Wang, L.: Smart manufacturing process and system automation – a critical review of the standards and envisioned scenarios. J. Manuf. Syst. **56**, 312–325 (2020)
10. Gotel, O., Cleland-Huang, J., Hayes, J.H., et al.: Traceability fundamentals. In: Software and Systems Traceability, pp 3–22. Springer-Verlag London Ltd. (2012)

11. Kumar, V., Hallqvist, C., Ekwall, D.: Developing a framework for traceability implementation in the textile supply chain. Systems (2017)
12. Kwon, S., Monnier, L.V., Barbau, R., Bernstein, W.Z.: Enriching standards-based digital thread by fusing as-designed and as-inspected data using knowledge graphs. Adv. Eng. Inform. (2020)
13. Shercliff, H.R., Lovatt, A.M.: Selection of manufacturing processes in design and the role of process modelling. Prog. Mater. Sci. **46**, 429–459 (2001)
14. Brovar, Y., Menshenin, Y., Knoll, D., Fortin, C.: Modelling of Engineering and Manufacturing Data Structures Through a DSM-Based Approach, pp 96–109 (2022)
15. Huet, G., Fortin, C., Mcsorley, G., et al.: The management of manufacturing processes using complementary information structures; integration of MBSE and PLM Approaches; View project Behavioural Product Lifecycle Management; View project INTERNATIONAL CONFERENCE ON ENGINEERING DESIGN; ICED11 (2011)
16. Soori, M., Arezoo, B., Dastres, R.: Internet of things for smart factories in industry 4.0: a review. Internet Things Cyber-Phys. Syst. **3**, 192–204 (2023)
17. Open Services for Lifecycle Collaboration (OSLC) (2023). http://open-services.net/wiki
18. Mcsorley G, De Génie D, École M, De Montréal P. A novel approach to product lifecycle management and engineering based on product in-use information (2014)
19. Sang, G.M., Xu, L., De Vrieze, P., et al.: Predictive maintenance in industry 4.0. In: ACM International Conference Proceeding Series; Association for Computing Machinery (2020)
20. Kirova, V., Kirby, N., Kothari, D., Childress, G.: Effective requirements traceability: models, tools, and practices. Bell Labs Tech. J. **12**(4), 143–157 (2008)

Potential of the Industrial Metaverse – A Taxonomic Approach

Manuel Holler[1](✉), Tuomo Eloranta[2], and Umut Demiriz[1]

[1] Zurich University of Applied Sciences, Theaterstrasse 17, 8400 Winterthur, Switzerland
{manuel.holler,umut.demiriz}@zhaw.ch
[2] Aalto University, Maarintie 8, 02150 Espoo, Finland
tuomo.eloranta@aalto.fi

Abstract. The metaverse is often regarded as the next successor of the current internet, and also for the industrial sector, the metaverse has the potential to enhance physical-digital approaches like the Industrial Internet of Things and facilitate the shift from Industry 4.0 to 5.0. To move away from context-specific value and comprehensively grasp its potential, we develop a classification model following the research question: "[RQ] Which dimensions and characteristics are suited to holistically describe use cases of the Industrial Metaverse?" To address, we pursue a taxonomic approach leveraging the reliable taxonomy development method by Nickerson et al. (2013) based on real-world uses cases of the Industrial Metaverse. Guided by the theory of the affordances, the resulting model classifies use cases by (1) IT artifact- (dominant building block, technology materiality, technology origin), (2) Organization- (lifecycle context, environmental context, role of the human) and (3) Affordances- (lifecycle value-add, sustainability pillar, value objective, value direction, impact) -related dimensions. Keeping review- and classification-related limitations in mind, we contribute a "theory for analysis" for science and a "tool to understand and innovate" for practice.

Keywords: Industrial Metaverse · Metaverse · Potential · Taxonomy · Use Case · Product Lifecycle Management · PLM

1 Introduction

The metaverse is often regarded as the next successor of the current internet [1–3]. Highly simplified, the concept describes a *"three-dimensional virtual world where avatars engage in political, economic, social and cultural activities"* [2:4211]. Driven by media-prominent company transformations of Facebook, Microsoft and Nvidia, the consumer perspective often stands in the center [4, 5]. For the industrial sector, the metaverse has the potential to enhance physical-digital approaches like the Industrial Internet of Things

Parts of this paper draw on modified, updated and rewritten content from the unpublished research proposal "Technology Affordances of the Industrial Metaverse" from the Swiss National Science Foundation.

Published by Springer Nature Switzerland AG 2025
P. Sureephong et al. (Eds.): PLM 2024, IFIP AICT 741, pp. 140–152, 2025.
https://doi.org/10.1007/978-3-031-93323-3_11

and facilitate the shift from Industry 4.0 to 5.0 [6–8]. Citing market projections by ABI Research, Waxer et al. [9] quantify the potential of this Industrial Metaverse: While the Consumer/Enterprise Metaverse is expected to deliver 30 resp. 50 BN USD by 2030, the Industrial Metaverse is forecasted to yield even 100 BN USD [9].

To leverage this substantial potential, several stakeholders have jumped on the bandwagon and realized standalone use cases: For instance, technology multinational Siemens has deployed immersive IoT-, AI- and XR-based factory planning and operations [10]. Another example is compressor manufacturer Burckhardt Compression which provides remote expert support leveraging similar combined technologies [11]. While these use cases certainly show context-specific value, a classification model to comprehensively grasp the potential is not available. Thus, we formulate the following research question: *"[RQ] Which dimensions and characteristics are suited to holistically describe use cases of the Industrial Metaverse?"* To address, we pursue a taxonomic approach leveraging the reliable taxonomy development method by Nickerson et al. [12] based on real-world uses cases of the Industrial Metaverse.

For that, the paper has been divided into five major parts: The first part deals with the Industrial Metaverse itself and related work. The second part then summarizes the methodological basis as well as the model development. The third part provides the taxonomy including in-depth description of dimensions and characteristics as major outcome. The fourth part follows up with a taxonomy application and reflection. The last part closes the paper with contributions, implications, limitations and an outlook.

2 Background

Industrial Metaverse. Nomenclatorally, the term is a combination of "meta" and "universe" whereby most sources refer to Neal Stephenson's novel "Snow Crash" [3]. While the notion of the Consumer Metaverse is meanwhile well and differentiated understood in scientific literature [2, 3, 13], the Industrial Metaverse is still in the discovery phase [8, 9, 14]. The most striking characteristic is the inherent tangibility and physicality within the industrial sector [6, 7, 9]. Another key property is the integration of the most advanced technologies such as Digital Twins, Internet of Things (IoT), Artificial Intelligence (AI) & Machine Learning (ML), Edge & Cloud Computing, Networks, Blockchain and Virtual/Augmented/Extended Reality (VR/AR/XR) allowing for great synergies [6, 7, 9].

Several stakeholders in academia and practice are working on their dynamic definitions of the Industrial Metaverse, whereby Meige and Eagar [10] provide a non-exhaustive overview. For this paper, we leverage the notion of Waxer et al. [9:5] as it integrates both practice and academia: *"The Industrial Metaverse – a metaverse sector that mirrors and simulates real machines, factories, cities, transportation networks, and other highly complex systems – will offer to its participants fully immersive, real-time, interactive, persistent and synchronous representations and simulations of the real world."* From a technical perspective, first reference architectures have been suggested [5, 7, 15]. Without diving too deep into computer science, the IEEE Decentralized Metaverse Initiative [15] proposes a system with the horizontal layers infrastructure, interaction and application, supported by the vertical layers industrial ecology and privacy/safety.

To address the "old wine in new bottles" discussion, two demarcations are of particular interest: For one, the Industrial Metaverse delimitates from other (consumer/enterprise) metaverses by increased relevance of interaction and persistence, yet reduced relevance of immersion [10]. For another, the Industrial Metaverse also dissociates from present Industry 4.0 approaches by the introduced technology convergence and enhanced importance of selected building blocks such as data visualization [10].

Related Work. To identify neighboring classification elements of Industrial Metaverse use cases, we used a systematic review according to vom Brocke et al. [16], as described in detail in section "3. Methodology". In this context, valuable contributions on the Industrial Metaverse equally have their roots in both academia (e.g., Zheng et al. [7]) and practice (e.g., Meige and Eagar [10]).

Beside more general metaverse taxonomies (e.g., Park and Kim [2]) in a wider sense, in the narrower sense especially one-dimensional, two-dimensional and three-dimensional models with decreasing numbers can be found: First, regarding one-dimensional models, Bohné et al. [17] classify applications by lifecycle stage (pre-production, production, post-production) and supported business functions (training, customer support etc.). Second, in terms of two-dimensional models, Zheng et al. [7] outline potential applications, such as R&D Virtual Crowdsourcing, within a framework that considers the dimensions of "industrial systems granularity" and "product life cycle." Third, referring to three-dimensional models, Meige and Eagar [10] group use cases by industry, category and business function as most comprehensive overview. Summarizing these findings in the language of Müller-Bloch and Kranz [18], a knowledge gap manifests as no comprehensive, empirically grounded and manufacturer-centric classification exists.

3 Methodology

Methodological Basis. The aim of this paper to classify use cases of the Industrial Metaverse. From a historical perspective, organizing fields and creating clarity is a fundamental need in research [12, 19]. Rooted in biology, classification steadily diffused to other disciplines [12, 19]. Thereby, the term refers to both: For one, classification (processes) are inductive ("empirical"), deductive ("conceptual") or intuitive ("ad hoc") [12, 19]. For another, classification (outcomes) are taxonomies (commonly developed empirically), typologies (commonly developed conceptually) or frameworks (general term) [12, 19]. Against this backdrop, our objects to classify are "business use cases", also defined as *"use cases describing goals and events, at the business level, that is, in a way that defines the work to be done without reference to the system to be built"* by Alexander and Neil [20:507].

The employed taxonomy development method by Nickerson et al. [12] (Fig. 1) has a number of advantages: The approach (1) is highly cited and successfully applied in many domains, (2) reverts to both induction and deduction and (3) ranges under the paradigm of DSR to produce useful artifacts [12]. To consider most recent guidance, Nickerson's original work is complemented by its successor [21]. Both author teams encourage taxonomy builders for rigor [12, 21], so we aim for diligence and transparency across the whole process.

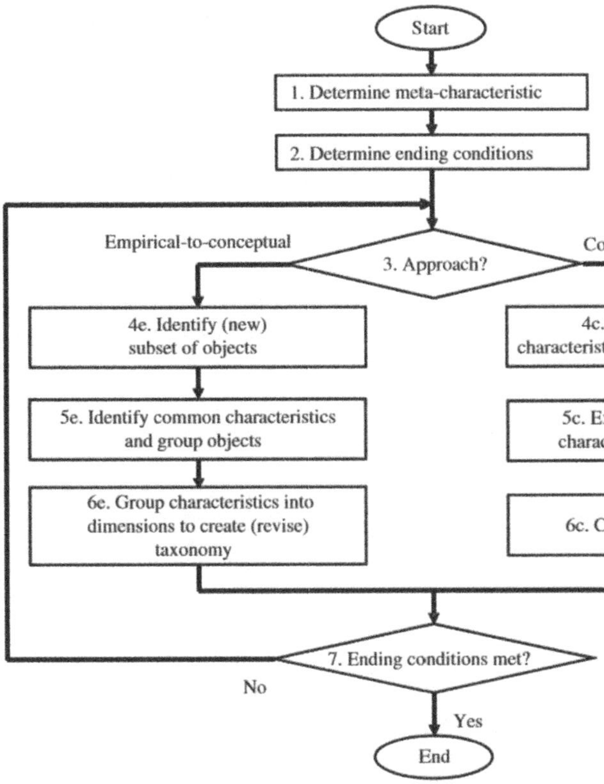

Fig. 1. Applied method for taxonomy development [12].

Taxonomy Development. Initially, we determined the meta characteristic in line with our goal ("scientific tool to understand and innovate") and audience ("product, service, business and IT executives in the industrial sector"). As theoretical basis for this most important characteristic, we applied the theory of affordances rooted in ecological psychology [22]. In short, affordances are *"the possibilities for goal-oriented action afforded to specified user groups by technical objects"* [23: 622]. An example: For the technical object "laptop", one person may perceive the affordance "write a paper", while another person may perceive "watch a movie". Or applied to our context: The Industrial Metaverse with its properties serves as technical object, to which the different industrial actors perceive different affordances (e.g., virtual product-market-fit testing). Recent theory models are multi-tiered: Affordances exist, are perceived, are actualized and have an effect [24]. This theory was selected as it *"provides a powerful lens for studying the co-constitutive relations between technology and people in organizations"* [25: 237]. Furthermore, the affordance theory has already proved successfully to study IT phenomena [26, 27].

In line with Nickerson et al. [12] ending conditions were defined. For one, these manifest in objective ending conditions with the most important *"mutually exclusive and collectively exhaustive"* [12: 343]. For another, in the sense of subjective ending conditions our taxonomy should be *"concise, robust, comprehensive, extendible, explanatory"* [12: 344].

For our empirical data basis, we adapted the systematic review according to vom Brocke et al. [16] as follows: Our review scope [16] was use case- and dimension-centric, neutral, general and representative. This was followed by topic amplification [16] around the core term "Industrial Metaverse", which led to our search in the database Google Scholar (2019–2024) including forward/backward search [16], but upon the early stage we also considered practice whitepapers. The analysis [16] was multi-stage from screening to analysis including duplicate removal. We included (1) real-world cases with relation to the industrial sector, (2) cases clearly designed by the authors as Industrial Metaverse cases and (3) cases showing high quality and depth. In contrast, we excluded (1) traditional Industry 4.0 cases, (2) brief cases without sufficient information and also (3) pure theory/marketing cases. With an overall amount of 64 cases as input for the classification, the top sources by far were Meige and Eagar [10] (27 cases) and Angerer et al. [11] (16 cases). Similarly, suitable dimensions and characteristics were analyzed. Use cases and potential dimensions were consolidated in a database. While use cases were described with an ID, title, description and source, dimensions were solely listed with the source.

This database served as foundation for the two limbs: In the "empirical-to-conceptual" stream, we analyzed the use cases for common characteristics (in line with the meta-characteristic) [12]. To the contrary, in the "conceptual-to-empirical" stream, we matched the identified dimensions and characteristics with the use cases [12]. As design processes are iterative [12, 21], we iterated the classification till all ending conditions were fullfilled. To guarantee its usefulness [12, 21], we also shaped the model with experienced product managers.

4 Results

In the progress of this paper, a taxonomy of use cases of the Industrial Metaverse was developed (Table 1). In line with our meta characteristic [24], the taxonomy is organized along the meta dimensions (1) IT artifact (technology), (2) organization (context) and (3) Affordances (potential). Within these, the actual dimensions (D_i) and characteristics (C_i) can be found. As we are particularly interested in understanding the potential of the Industrial Metaverse, these dimensions and characteristics are overweighted (3-3-5).

D1: Dominant Building Block. While the Industrial Metaverse per definition involves technology integrations, the analyzed data show the occurrence of one or more dominating building block(s). Accordingly, this dimension features the characteristics (C1.1) Digital Twins, (C1.2) IoT, (C1.3) AI & ML, (C1.4) Blockchain, (C1.5) XR and (C1.6) Combinations, which also reflect in literature (e.g., Waxer et al. [9]). For example, for Bosch's case "Design Review of Products in the Metaverse" [11] certainly XR stands out, while Siemens's more holistic case "Factory Planning and Operations" [11] shows integrations of IoT, AI and XR. In this context, the different relevance is discernable, while IoT, AI and XR can be considered as mainstream, blockchain clearly must be allocated to niche building blocks.

Table 1. Taxonomy of use cases of the Industrial Metaverse.

IT artifact (technology)							
D1: Dominant building block	C1.1: Digital Twins	C1.2: IoT	C1.3: AI & ML	C1.4: Blockchain	C1.5: XR	C1.6: Combinations	
D2: Technology materiality	C2.1: Digital materiality (only software)	C2.2: Digital and physical materiality (software & hardware)					
D3: Technology origin	C3.1: Off-the-shelf platform	C3.2: Customized platform	C3.3: Self-developed platform	C3.4: Co-developed platform			
Organization (context)							
D4: Lifecycle context	C4.1: Beginning-of-life	C4.2: Middle-of-life	C4.3: End-of-life	C4.4: Whole lifecycle			
D5: Environmental context	C5.1: Product/ service	C5.2: Process	C5.3: Machine/ line/factory	C5.4: Human	C5.5: Ecosystem		
D6: Role of the human	C6.1: Human in the lead	C6.2: Human in the loop	C6.3: Full automation				
Affordances (potential)							
D7: Lifecycle value-add	C7.1: Improved design	C7.2: Enhanced manu-facturing	C7.3: Optimized distribution	C7.4: Augmented use	C7.5: Amelio-rated support	C7.6: Improved retire	C7.7: Amended whole lifecycle
D8: Sustainability pillar	C8.1: Economic sustainability	C8.2: Social sustainability	C8.3: Ecological sustainability	C8.4: Combinations			
D9: Value objective	C9.1: Efficiency	C9.2: Effectiveness					
D10: Value direction	C10.1: Enterprise-internal	C10.2: Enterprise-external					
D11: Impact	C11.1: Low impact	C11.2: Medium impact	C11.3: High impact				

D2: Technology Materiality. From both empirically clustering the use cases and conceptual literature work (e.g., Jarrahi [28]) two fundamental types of technology materiality became evident, namely (C2.1) Digital materiality (only software) and (C2.2) Digital and physical materiality (software & hardware). For one, digital materiality often appears in early lifecycle stages where not yet a physical product exists (e.g., "Product-Feature-Testing and Customer Feedback" by digital.auto [11]). For another, digital and physical materiality occurs in later stages, where the physical product exists and is connected by IoT components (e.g., "Digital Rail Operations" by Deutsche Bahn [11]). In this dimension, the ambiguity of the virtual metaverse and the tangible industry becomes striking.

D3: Technology Origin. During the classification, it become apparent that the employed metaverse platforms have multiple origins. Also informed by existing literature (e.g., Li et al. [29]), this dimension comprises the characteristics (C3.1) Off-the-shelf, (C3.2) Customized, (C3.3) Self-developed and (C3.4) Co-developed platform. While "off-the-shelf" (use of standard platforms of the large metaverse players) can be applied for mainstream cases (e.g., "Metaverse Training" by Deutsche Bahn [11]), industrial necessities often lead to "customization" (e.g., "Virtual Safety Street" by RWE [11]). "Self-developed" solutions especially appear at companies with dual roles that develop metaverse software for their customers and also apply it for their own factories (e.g., "Digital Native Factory" by Siemens [10]), whereby high complexity often requires "Co-development" (e.g., Siemens, Nvidia and Sony).

D4: Lifecycle Context. With regard to the organization, various lifecycle contexts are discernable. Conceptual work (e.g., Terzi et al. [30]) and empirical clustering resulted in the characteristics (C4.1) Beginning-of-life, (C4.2) Middle-of-life, (C4.3) End-of-life and (C4.4) Whole lifecycle. To illustrate, BP's use case "Digital Twins for Asset Optimization" [10] is embedded in typical use, support and maintenance activities accomplished by corresponding operators and MRO experts. Similarly, other cases have their organizational context in design/manufacturing (e.g., marketing managers, product developers, factory planners) respectively retiring settings (e.g., sustainability managers). Thereby, a clear focus on beginning-of-life and middle-of- life contexts is striking, whereas end-of-life settings are rather rare.

D5: Environmental Context. In terms of environmental context, the cases also demonstrated a wide spectrum. Aligning with product lifecycle management literature (Terzi et al. [30]), the cases center around (C5.1) Product/service (e.g., "Passenger Welcome Sequence" by digital.auto), (C5.2) Process (e.g., "AI-supported Manufacturing Processes" by Tesla), different levels of production systems (C5.3) Machine/line/factory (e.g., "Digital Factory" by BWM) and (C5.4) Human (e.g., "Spatial Motion Analysis" by Mercedes) [11]. Beside these clearly assignable contexts, a substantial share of cases touches multiple items in the sense of an ecosystem (C5.5) (e.g., Renault's "Industrial Metaverse Factory" [10]). With production activities in the heart of industrial value chains, most cases can be categorized to (C5.3) Machine/line/factory.

D6: Role of the Human. Another important dimension to characterize the organizational set up, is the relationship between technology and the human. This role manifested in a spectrum from "only tech" to "only human", where existing automation levels (e.g., Vagia et al. [31]) were adapted to the required granularity of this work. A first set such as "Immersive Product Design Improvement" (Boeing [10]) is clearly human-dominated (C6.1: Human in the lead). A second group like "Real Estate Twin" (Deutsche Bahn [11]) shows a set-up where most tasks are already taken over by the machine with human surveillance (C6.2: Human in the loop). A third category such as "AI-Optimization of Quality Controls" (Pegatron [11]) shows full automation without any human involvement (C6.3: Full automation). While this last form certainly is the ultimate goal, most use cases still show a hybrid mode.

D7: Lifecycle Value-add. While the Industrial Metaverse ideally unfolds its potential across the whole product lifecycle, the added-value is often bound to a specific stage. Coding the data while continuously comparing with the reviewed literature (e.g., Terzi et al. [30]), this important dimension manifests in the characteristics (C7.1) Improved design, (C7.2) Enhanced manufacturing, (C7.3) Optimized distribution, (C7.4) Augmented use, (C7.5) Ameliorated support, (C7.6) Improved retire and (C7.7) Amended whole lifecycle. As illustrative use case Ford's "VR-based Automotive Service Training" [10] offers the OEM ameliorated support. Here it should be noted that these values range from perception up to real effects [24], depending on the implementation maturity of the use case.

D8: Sustainability Pillar. Derived both inductively and deductively, the Industrial Metaverse can address various macro goals. In line with the three-pillar model [32], we differentiate (C8.1) Economic, (C8.2) Social, (C8.3) Ecological sustainability and (C8.4) Combinations. While certainly most cases are initiated for economic value (e.g., saving costs, opening new revenues), interestingly several also have social (e.g., improving ergonomics, reducing injury risks) and ecological (e.g., decreasing carbon emissions, enhancing circularity) motives. Here, "Remote Expert Support by Industrial Metaverse" by Burckhardt Compression [11] can be brought up where issues threating productivity can be fixed short-term (economic), where intercontinental flights can be avoided (environmental) and where the remote expert can stay with his family over night (social).

D9: Value Objective. Organizations perceive value of the Industrial Metaverse in two fundamental ways: For one, a first group of use cases (e.g., "Industrial Metaverse Factory" by Renault [10]) aims for efficiency-driven optimizations under the Industry. 4.0 paradigm to stay competitive (C9.1: Efficiency). For another, a second group (e.g., "Product-Feature-Testing and Customer Feedback" by digital.auto [11]) targets to achieve new things, such as creating new products, revenue streams or customers (C9.2: Effectiveness). While the former might be decisive for industrial executives to start the metaverse journey in the short term, the latter probably may be even more relevant in the long-term.

D10: Value Direction. Besides this value objective, the value direction was identified as equally important dimension from use case grouping and literature review (e.g., Barth et al. [33]). Following this line, this dimension manifests in the key characteristics (D10.1) Enterprise-internal and (D10.2) Enterprise-external value creation. Whereas internally directed value creation targets own employees and processes (e.g., Pegatron's case "AI-Optimization of Quality Controls" [10]), externally oriented value creation goes beyond the company borders directed towards customers and suppliers (e.g., Burckhardt Compression's case "Remote Expert Support by Industrial Metaverse" [11]).

D11: Impact. Finally, during the research it become clear that another dimension is necessary to describe the value for the industrial enterprises. The potential enabled by the Industrial Metaverse manifests in (D11.1) Low, (D11.2) Medium and (D11.3) High impact. Depending on the context, parameters such as scope (local vs. global), duration (temporary vs. permanent) and urgency (normal situation vs. emergency) can be used to operationalize this impact. Accordingly, low impact cases show local, temporary or normal character, whereas high impact cases feature global, permanent or emergency value. Starting small and growing, manufacturers ideally seek for high value cases (e.g., large digitally planned and operated factories, "Digital Native Factory" by Siemens [10]).

5 Discussion

Taxonomy Application. To comprehensively grasp the potential of the Industrial Metaverse, we developed a classification model. To exemplify the taxonomy, we demonstrate it by the aid of an expressive use case. To illustrate it as good as possible, we selected the narrow and not too ample case "Remote Expert Support by Industrial Metaverse" [11, 34] (Table 2, relevant characteristics in boldface) by Burckhardt Compression, a Swiss manufacturer of large compressors.

Table 2. Classified use case "Remote Expert Support by Industrial Metaverse".

IT artifact (technology)							
D1: Dominant building block	C1.1: Digital Twins	C1.2: IoT	C1.3: AI & ML	C1.4: Blockchain	C1.5: XR	**C1.6: Combinations**	
D2: Technology materiality	C2.1: Digital materiality (only software)	**C2.2: Digital and physical materiality (software & hardware)**					
D3: Technology origin	C3.1: Off-the-shelf platform	**C3.2: Customized platform**	C3.3: Self-developed platform	C3.4: Co-developed platform			
Organization (context)							
D4: Lifecycle context	C4.1: Beginning-of-life	**C4.2: Middle-of-life**	C4.3: End-of-life	C4.4: Whole lifecycle			
D5: Environmental context	**C5.1: Product/ service**	C5.2: Process	C5.3: Machine/ line/factory	C5.4: Human	C5.5: Ecosystem		
D6: Role of the human	**C6.1: Human in the lead**	C6.2: Human in the loop	C6.3: Full automation				
Affordances (potential)							
D7: Lifecycle value-add	C7.1: Improved design	C7.2: Enhanced manu-facturing	C7.3: Optimized distribution	C7.4: Augmented use	**C7.5: Amelio-rated support**	C7.6: Improved retire	C7.7: Amended whole lifecycle
D8: Sustainability pillar	C8.1: Economic sustainability	C8.2: Social sustainability	C8.3: Ecological sustainability	**C8.4: Combinations**			
D9: Value objective	C9.1: Efficiency	**C9.2: Effectiveness**					
D10: Value direction	C10.1: Enterprise-internal	**C10.2: Enterprise-external**					
D11: Impact	C11.1: Low impact	C11.2: Medium impact	**C11.3: High impact**				

In aggregate form, the Industrial Metaverse is used to "teleport" experienced technicians to remote locations somewhere in the ocean. The ship personal uses a mobile device with a LIDAR sensor to create a scan (digital twin) of the compressor, which is sent to the remote expert, where he has the possibility to overlay it with other CAD, PLM and IoT data. In combination with the live video from the ship, this persistent metaverse serves as foundation for interactive and immersive troubleshooting, whereby the field service technician finally implements the recommended working steps [11, 34].

Regarding the dominant building block, we can see a very compelling integration of the most advanced digital technologies (D1). Furthermore, the case shows digital and physical materiality (D2), combining digital elements (o.a. CAD, PLM data) with the physical compressor (o.a. scan, IoT data). To realize, Burckhardt Compression leverages a customized metaverse platform (D3) from Boston-based provider PTC. In terms of the lifecycle context, the case clearly addresses Middle-of-life (D4) with the corresponding MRO tasks and personnel. Beyond, the use case is embedded in an environmental product-/service-centric context (D5) minimizing compressor downtime. With a large proportion of manual, labor-intense repair activities, the human is in the lead (D6) in this case. Regarding the outcome, ameliorated support is the lifecycle value-add (D7). As introduced, Burckhardt Compression thereby contributes to economic, social and ecological sustainability (D8). While the so far not doable "expert beaming" is made possible, the value objective is effectiveness (D9). Upon the interaction with the external ship operator, the value direction can be classified as enterprise-external (D10). Lastly, despite its local and temporary character, but high level of urgency, we classify this case as high impact case (D11).

Taxonomy Reflection. This reflection follows the line of (1) conspicuous insights and (2) comparison with literature. First, the nascent definition of the Industrial Metaverse and use cases is striking. Not unexpected upon the early stage, the works referred show various nations and also the use cases are often difficult to distinguish from existing Industry 4.0 approaches. Beyond, with regard to the distribution of the use cases, a significant insight is the breath in combination with frequently occurring manifestations. While we can detect a plethora of rather "exotic" cases (e.g., blockchain-based traceability), also clear foci (e.g., virtual factory planning) are discernable. Despite some lifecycle-oriented, integral use cases, most industrial companies approach the Industrial Metaverse in a very focused manner, addressing a specific lifecycle stage, technology or business function.

Second, comparing the identified dimensions and characteristics with the introduced one-, two- and three-dimensional models [7, 10, 17], but also related smart use case taxonomies (e.g., Smart service systems by Herterich et al. [35], Smart product development scenarios by Holler et al. [36]), common but also novel elements can be found. While our model also reverts to established dimensions like the "Lifecycle phase" [7, 35], our taxonomy also features unique descriptors such as the "Dominant building block".

6 Conclusion

This paper set out to develop a taxonomy of use cases of the Industrial Metaverse. For academics, up to the best of the authors' knowledge, we make a contribution by a first comprehensive, empirically grounded and manufacturer-centric classification. Leveraging the framework of Gregor [37], we therefore provide a "theory for analysis" of the Industrial Metaverse. Against the backdrop of the early stage and calls-for-research, we make an entry point of the Industrial Metaverse to the community of product lifecycle management [30, 38, 39]. For practitioners, the classification similarly can be leveraged for technology management: First, it gives managers the possibility to assess and understand existing use cases ("How can the latest case brought up in a meeting or posted on LinkedIn be categorized?"). Second, it also enables them to cogitate novel and unconventional applications ("How could a case with the dominating building block "Blockchain" in the lifecycle context "End-of-life" look like?").

Only the last months of research on the Industrial Metaverse has shown substantial progress. Manufacturers should embrace this new technology, and balance technology push ("Where could I use the metaverse?") and market pull ("Where does the metaverse solve my problems?"). Beside the well-discussed focus use cases, also the investigation of niches makes sense. Steadily keeping existing digitalization activities and privacy/security aspects in mind, the maturity towards a holistic Industrial Metaverse should be continuously increased.

Notwithstanding these contributions, our paper also must critically deal with restrictions: First, review-related, upon the representative review, we might have not considered all use cases and dimensions/characteristics [16], also of course there may be further unknown cases with potentially novel manifestations in the secret industrial research & development departments. Second, classification-related, by the sometimes unclear description and ambiguities of the coded cases and the qualitative approach in general [19], the use cases were classified to the best of our knowledge. Hence, further empirical data are needed to further validate and refine the taxonomy, but also to enhance the understanding and relevance of this complex model.

Scientific research on the Industrial Metaverse is in its infancy: To dive deeper into the potentials and challenges of the Industrial Metaverse [7], more in-depth studies collecting primary data within different industry branches and departments add value. Furthermore, to pave the way for industrial companies to successfully adopt the metaverse and systematically increase its maturity, the development of a capability model [40] is also worth mentioning.

Acknowledgments. The authors of this article were supported by the Swiss National Science Foundation "Spark" program (grant number 220855).

Disclosure of Interests. The authors have no competing interests to declare that are relevant to the content of this article.

References

1. Dwivedi, Y.K., et al.: Metaverse beyond the hype: multidisciplinary perspectives on emerging challenges, opportunities, and agenda for research, practice and policy. Int. J. Inf. Manag. **66**, 102542 (2022)
2. Park, S.M., Kim, Y.-G.: A metaverse: taxonomy, components, applications, and open challenges. IEEE Access **10**, 4209–4251 (2022)
3. Dolata, M., Schwabe, G.: What is the metaverse and who seeks to define it? Mapping the site of social construction. J. Inf. Technol. **38**(3), 239–266 (2023)
4. Wang, H., et al.: A survey on metaverse: the state-of-the-art, technologies, applications, and challenges. IEEE Internet Things J. **10**(16), 14671–14688 (2023)
5. Bhattacharya, P., et al.: Towards future internet: the metaverse perspective for diverse industrial applications. Mathematics **11**(4), 941 (2023)
6. Yao, X., Ma, N., Zhang, J., Wang, K., Yang, E., Faccio, M.: Enhancing wisdom manufacturing as industrial metaverse for industry and society 5.0. J. Intell. Manuf. **35**, 235–255 (2022)
7. Zheng, Z., et al.: Industrial metaverse: connotation, features, technologies, applications and challenges. In: Fan, W., Zhang, L., Li, N., Song, X. (eds.) Methods and Applications for Modeling and Simulation of Complex Systems, CCIS, vol. 1712, pp. 239–263. Springer, Singapore (2022)

8. Manikandan, S., Elakiya, E., Punitha Devi, C., Khasim, S.: Industrial Revolution and Metaverse: Industry 5.0. Quing Publications, Puducherry, India (2023)
9. Waxer, C., et al.: The emergent industrial metaverse. MIT Technology Review Insights Whitepaper (2023)
10. Meige, A., Eagar, R.: The industrial metaverse - Making the invisible visible to drive sustainable growth. Arthur D. Little Whitepaper (2023)
11. Angerer, H., et al.: Metaverse Use Cases, Mehrwerte und Potenziale für den Wirtschaftsstandort Deutschland. Bitkom Whitepaper (2023)
12. Nickerson, R.C., Varshney, U., Muntermann, J.: A method for taxonomy development and its application in information systems. Eur. J. Inf. Syst. **22**(3), 336–359 (2013)
13. Ritterbusch, G.D., Teichmann, M.R.: Defining the metaverse: a systematic literature review. IEEE Access **11**, 12368–12377 (2023)
14. Lee, J., Kundu, P.: Integrated cyber-physical systems and industrial metaverse for remote manufacturing. Manuf. Lett. **34**, 12–15 (2022)
15. Chi, C., et al.: The industrial metaverse report. IEEE Decentralized Metaverse Initiative Industry Report (2023)
16. Vom Brocke, J., Simons, A., Riemer, K., Niehaves, B., Plattfaut, R., Cleven, A.: Reconstructing the giant: on the importance of rigour in documenting the literature search process. In: 17th European Conference on Information Systems (ECIS), Verona, Italy (2009)
17. Bohné, T., Li, C., Triantafyllidis, K.: Exploring the industrial metaverse: a roadmap to the future. University of Cambridge & World Economic Forum Whitepaper (2023)
18. Müller-Bloch, C., Kranz, J.A.: Framework for rigorously identifying research gaps in qualitative literature reviews. In: 36th International Conference on Information Systems (ICIS), Fort Worth, United States (2015)
19. Bailey, K.D.: Typologies and Taxonomies: An Introduction to Classification Techniques. Sage, Thousand Oaks, California (1994)
20. Alexander, I., Neil, M.: Scenarios, Stories, Use Cases: Through The Systems Development Life-cycle. John Wiley & Sons, Chichester, United Kingdom (2004)
21. Kundisch, D., et al.: An update for taxonomy designers. Bus. Inf. Syst. Eng. **64**(4), 421–439 (2022)
22. Gibson, J.J.: The ecological approach to visual perception. Lawrence Erlbaum Associates, Hillsdale, United States (1986)
23. Markus, M.L., Silver, M.S.: A foundation for the study of IT effects: a new look at De- Sanctis and Poole's concepts of structural features and spirit. J. AIS **9**(10/11), 609–632 (2008)
24. Pozzi, G., Pigni, F., Vitari, C.: Affordance theory in the IS discipline: A review and synthesis of the literature. In: 20th Americas Conference on Information Systems (AMCIS), Savannah, United States (2014)
25. Fayard, A.-L., Weeks, J.: Affordances for practice. Inf. Organ. **24**(4), 236–249 (2014)
26. Dincelli, E., Yayla, A.: Immersive virtual reality in the age of the metaverse: a hybrid- narrative review based on the technology affordance perspective. J. Strat. Inf. Syst. **31**(2), 101717 (2022)
27. Herterich, M., Dremel, C., Wulf, J., vom Brocke, J.: The emergence of smart service ecosystems – the role of socio-technical antecedents and affordances. Inf. Syst. J. **33**(3), 524–566 (2023)
28. Jarrahi, M.H.: Digital and physical materiality of information technologies: the case of Fitbit activity tracking devices. In: 48th Hawaii International Conference on System Sciences (HICSS), Kauai HI (2015)
29. Li, J., Conradi, R., Bunse, C., Torchiano, M., Slyngstad, O.P.N., Morisio, M.: Development with off-the-shelf components: 10 facts. IEEE Softw. **26**(2), 80–87 (2009)
30. Terzi, S., Bouras, A., Dutta, D., Garetti, M., Kiritsis, D.: Product lifecycle management - From its history to its new role. Int. J. Prod. Lifecycle Manag. **4**(4), 360–389 (2010)

31. Vagia, M., Transeth, A.A., Fjerdingen, S.A.: A literature review on the levels of automation during the years. What are the different taxonomies that have been proposed? Appl. Ergon. **53**, 190–202 (2016)
32. Brundtland Commission: Report of the world commission on environment and development: Our common future. Oxford University Press, Oxford, United Kingdom (1987)
33. Barth, L., Ehrat, E., Fuchs, R., Haarmann, F.: Systematization of digital twins: ontology and conceptual framework. In: 3rd International Conference on Information Science and Systems (ICISS), Cambridge, United Kingdom (2020)
34. Burckhardt Compression: Burckhardt Compression Industrial Metaverse by PTC, https://www.youtube.com/watch?v=lANFZqErVLM, last accessed: 2024/03/13
35. Herterich, M.M., Buehnen, T., Uebernickel, F., Brenner, W.: A taxonomy of industrial service systems enabled by digital product innovation. In: 49th Hawaii International Conference on System Sciences (HICSS), Koloa HI (2016)
36. Holler, M., Uebernickel, F., Brenner, W.: Digital product innovation in manufacturing industries – towards a taxonomy for feedback-driven product development scenarios. In: 50th Hawaii International Conference on System Sciences (HICSS), Waikoloa HI (2017)
37. Gregor, S.: The nature of theory in information systems. MIS Q. **30**(3), 611–642 (2006)
38. Peukert, C., Weinhardt, C., Hinz, O., van der Aalst, W.M.P.: Metaverse: how to approach its challenges from a BISE perspective. Bus. Inf. Syst. Eng. **64**(4), 401–406 (2022)
39. Rivest, L., Braesch, B., Nyffenegger, F., Danjou, C., Maranzana, N., Segonds, F.: Identifying PLM themes and clusters from a decade of research literature. Int. J. Prod. Lifecycle Manag. **12**(2), 81–106 (2019)
40. Becker, J., Knackstedt, R., Pöppelbuß, J.: Developing maturity models for IT management. Bus. Inf. Syst. Eng. **1**(3), 213–222 (2009)

On the Integration of the SAPPhIRE Model in the Digital Twin Development Process: A Train Braking System Use Case

Dmitrii Ershenko(✉), Saina Sadeghzadeh, Clement Fortin, and Andreas Panayi

Skolkovo Innovation Center, Skolkovo Institute of Science and Technology (Skoltech), Bolshoy Boulevard 30, bld. 1, Moscow, Russia
Dmitrii.Ershenko@skoltech.ru

Abstract. Advances in digital technologies over the past decade have made Digital Twins an important ingredient in the development of new products and tracking their performance over the lifecycle. Nonetheless, the definition and structure of Digital Twins are still under debate. The SAPPhIRE model initially developed to represent the causality of a system offer significant advantages, such as the product behavior and its evolution, and most importantly, how different parts of the system interact. In this paper, the concept of representation of the virtual part of a Digital Twin based on the SAPPhIRE model is proposed as a solution for defining and developing the Digital Twin. The synergy between the Digital Twin model and the extended SAPPhIRE model provides an opportunity to conveniently build the virtual part of the Digital Twin. To substantiate the concept proposed in this paper, a Digital Twin of a train braking system is demonstrated.

Keywords: Digital Twin · SAPPhIRE model · Behavior · 1D simulation

1 Introduction

1.1 Digital Twin

Nowadays, the multidisciplinary product development processes and the complexity of systems has significantly increased. Oliver L. de Weck [1], defined complex systems as a system with components and interconnections, interactions, or interdependencies that are difficult to describe, understand, predict, manage, design, or change. This complexity motivated specialists to develop new tools and new approaches. The Digital Twin (DT) is a breakthrough approach and solution to properly manage the complexity in product development.

The concept of building a digital representation of a system that is identical to the physical one has been around since the last century. Initially the goal was to explore the design space in the virtual world in order to drive well-informed engineering decisions in prototype development while limiting the amount of expensive physical tests. Since

Published by Springer Nature Switzerland AG 2025
P. Sureephong et al. (Eds.): PLM 2024, IFIP AICT 741, pp. 153–167, 2025.
https://doi.org/10.1007/978-3-031-93323-3_12

then, computing power has increased dramatically and building a virtual copy of a real system is no longer impossible. The advancements pushed people to find new ways of imitating real products. In 2003, Grieves, in his Executive Course on Product Lifecycle Management (PLM) at the University of Michigan presented the concept of DT for the first time [2]. He proposed a rather simple and clear concept, comprised of three main parts: a) physical product in real space, b) digital product in virtual space, and c) the information links that connect the digital and real products for data exchange. From his definition, the DT could be defined not only as a copy of the real system, but as a bidirectional flow of information: from real to virtual (feedforward) and from virtual to real (feedback). A simple illustration of this concept can be seen in Fig. 1.

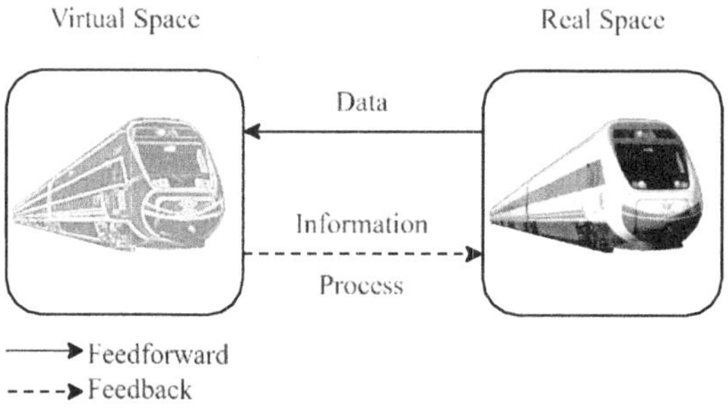

Fig. 1. DT as a three parts concept

Following this, the embrace and implementation of the DT concept has only grown over the years. Various scientists and even different companies put forward their own understanding and concept of the DT implementation. F. Tao extended the three-part concept proposed by Grieves. His proposed DT structure is shown in Fig. 2, where "PE" represents the physical entity; "VE" represents the virtual entity; "SS" represents the services for both "PE" and "VE"; "DD" represents the "DT" data; "CN" represents the connection of different parts. Here, as in the previous concept, all five parts of the DT are equally important [3].

There is still a lot of debate about how to define a DT, what a DT is in specific cases and where this DT occurs. As such, the definition of DT is still evolving. All previous works expressed the importance of DT, however, there is no consensus on how to build the model in a standardized way. In this paper, the DT concept developments are intended to form the foundation for a predictive/prescriptive maintenance analytics for a train system. The SAPPhIRE model is proposed as an approach to introduce standardization in DT developments.

1.2 SAPPhIRE Model of Causality

The SAPPhIRE model was first created to represent a more comprehensive model of the system's behavior, structure and function at different levels of granularity [4]. While previous models provide only a partial view, the SAPPhIRE model represents a causal and understandable view of the system with different information flows. The aim of the

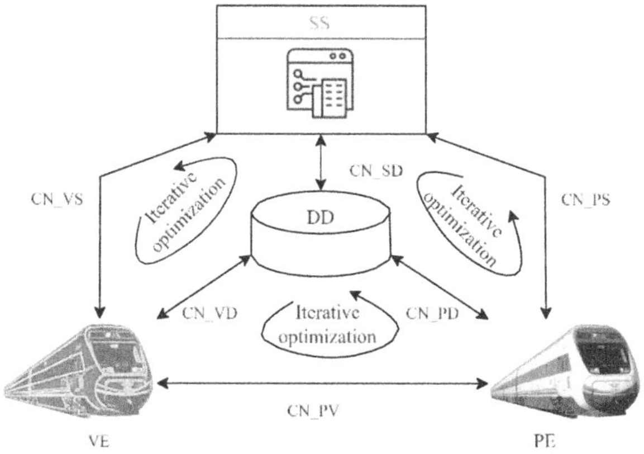

Fig. 2. DT as a five parts concept

SAPPhIRE model is to demonstrate the causality of the system [5] where the output depends on the previous and current input of the system, to get results. It can represent the structure of a complex system through multiple models linking to each other [6]. The SAPPhIRE model comprises of seven main constructs including State of change, Action, Part, Phenomenon, Input, oRgan and Effect, which are connected to each other by their definition [7]. A schematic is shown in Fig. 3.

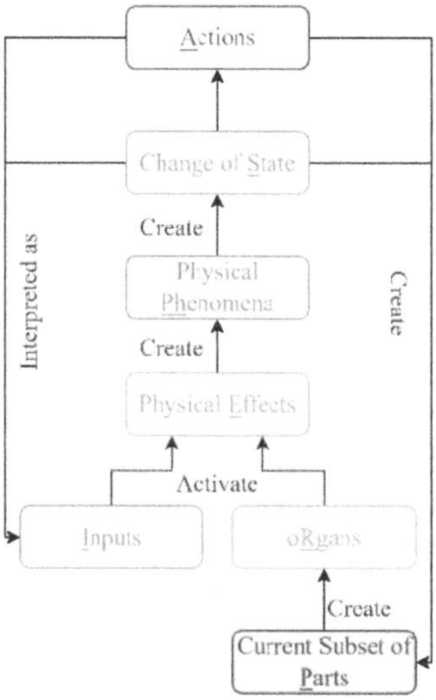

Fig. 3. The SAPPhIRE model of causality

An important part of the SAPPhIRE model is the connection between constructs. The SAPPhIRE model has been transferred from a simple series of actions to the domain of logical process [4]. As it can be seen in Fig. 3, each construct is linking with each other in order to indicate the Causal chain of the system, where the output of the Action is dependent on the Input and Subset of Parts which can also change effectively the configuration of a system. As a result, the SAPPhIRE model represents almost all aspects of the system including physical aspects (Parts and oRgans), system behavior (Physical Effects and Phenomena) and changes of the system over time (Change of State, Inputs, Actions). Therefore, all these categories are vital not only from the causality side, but also from the product lifecycle information and management side.

Extended SAPPHIRE Model

The SAPPhIRE model has been developed by Chakabarti, illustrating that it can be applicable as a model of analysis in complex system. However, certain shortcomings have been identified [4], which have led to the proposal of a modified SAPPhIRE model, the Extended SAPPhIRE model, as shown in Fig. 4. Like the standard SAPPhIRE model, Actions are an interpretation of the Change of State. In turn, interpretations should be processed by someone or something outside the system. It can be a customer, a designer, an engineer or just an external observer. In addition, all these interpretations could have their own opinions which could suggest changes to the system or not. In essence, the standard SAPPhIRE model is not open to any observation and interpretation. McSorley, suggested that there should be another construct between actions and state: Observation. Consequently, the new construct Observation interprets whether the Change of State requires any Action, subsequently, corrective actions should be made which can lead to re-evaluation of Constitutive Law. This brings more value to the system by not performing any actions when it is not required [4]. This process essentially enables robust control over any changes related to the product system and defines an accurate relationship between system's elements.

The different constructs of the extended SAPPhIRE model of causality are explained in the list below:

State: Represents the state of a system (and environment) at a particular time.
Action: Represents an abstract description or high level interpretation of a change of state, or the introduction of new input.
Parts: Represents a set of physical components or interface composing the system and its interaction with environment.
Phenomenon: Represents the exchange between a system and its environment in which the two entities have an influence on each other.
Inputs: Represents a physical aspect that transcends the boundaries of a system and plays a critical role in how the system interacts with its environment.
oRgans: Represents a structural context necessary for a physical effect to be activated including attributes, subparts and conditions of a system and environment.
Effect: Represents natural principles that govern interactions within the system and between the system and the environment.
In the extended model, several modifications have been made to the pre-existing constructions, in addition to the Observations and Actions. Moreover, the extended SAPPhIRE model of causality can be used as a PLM framework which in turn supports the management of product information throughout a product's lifecycle (from conception, through design and manufacturing, to operation, service, and disposal). Furthermore,

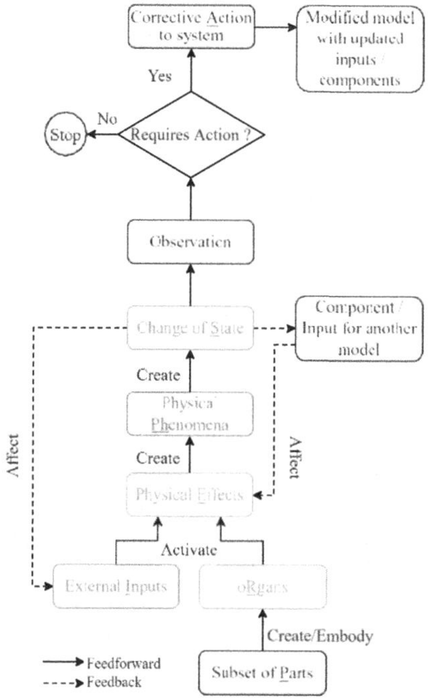

Fig. 4. The extended SAPPhIRE model of causality

SAPPhIRE model has been shown to address each element of the product including spatial, behavioral, and temporal which consolidates the development of real product life cycle information and engineering management systems [4]. Furthermore, this model provides a better understanding of service and maintenance information of a product during the operational phase. It will make in-service and maintenance information more organisable since the decision should be made whether the corrective action is required or not to the system.

2 Literature Review

A review of literature shows that the growth of demand for Digital Twins (DTs) is rapidly increasing every year both in academia and industry. This statement is substantiated by the rapid growth in the number of publications on the topic. For example, in the period from 2009 to 2018, the number of papers dedicated to DT over 10 years dramatically rose from almost zero to almost 50 papers per year [8].

Even with such a large number of publications on DT, it can still be seen that there is no consensus on what a DT should look like. Different authors brought different ideas to the table not only in the context of building DT, but also in the context of defining DT. Some researchers mentioned that it should focus on simulation, however others suggested that DT consists of three main parts [2]. Others proposed about five parts [9]. Meanwhile, there are still some propositions that each connection should be bi-directional in DT [8]. Hence, NASA gives their own complicated definition that DT is integrated multiphysics, multiscale, probabilistic simulation [10]. These differences lead to an underestimation of the importance of the DT concept. At the same time, all the papers represent that the DT has a great capacity for implementation. It is widely used in various disciplines from medical to the aerospace industry.

The DT is not only the next generation of simulation, but also a powerful tool for comparing the physical space with the virtual space to drive well-informed decisions. This means that the DT model should be properly evaluated to ensure its accuracy in reflecting the physical and virtual spaces. At the same time, without considerable effort to describe and quantify the benefits, it is difficult to prove that the DT concept is an appropriate and binding solution. Therefore, key questions remain for the DT concepts. For example, the definition of the DT and its level of fidelity. These parts are still not clarified despite all the years of research.

The SAPPhIRE model was first introduced in 2005 by Chakrabarti [5]. After this publication, he and his research team continued investigations to reach the final structure of the SAPPhIRE model [7]. In 2014, G. McSorley, C. Fortin and G. Huet, introduced the extended SAPPhIRE model [4]. However, all these works have not implemented the SAPPhIRE model in the DT domain. The most common use cases are product in use information, in service scenarios, or in other stages of product life cycle [11].

Consequently, the combination of all these literatures allows for the understanding of the different views and opinions on the research problem. The extended SAPPhIRE model can be readily integrate with the DT development process to make the creation process structured and eliminate errors related to misunderstanding of how to create a DT.

3 Paper Objectives and Structure

The aim of this paper is to highlight the potential impact of the creation of a DT based on the SAPPhIRE model which can enhance the development of a product-system at the early stage of its life cycle. This approach could also support predictive maintenance by providing better quality to operation. To demonstrate this approach, a model of a train braking system has been chosen as a use case. The SAPPhIRE model has been employed to create the DT and outline the process of building the DT as predictive model for the train braking system. This creates a unified digital system which can be validated based on bi-directional relations (known as Twining which an essential part of the DT process). The SAPPhIRE model can be used to transfer information, providing relationships between different parts of a product which in turn gives the ability to facilitate information transfer between different stages of the product life cycle in an efficient manner.

This paper has the following structure. Section 1 introduces the problem addressed in this paper and gives a brief introduction to the DT and the extended SAPPhIRE model. The literature review is presented in Sect. 2. The objective and structure of the paper are presented in Sect. 3. Section 4 is dedicated to the description of a proposed approach based on the integration of the SAPPhIRE model in the DT development process. Section 5 is dedicated to the validation of the proposed approach by building a virtual part of the train braking system as a use case. The discussion and conclusion are in Sects. 6.

4 Integrated Scheme of SAPPhIRE and DT Models

4.1 Relation Between SAPPhIRE, 1D Simulations and DT

The SAPPhIRE model has various parts that are similar to the DT, including bi-directional relationships between system elements that are the main criteria for DTs. Therefore, the extended SAPPhIRE model which is the basis for potential impact of the development algorithm for DT, can determine which components are needed to create the DT, which external inputs crosses boundary of the system, what potential physical phenomena will occur at the virtual part of DTs and what change of state will be done after simulation.

In addition, the results of the simulations can improve the performance of the system, and anticipate potential problems that could be addressed to improve the efficiency of the system. This approach can reduce the possibility of unexpected problems in the later stages of the product life cycle. In other words, one of the use cases of DTs built with the contribution of SAPPhIRE model is to use it for predictive analysis.

The following example illustrates the relationship between the SAPPhIRE model, 1D simulation and DT in a simplified way. Suppose there is a sliding block that is descending from an inclined slope. At the beginning of the descent, an additional force acts the block for one second. Friction is assumed to be negligible and the only resistive force that has an impact on the block is wind resistance. The displacement of the block is the parameter to be studied. A simple sketch is shown in Fig. 5.

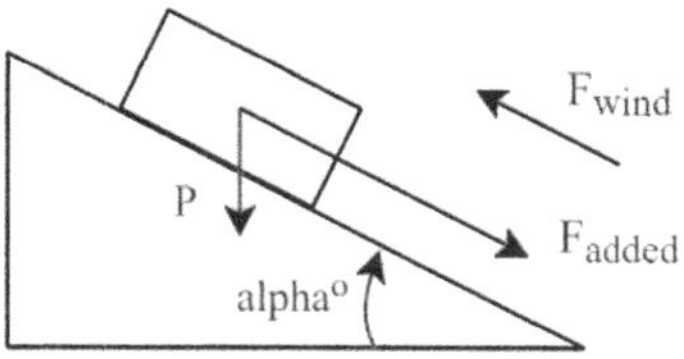

Fig. 5. Sliding block from an inclined slope

The extended SAPPhIRE model proposed by G. McSorley could be built based on the simplified example of the sliding block on an inclined slope (Fig. 6).

From the extended SAPPhIRE model, it is evident that Subsets of Parts (Downhill and Block) create oRgans for the system Block Mass and Downhill Inclination. Concurrently, there are External Inputs that cross the boundary of the system: external force, gravity and windage coefficient. As the next step, oRgans and External Inputs activate the physical law governing wind resistance (Physical Effect). This in turn creates another block Physical Phenomena which describes a change of displacement due to wind resistance. The Change of State tracks whether there is a deviation of displacement from the required one. Hence, there is Observation of results and corrective required Actions are

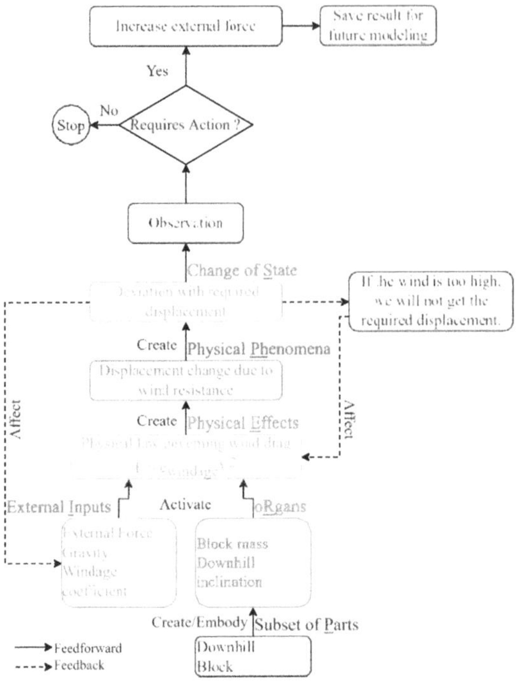

Fig. 6. The extended SAPPhIRE model of the sliding block

implemented. If the deviation from required displacement is in threshold, there are no actions. If the deviation is too high some action is required, for example the external force could be increased. A schematic of this 1D simulation model in Siemens Amesim software with the different constructs of the extended SAPPhIRE model is shown in Fig. 7.

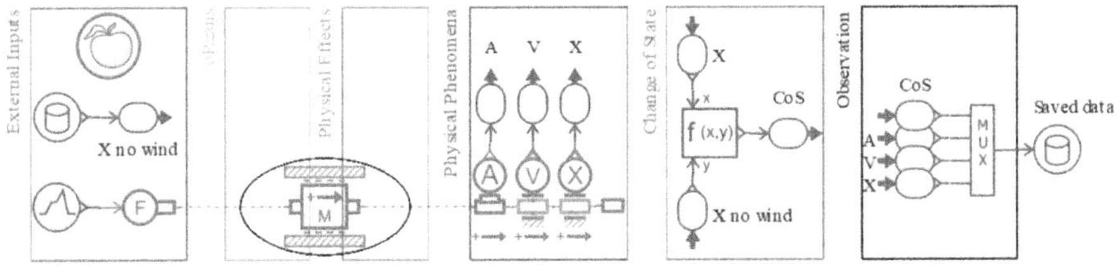

Fig. 7. 1D simulation model of sliding block

This simulation model is grouped in the same construct blocks as the extended SAPPhIRE model. Each block has external Inputs that cross the boundary of the system and are needed for calculations. However, every single block of the model combines oRgans and Physical Effects. For example, in Fig. 7, the block which is circled in black oval represents both the sliding block and slope inside it. At the same time this block has physical laws inside it, which evaluate the movement of the block on the slope. During the simulation Physical Phenomena occurs, e.g. how acceleration (A), velocity (V) and displacement (D) change. This leads to deviation from real data (label: X no wind). As a result, all outputs can be visualized and Observed, save the data and make a conclusion about the Change of State, whether this system needs any Action or not.

These similarities between the organization of simulation and SAPPhIRE model make it possible to integrate the SAPPhIRE model in the DT development process. Therefore, this approach will be used as a tool for building the virtual parts of DT in the current work.

4.2 Integration Between DT and SAPPhIRE Model

The DT model is quite similar to the extended SAPPhIRE model. The similarity is not only in the area of components, which are the same for both models, but also in the area of information flow, which is important for DT. There will be no DT if there is no information flow between the physical entity and the virtual entity. Thus, the word "digital" means that the virtual part should be built using digital technologies. Such a tool was defined at the previous Sect. 4.1. As a result, this paper proposes integration between the DT model and the extended SAPPhIRE model. The proposed integrated scheme is shown in Fig. 8. All of the SAPPhIRE model constructs are integrated to the virtual part of the DT model.

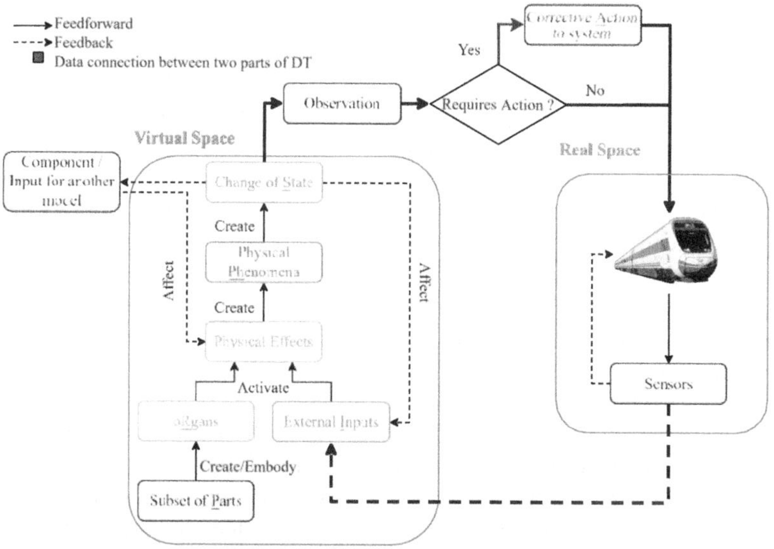

Fig. 8. Integrated scheme of SAPPhIRE and DT models

The process begins with a Subset of Parts. On the basis of this, the components that will be modeled in the virtual part of DT, are defined. Then External Inputs are defined which have been taken from sensors on the physical object. Next, the Subset of Parts creates parts from which the virtual part of DT will be created. Hence, as for the SAPPhIRE model, oRgans and External Inputs activate Physical Effects. This is where the laws that govern the physical system are defined. After that, the Physical Phenomena where changes in parameters can be observed. At this stage, identical sensors exist in the digital and physical entities; based on phenomena data from physical system sensors and from modeling a comparison is made to determine whether there is a Change of State. Next is the Observation of such deviations. If the deviation is within the required threshold, then no corrective actions are needed. If not, then something should be corrected. The information about required corrective actions or lack of them flows to the physical space through the red lines in Fig. 8. The real space takes into account the received information and reacts accordingly. The sensors on the real entity track the necessary information about the system behavior. This information flows back to the virtual space to enable observations and consequent actions.. All these form a continues loop during system operation for data exchange between the DT and the physical system for well-informed data driven decisions for efficient operation.

As a result, there is an understandable scheme on which the DT can be built. In turn, it indicates how virtual space should be organized and modeled effectively. Based on this approach, a virtual part of the train braking system is built as a use case.

5 Use Case

The train system structure is complex and comprises different subsystems including Braking system, Control system, Network system, etc. Therefore, instead of building a DT of the entire train system, the train braking system has been chosen as a representative use case to properly demonstrate the concept proposed in this paper.

At the functional level, the train braking system is strongly reliant on compressed air to operate on the majority of trains. Thus, compressed air is a crucial component in the train braking system. There are two main lines for air supply: the Feed Line (FL) and the Brake Line (BL). Considering the braking systems, the pressure at these two lines should be monitored, how it increases or decreases. There are three different modes of braking: emergency braking, pneumatic braking and electro-pneumatic braking. These modes consume air at the Brake Line. The pressure at the Brake Line is supported by the main Feed Line. Therefore, the pressure of air at the Feed Line drops when braking occurs. However, if the Brake Line is required only for braking, the Feed Line also transmits compressed air for other train systems including Sand, Whistle and Horn. Sand is needed to increase friction between train wheels and railroad tracks. Whistle and Horn are necessary to generate warning sounds or sound signals to traffic controllers at a railroad crossing. At the same time, air should be supplied to the Feed Line. For this purpose, there are compressors on the train that add air to the system at a certain pressure. For developing the DT of the train braking system, the integrated scheme of SAPPhIRE and DT models was created as shown in Fig. 9:

In total there are 15 different sensors on the train for 5 cars. At the first car there are 5 of them: pressure at the FL and BL, signal at the activation of Horn, Whistle and Sand. On cars two, four and five there are two sensors on each car: pressure at FL and BL. At car three there are four sensors: pressure at FL and BL together with sensors of starting of two compressors (C1 and C2). The sensors on the different cars are shown in Fig. 10.

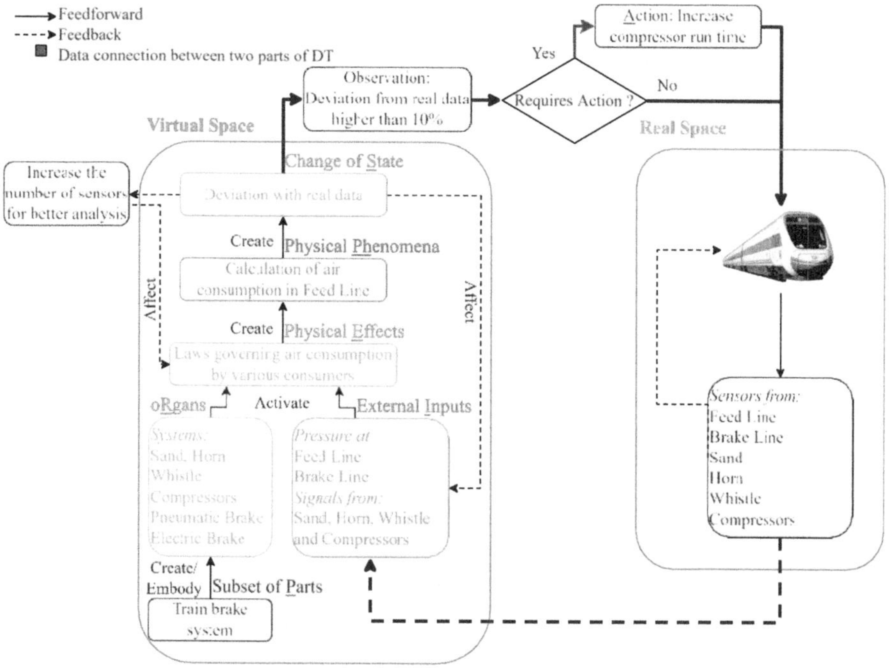

Fig. 9. Integrated scheme for the train

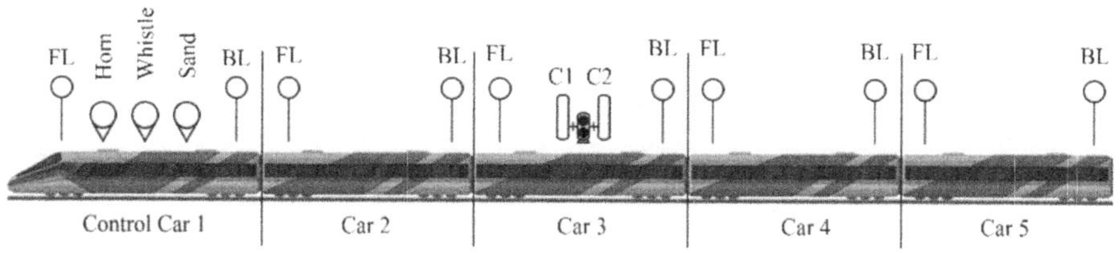

Fig. 10. Train sensor distribution

This scheme is the basis for building the DT. The next step is to use the software as a tool for creating virtual part of DT. Therefore, by considering Fig. 9, the virtual part of DT was built as shown in Fig. 11:

Here the first construct is External Inputs. This data comes from the physical train. Firstly, there is pressure on the Feed Line and the Brake Line at the five cars. The next data are the signal of switching on of Sand, Horn and Whistle systems. The last data are the signal of switching on the two compressors. All these data from the physical train goes to the train block, which combines both the physical effects that govern the braking system and the organs. After that the physical phenomena block calculates the pressure at the Feed Line. The result of this calculation is presented in Fig. 12.

Decreasing pressure in the Feed Line defines that air is being consumed by brakes or other systems such as Sand, Whistle or Horn. Increasing pressure on the Feed Line

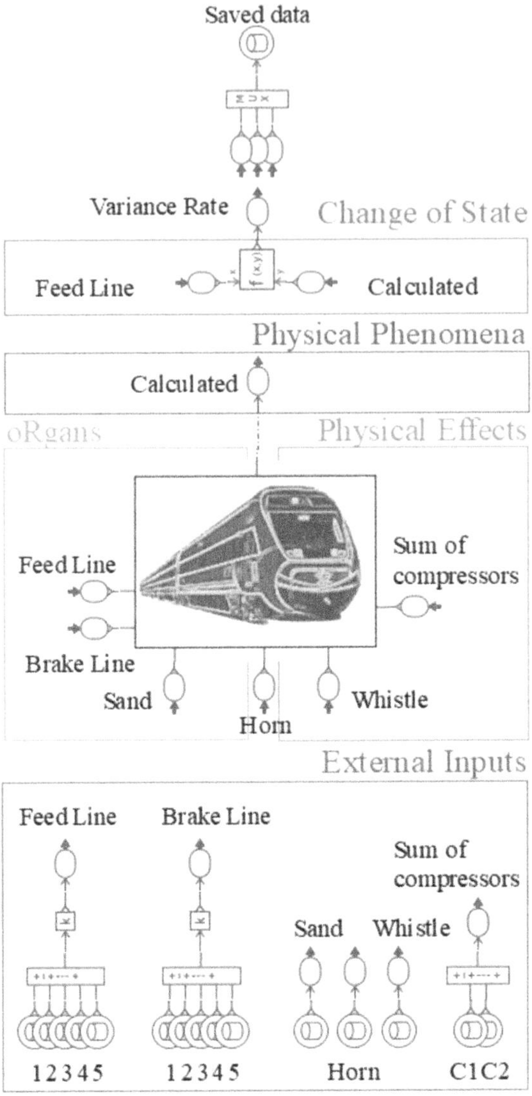

Fig. 11. Virtual part of DT of the train braking system

defines that two compressors are working to add air to the system. Next is the Change of State deviation from physical data (Fig. 13). Based on the Change of State an Observation is be made considering the variance rate.

If the variance is within the threshold, then no action is needed, if the variance is too high, then an action is required. For example, compressors should operate more often. This information is transferred to the physical system. The sensors on the train cache the pressure on the Feed Line and transfer this information back to the virtual part of the

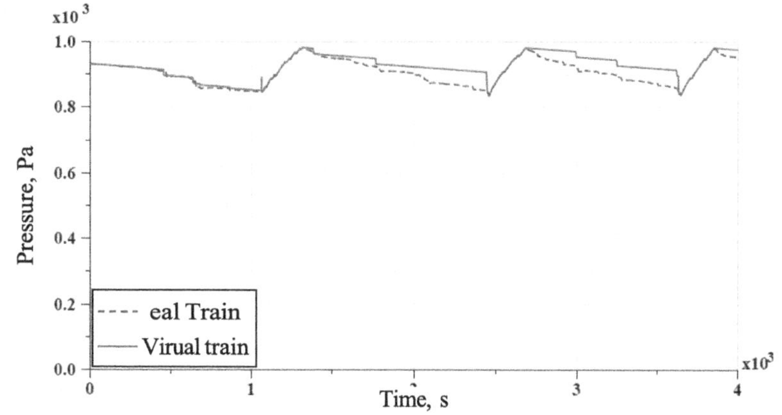

Fig. 12. Feed Line pressure

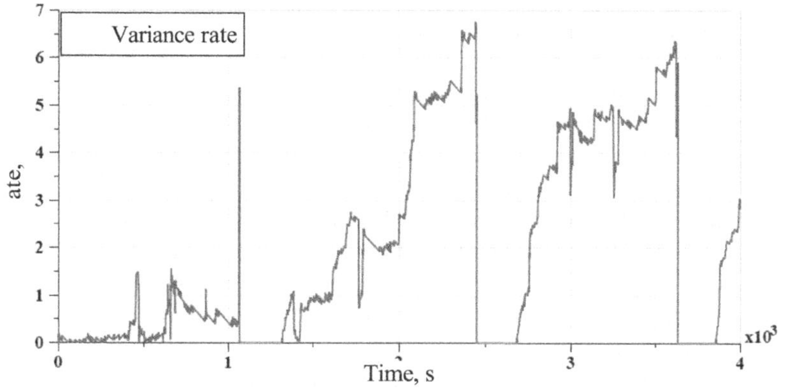

Fig. 13. Variance rate between physical data and calculated

DT. This process is continuous and goes through all the stages proposed in the integrated model (Fig. 9).

6 Discussion and Conclusion

Most often, models to build DT are unstructured and difficult to comprehend for external users. The integrated scheme of the extended SAPPhIRE model and the Digital Twin model provides a well structured methodology to define any DT. The resulting model sets the foundation for using this integrated methodology to build well-defined Digital Twins, which in turn will enable the structured data flow with the physical system to drive well-informed decisions. It is interesting to note, that although the SAPPhIRE model was originally defined for the conceptual design stage of products and system, it was clearly demonstrated that it can be used effectively to define Digital Twins. Its value lies in its power of expression of the behavior of the system which is the most important characteristic required for Digital Twins. Its internal structure with External Inputs, oRgans, Change of State and so on provides a very rich construct to express the profound nature of the DT concept.

The DT itself could be applied at different stages of the product life cycle including design, production, operation and recycling. In this case, they could be cascaded into various DT to the different phases of life cycle as was already demonstrated by McSorley [12]. The proposed model has a construction that comes from the SAPPhIRE model. With the contribution of Observation or Input for another model construct, the connection between different models at different lifecycle stages could be implemented. This extends the use of the integrated model in terms of different stages of the product life cycle. Furthermore, an important part of the whole DT concept is the information flow between the virtual entity and the physical entity. The proposed integrated model allows for information flow traceability not only between physical and virtual entities, but also within each of them. Previous concepts did not provide such a level of information traceability. In the proposed methodology, the observation construct is not only part of the information flow from virtual to physical part of DT, but also part of the reverse information flow: from physical to virtual. This gives the ability to readily understand the need for any actions that will change entities. Therefore, the integrated scheme of the SAPPhIRE model and DT is suitable for a variety of use cases with a proper representation of the information flow.

The proposed methodology sets the foundation for a standardized approach to build Structured, Traceable, Efficient and Manageable (STEM) Digital Twins.

In conclusion, while the proposed STEM DT methodology ensures seamless integration for a unified digital system, the integrated scheme of the extended SAPPhIRE model and the DT model could be used to support predictive maintenance or predictive analytics. It could occur at the stage of information flow from virtual to physical and vice versa. Moreover, the methodology could include information coming not only from the physical entity itself, but also from other sources, such as maintenance work or technical reports. The use case presented in this paper serves as a demonstrated example for the change of state which could be perceived as the way of leakage of air from the Feed Line and it will serve as the foundation for developing the predictive maintenance techniques for this system.

References

1. De Weck, O.L., Roos, D., Magee, C.L.: Engineering Systems: Meeting Human Needs in a Complex Technological World (2011)
2. Grieves, M.: Digital Twin: Manufacturing Excellence through Virtual Factory Replication (2015)
3. Tao, F., Zhang, M.: Digital twin shop-floor: a new shop-floor paradigm towards smart manufacturing. IEEE Access **5**, 20418–20427 (2017)
4. McSorley, G., Fortin, C., Huet, G.: Modified SAPPhIRE model as a framework for product lifecycle management. In: DS 77: Proceedings of the DESIGN 2014 13th International Design Conference, pp. 1843–1852 (2014)

5. Chakrabarti, A., Sarkar, P., Leelavathamma, B., Nataraju, B.S.: A functional representation for aiding biomimetic and artificial inspiration of new ideas. AI EDAM. **19**, 113–132 (2005)
6. Bhattacharya, K., Chakrabarti, A.: Application of SAPPhIRE Model of Causality in the Design of Product-Service Systems, pp. 527–539 (2023)
7. Srinivasan, V., Chakrabarti, A.: SAPPhIRE an approach to analysis and synthesis. ICED **2**, 417–428 (2009)
8. Jones, D., Snider, C., Nassehi, A., Yon, J., Hicks, B.: Characterising the digital twin: a systematic literature review. CIRP J. Manuf. Sci. Technol. **29**, 36–52 (2020)
9. Tao, F., Zhang, H., Liu, A., Nee, A.Y.C.: Digital twin in industry: state-of-the-art. IEEE Trans. Ind. Inf. **15**, 2405–2415 (2019)
10. Glaessgen, E.H., Stargel, D.S.: The digital twin paradigm for future NASA and U.S. Air force vehicles. Collection of Technical Papers - AIAA/ASME/ASCE/AHS/ASC Structures, Structural Dynamics and Materials Conference (2012)
11. Peters, S., Fortin, C., McSorley, G.: A novel approach to product lifecycle management and engineering using behavioural models for the conceptual design phase. IFIP Adv. Inf. Commun. Technol. **565**, 159–169 (2019). IFIP
12. McSorley, G.: A novel approach to product lifecycle management and engineering based on product in-use information (2014)

Models for Manufacturing (MfM) Methodology Applied to the Automated Characterization of Aerospace Sheet Metal Parts Manufacturing Process

Manuel Roldán[1], Murillo Skrzek[1,2], Domingo Morales-Palma[1], Anderson Luis Szejka[2], and Fernando Mas[1,3](✉)

[1] University of Seville, 41092 Seville, Spain
{manrolrey,dmpalma,fmas}@us.es
[2] Industrial and Systems Engineering Graduate Program, Pontifical Catholic University of Parana, Curitiba, Brazil
{murillo.skrzek,anderson.szejka}@pucpr.br
[3] CT Engineering Group, Madrid, Spain

Abstract. The evolution of aerospace products, with an increasing number of components, more demanding delivery times, and reduced costs, is leading companies to look for integrated management systems. The future of the aerospace industry is linked to the development of new methodologies, such as Model-Based Engineering (MBE). In the last two decades, the use of semantic structures, such as ontological models, to improve information systems has proven to be very effective in the industry. Models for Manufacturing (MfM) is a recent OBE (Ontology-Based Engineering) methodology for defining complex manufacturing systems' data, functions, and behaviours using graphical models. It is based on agnostic tools and is structured in an architecture of three independent layers. In this paper, the MfM methodology is applied to describe the manufacturing process of Aerospace Sheet Metal (ASM) parts. The implementation is described through a practical example for a part manufactured with one of the most used sheet metal forming processes in the aerospace industry: hydroforming. A study analysed 26 typical sheet metal parts, finding bending and hydroforming to be the main processes. The model aims to characterize the forming process, i.e., to recognize the most appropriate manufacturing process, and subsequently to define the necessary operations and resources, based on the part's Computer Aided-Design model. In addition, the model is complemented by other existing MfM tools for sheet metal forming.

Keywords: Model-Based Engineering (MBE) · Aerospace Sheet Metal (ASM) · Models for Manufacturing (MfM) · hydroforming

Published by Springer Nature Switzerland AG 2025
P. Sureephong et al. (Eds.): PLM 2024, IFIP AICT 741, pp. 168–183, 2025.
https://doi.org/10.1007/978-3-031-93323-3_13

1 Introduction

In recent years, there has been a notable increase in the complexity of aerospace components, necessitating the establishment of sophisticated manufacturing facilities and inventive methodologies to effectively tackle the obstacles associated with cost reduction, production time, and the assurance of superior quality. Furthermore, the design and production of an aircraft requires concerted efforts of engineers of different nationalities and backgrounds, who are required to participate in the exchange of information and expertise related to multiple elements during the various stages of product development or production planning [1].

During the development of an aircraft project, it is crucial to avoid any kind of error as it can result in a significant increase in costs and project timeline. Consequently, engineers allocate a substantial amount of time to all industrial processes of a particular part, considering a multitude of variables and necessary establishments, all of which depend on the level of complexity involved [2].

Ontology-Based Engineering (OBE) arises as a novel methodology, originating from prior techniques, due to technological progress in the field of computer science, to achieve these objectives and enhance the multidisciplinary design and authentic simulation of intricate systems [3].

Models for Manufacturing (MfM) is a recent Ontology Based Engineering (OBE) methodology for defining and reproducing data, functions, and behaviours of complex manufacturing systems through computer-aided graphical modelling tools. It comes from the concept of Model-Based Engineering (MBE) methodology [4]. This methodology uses models and its abstract behavioural approach to enable a more robust engineering definition. The MfM methodology proposes a three-layer model architecture (3LM) [5], which will be described in detail in this paper.

Figure 1 shows the aerospace sheet metal (ASM) manufacturing process. In the first part (a), the features of the 3D model are extracted as proposed by [6], followed by (b) the process planning. This stage of process planning is currently conducted by a team of experienced engineers who will determine which types of machine to use, the sequence of operations, infrastructure and tools, as well as manufacturing strategies. This stage represents the research gap addressed by this article, as companies currently lack knowledge capture systems to optimize manufacturing processes. In the final step (c), the part manufacture is carried out according to the process planning.

However, it takes a long time to plan and manufacture new aerospace parts and requires extremely qualified labour for such a task. Additionally, with the amount of information involved, the probability of an error occurring during this process is very high. Additionally, conventional planning processes are verified to have higher expenses and consume engineering time. In this context, *is it possible to automate the process of planning aerospace parts based on knowledge representation using OBE and semantic web-based data provided by the 3D model*?

The main objective of this research is to propose the creation of an automated knowledge system to support the ASM parts planning process. This system can support industry planning departments, capturing not only explicit domain knowledge, but also implicit and tacit domain experts' knowledge, providing processes, methods, and tools to support the capture and reuse of manufacturing knowledge in a model-based context. Section 2

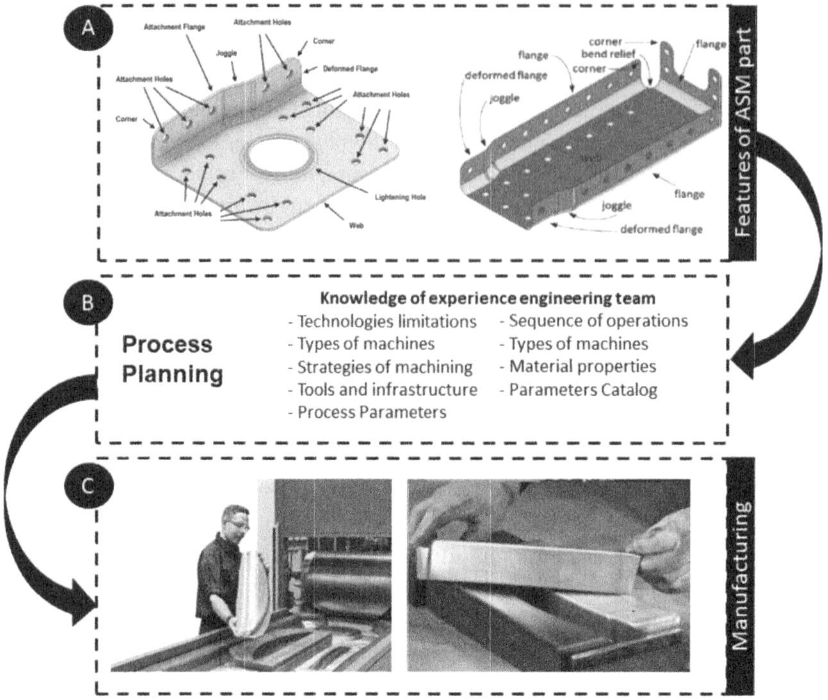

Fig. 1. Steps of ASM manufacturing.

of this article presents the literature review, followed by the conceptual approach of the research in Sect. 3, Sect. 4 presents an application case, and Sect. 5 presents the conclusions and ideas for future work.

2 Literature Review

For this initial content analysis, the authors followed the recommendations outlined by [7] and adopted the research methodology proposed by [8]. The PRISMA method (Preferred Reporting Items for Systematic Reviews and Meta-Analyses) was utilized for this review.

The literature review methodology is outlined in four steps: (a) the keyword selection by combining keywords such as Aerospace Industry, Cyber-Physical Production Systems (CPPS), Reconfigurable Manufacturing Systems (RMS), and Semantic Interoperability. Several combinations associated with the origins and synonyms of these words were explored; (b) The definition and selection of databases such as Scopus, Google Scholar, and ScienceDirect were designated as the chosen databases for the research; (c) the inclusion and exclusion criteria; and (d) the selection of the most relevant articles within the theme. The search yielded a total of 2,457 papers.

A first exhaustive review of the main bibliographic sources resulted in a total of 113 papers that could be relevant to the present work categorizing them based on three distinct research methodologies. The first group (a) is the papers that bring a general approach to industry knowledge systems concepts followed by the second group (b), which is the papers aboard a specific application of a knowledge system in the aerospace industry, and last group (c) is the papers that address methodologies for manufacturing aerospace parts.

Figure 2 shows the number of articles published per year identified in this review of the literature according to each group.

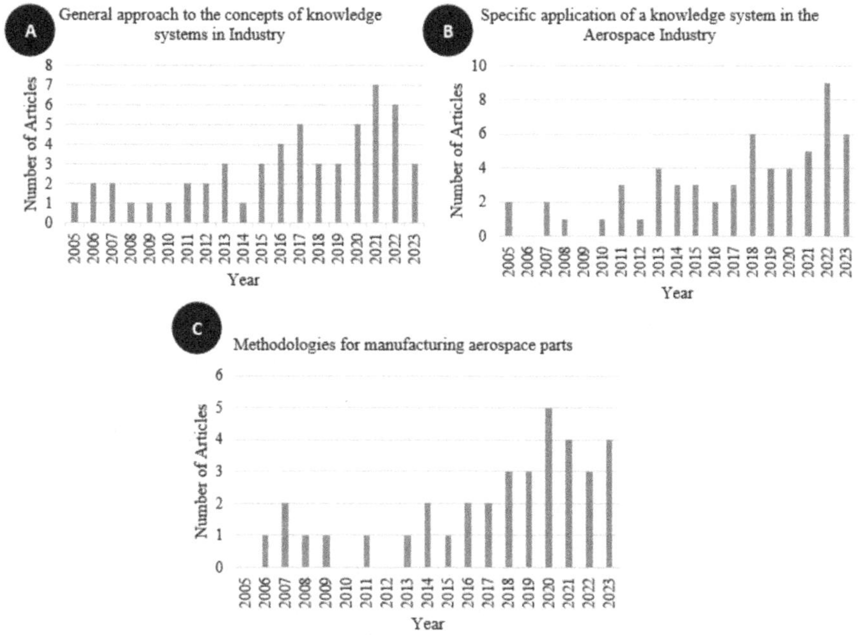

Fig. 2. Graph of publications by year according to each group.

Of the 113 chosen contributions, (a) 55 delineate general approaches to knowledge systems in the industry, (b) 59 concentrates on applying a knowledge system in the Aerospace Industry. And (c) 39 address methodologies for manufacturing aerospace parts. It is possible to observe a growing number of research studies within the three areas addressed, thus the topic is gaining prominence and visibility in academic and industrial circles.

Among the most relevant works [9] addresses the use of Knowledge Engineering for parametric aircraft modelling toward optimization, employing a combination of object-oriented programming. It focuses more on the methodology and approach of utilizing knowledge engineering techniques. The main result of [10] is the development and application of a Knowledge Engineering system to support the engineering design process in the aerospace industry, specifically for the design of fuselage skin panels in Metal Fiber Laminate for the Airbus A380 aircraft. The results of applying KBE are highly accurate, with a 50% reduction in pre-impregnated laminate basic panel cutting waste.

The paper [11] proposes a knowledge-based prototype application to assist and guide designers in defining and evaluating conceptual alternatives for the Final Assembly Line (FAL). The main limitation of the article is that the proposed application is highly based on the knowledge and experience of the personnel. Although the application aims to improve the design of aircraft assembly lines, it still requires the interaction of specialized technical teams with the system to define alternatives. Furthermore, the [12] developed a Semantic Interoperable Intelligent Manufacturing System (SIIMS) that promotes semantic interoperability throughout the manufacturing system, integrating information and concepts from various domains and sharing them between different phases of the manufacturing process. The limitation of the article is that the proposed SIIMS requires further evaluation in various industry areas, considering more complex processes in the analysis.

The main result of [13] is the application of data-driven modelling and model learning procedures in various fields of science and engineering. The article mentions the difficulties in addressing all aspects of engineering and scientific modelling with data-driven procedures. [14] discusses the successful application of reference ontologies to support semantic interoperability during the product development process, which contributes to improving product design and manufacturing. Additionally, there is potential to explore more complex and diverse products, such as those with complex geometry, tight tolerances, or multiple assemblies. Finally, [15] addresses the development of a Knowledge-Based Engineering (KBE) application for the design and manufacture of high-speed milling (HSM) accessories. The article suggests that automating the design process should be approached in partial stages, combining designer decisions with automatic expert-based knowledge judgments.

All articles present positive results regarding the application of ontologies in the industry. However, they still explore issues such as essential commercial tools for application implementation and the extraction of knowledge from professionals in various fields to acquire a comprehensive system.

3 MfM Methodology

The MfM methodology is based on a 3-Layer Model (3LM) [5]: data layer, ontology layer and service layer shown in Fig. 3. The 3LM is supported by user-friendly modelling tools and guarantees independence between the three layers, keeping the data layer, the ontology layer, and the service layer isolated.

The data layer gathers all information in terms of databases: legacy software databases, commercial software databases, or databases available in the cloud. The ontology layer is the core of the 3LM framework, as it contains the knowledge of the enterprise itself, the data and semantic models, as well as the associated simulation or behavioural requirements. Finally, the services layer comprises software services, simulation tools, visualizers, data analysis services, dashboards, etc.

The MfM methodology establishes a set of golden rules for the development of ontologies:

- The MfM methodology provides a set of metamodels to define ontologies.
- The MfM methodology is tool-agnostic and does not rely on any particular tool.

- The MfM methodology promotes a mechanism for lifecycle management and configuration control using MLM.

The ontology layer becomes the most important layer of the model because it collects, stores, and manages the knowledge. This layer is made up of four models: scope, data, behaviour, and semantics.

The scope model is the first step toward the creation of the ontology and defines the boundaries on which the ontology works. The data model defines the information managed in the scope in scope. In the proposed methodology, the data modeler organizes the data objects into a concept map, creates relationships between objects, and enriches the data model by adding properties to the objects.

The behaviour model defines the internal activity for each of the activities or processes described in the scope model, i.e., it depicts the behaviour of the simulation or the requirements to be met by the software.

The semantic model aims to maintain the connection between ontologies throughout their life cycle. Thus, to maintain traceability and facilitate the automation and processing of semantic information, the data information is managed using a key concept-value relationship.

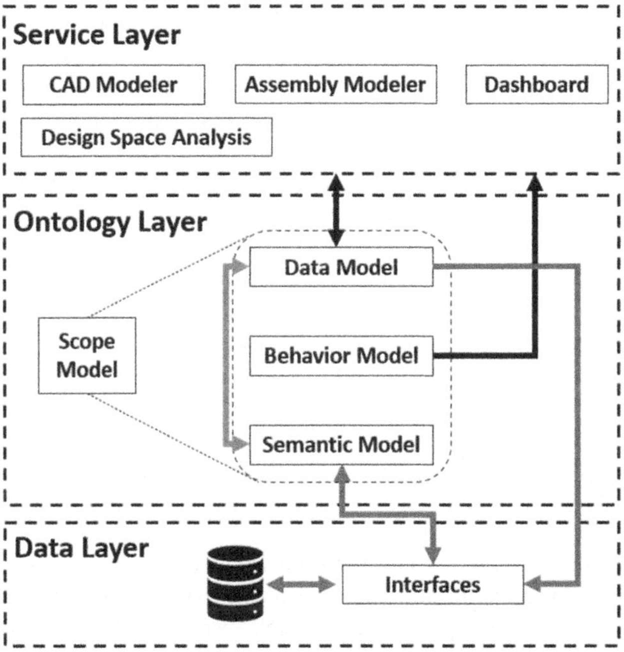

Fig. 3. MfM architecture: 3 Layers Model.

4 Use Case: Automated Characterization of an ASM Part Manufacturing Process

Sheet metal parts represent a high rate of the metallic structural components of an aircraft, and consequently, sheet metal forming comprises a large part of industrial metal production. This article aims to provide a new tool to facilitate industrialisation tasks in this sector. For example, when a company faces the need to quote the development costs of a new project, it shall analyse each of the parts involved, identifying the necessary processes, their complexity, the resources needed, or the hours incurred.

On the other hand, when industrialising a new project, a great deal of documentation is generated, in addition to selecting the necessary resources.

In such cases, usually, the only knowledge available is the technical documentation or the 3D model, and additional information such as the material or the number of parts to be manufactured.

4.1 MfM Ontology Model

The Scope Model consists of a series of IDEF0 diagrams, each of them defining the decomposition of an "activity" into sub-activities. To develop the scope model, it was decided to use IDEF0 diagrams with the Ramus tool, since it is a tool widely used in the aerospace sector for decades and it fulfils one of the requirements of the MfM methodology, being "easy to write" and "easy to read". In the Scope Model are defined all the activities, means, and data object elements and their relationships. On the other hand, for each elementary activity of the scope model, i.e., for each activity that has no sub-activity, a behaviour diagram is required. For the use case, seven behaviour diagrams are defined. For the graphic representation of the behaviour diagrams, tools such as GraphViz are usually used.

The main activity is A0 "Automated characterization of a sheet metal part manufacturing process". Figure 4 represents how the A0 is decomposed into four sub- activities: A1 "Manufacturing process identification", A2 "Identifying Process", A3 "Recognizing Flat pattern" and A4 "Manufacturing process description". The two main activities are A1 and A4. In the first one (A1), the manufacturing process (only bending and hydroforming process will be considered in this study case) is decided by recognizing and processing the part manufacturing features. Then, the manufacturing process is described in A4, by grouping, sorting, and allocating the necessary resources and parameters to each operation.

It is considered as input the CAD 3D model, "Part Model", and their properties, "Part Properties", including information such as the material. However, additional data are required, such as material properties, available resources, manufacturing parameters for the applicable operations, and the operations associated with each feature of the part. With this information, it is possible to estimate the operations and resources needed to manufacture the desired part, based on the features.

The Scope Model starts with activity A1, where an automated features recognition software is needed, A11 [6]. Then, as it is described in the behaviour diagram (see Fig. 5),

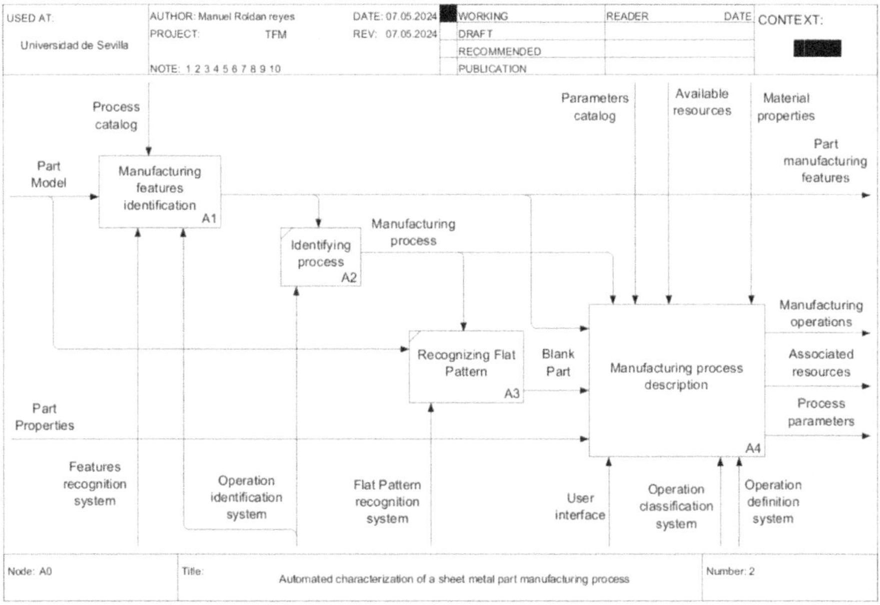

Fig. 4. A0, "Automated characterization of a sheet metal part manufacturing process".

the data object "Process Catalog" (see Table 1) contains the information needed to allocate to each feature two additional attributes (A12): *Feature Type (trim or deformation)* and *Operation T*ype (cutting, drilling, bending, or hydroforming).

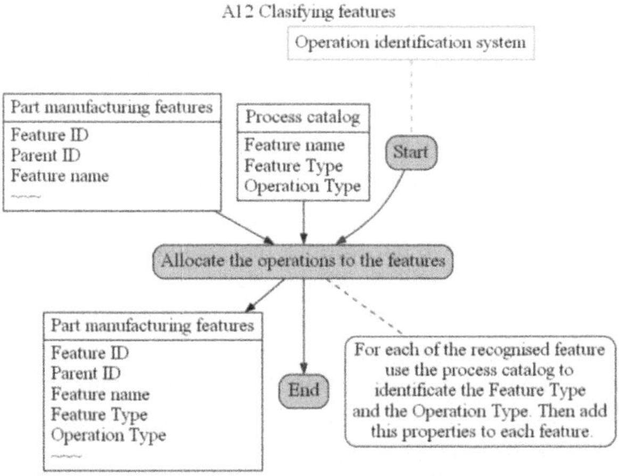

Fig. 5. Behaviour diagram of activity A12, "Clasifying features".

This information is used in A2 to determine if the part will be manufactured by bending or hydroforming process, which is one of the most valuable types of information

Table 1. Process Catalog.

Feature name	Type	Feature Type	Operation Type
cutout		trim	cutting
stringer cutout		trim	cutting
bend relief		trim	cutting
corner		trim	cutting
tooling hole		trim	drilling
attachment hole		trim	drilling
Lightening cutout		trim	cutting
Lightening hole		trim	cutting
Attachment flange	planar	deformation	bending
Attachment flange	curve	deformation	hydroforming
Stiffening flange	planar	deformation	bending
Stiffening flange	curve	deformation	hydroforming
Lip		deformation	hydroforming
Joggle		deformation	hydroforming
Deformed flange		deformation	hydroforming
Deformed web		deformation	hydroforming
Bead		deformation	hydroforming
Lightening cutout		deformation	hydroforming
Lightening hole		deformation	hydroforming

for the intended application. The manufacturing process is defined based on the types of features, that is, if there is any 'Operation Type = Hydroforming', the part will be hydroformed. The decision is supported by the rationale of the process itself. Hydroforming is a challenging process in which the sheet metal is deformed over a tool while being pushed by a rubber membrane, generating all the deformations in a single step. It would not make sense to leave the features that can be made by bending for a later operation since higher operating costs will usually be incurred (see Fig. 6).

Once the manufacturing process is established, the flat pattern is recognized in A3 through the information provided by the 3D model and the features. Many commercial tools can be used to obtain this information. However, keeping in mind the agnostic concept of the MfM methodology, the software is suggested to extract the geometry of the blank part from the 3D model and the features.

The method proposed in this paper depends on the manufacturing process selected in A2. For bending, all the features type = trim may be considered (note that holes are included). However, only operation type = cutting is kept in the case of hydroforming (holes must be done in subsequent operations due to potential deformations).

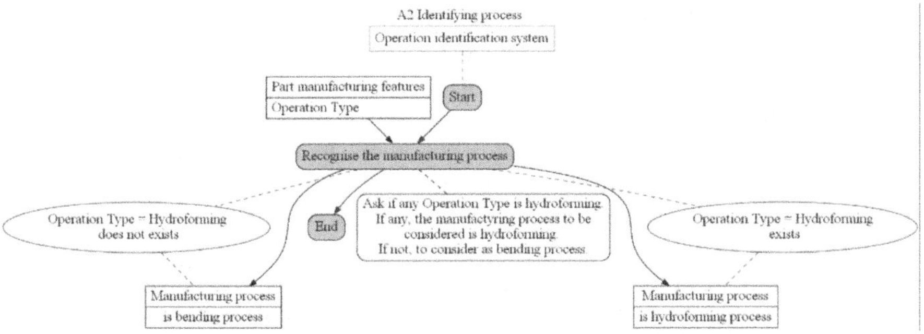

Fig. 6. Behaviour diagram of activity A2, "Identifying Process".

Finally, the manufacturing process is described in A4, which comprises three internal activities: A41 "Grouping operations of the process", A42 "Sorting operations of each group" and A43, "Defining Operation". These activities describe and characterize the operations of the process. Similarly, to the previous activity, at this late stage, the grouping and ordering of the operations depend on the manufacturing process. For bending, only two groups are considered: blank cutting and bending. For hydroforming, a drilling group is needed at the end of the process according to the previous statement. Finally, to order the operations, logical optimization criteria must be established. In the literature, optimization models of parameters and operations can be found in the sheet metal forming process [16]. In addition, the sequence of each operation must take into account the feasibility of the subsequent one, since an incorrect execution sequence can result in geometries that would not allow the punch to enter, making the part unusable.

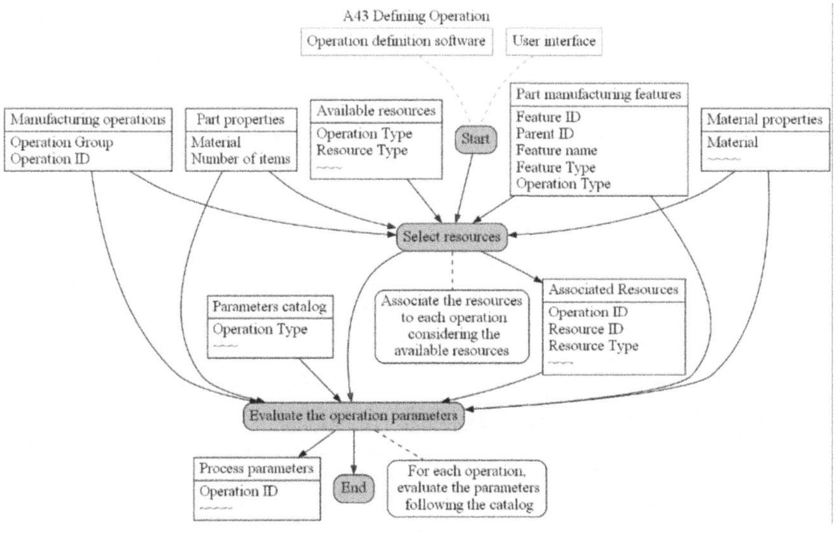

Fig. 7. Behaviour diagram of activity A43, "Defining Operation".

The last sub-activity concerns the allocation of resources and parameters to each operation. For this purpose, two data objects are considered necessary: *Available resources* and *Parameters catalog*. In both cases, there is a univocal relationship between the type of operation and the required resource type, as well as between each operation and the applicable process parameters. In the literature, there are some models to estimate the process parameters from the features (Fig. 7).

The Ontology Data Model in Fig. 8 shows the data objects contained in the Scope and Behavior models, which are the data treated in the defined ontology. This Data model complains about the structure of the needed databases in the implementation of the ontology. To realize the data model, concept maps are used, as they are a simple and useful graphical tool to structure and represent knowledge. Some of the applications are ConceptMap or Graphviz.

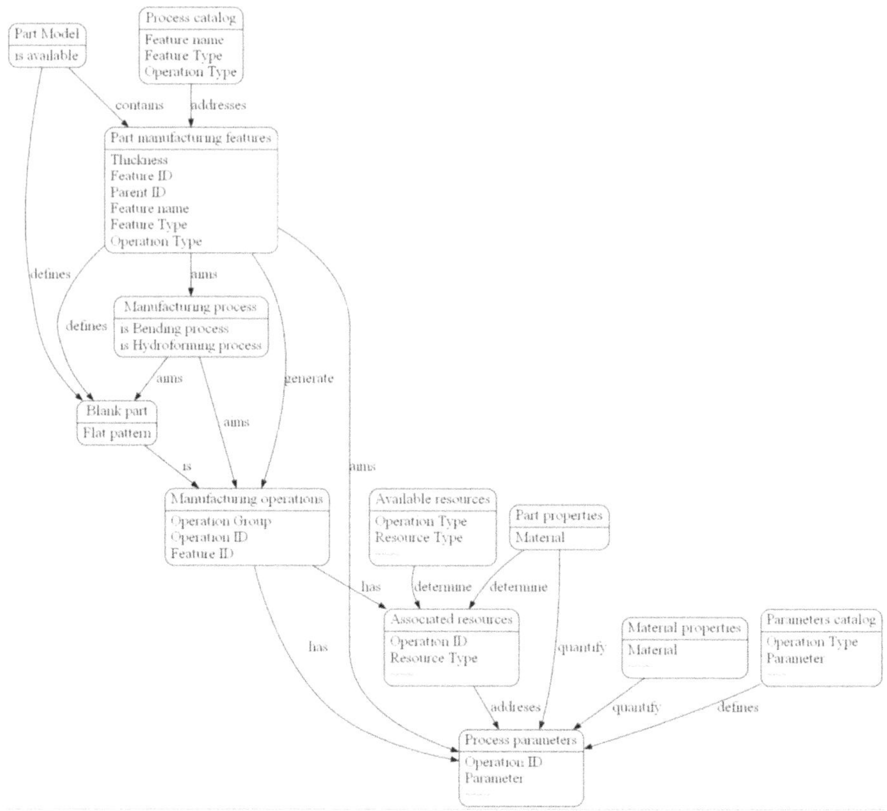

Fig. 8. Ontology Data Model.

4.2 Implementation

To facilitate understanding of the model, one of the parts used by Ghaffarishahri and Rivest in [17] will be used. They developed their model based on a total of 26 parts belonging to the frames, bulkheads, floor, and cockpit of a Bombardier DHC-8-102.

This is a typical design of an ASM hydroformed part that has two tooling holes in the web and a flange on each side. Each flange has a different configuration, although they represent three of the most common designs for this feature. A simple flange, perpendicular and immediate to the web and with no holes, another flange with 3 holes, and finally a flange with a joggle, which will be decisive in the choice of the manufacturing process.

The first step is to recognize the manufacturing features of the instance. The features obtained [6] and the allocation of the properties "Feature Type" and "Operation Type" are shown in Fig. 9 (activities A11 and A12 of the scope model). Note that the features with curvature (i.e., lips, lightening, beads, and joggle) are those that require the use of hydroforming. For this instance, the joggle and the deformed flange are the features that condition the part to be hydroformed.

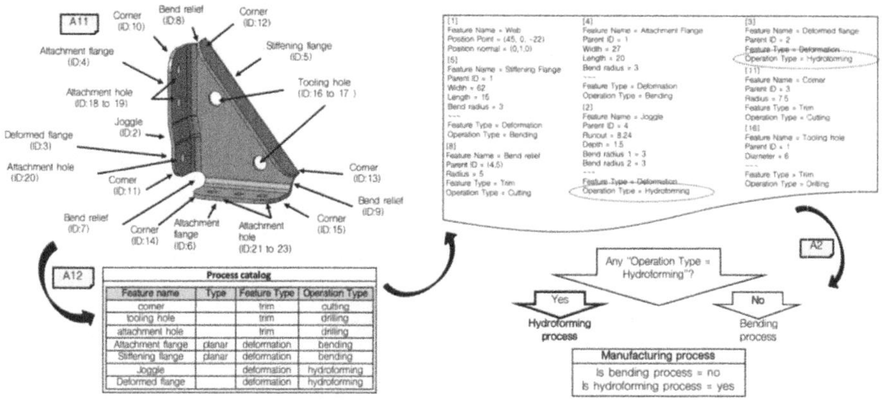

Fig. 9. Activities A11 "Recognizing features" to A2 "Identifying Process".

Once the features are available, flat pattern recognition is performed following the steps described in the behaviour model. In this case, as this part is going to be hydroformed, the blank part is defined through the 3D model and the features with "Operation Type = cutting", so drilling holes are not considered, differing from a part manufactured with bending operations only. The reason is based on the deformation suffered by the sheet during the hydroforming process, which recommended keeping the holes for subsequent operation (Fig. 10).

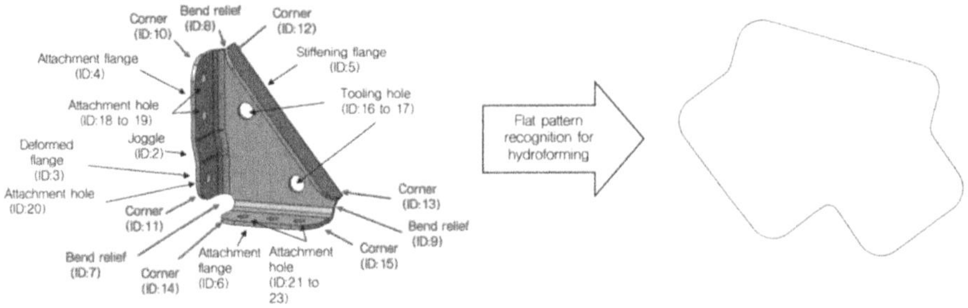

Fig. 10. Flat pattern recognition. (Features with *Operation Type* = *cutting* marked in red).

Once the blank part is recognized, the features may be grouped in terms of the operation required to generate them. For hydroforming, 3 groups of operations are defined: blank cutting, hydroforming, and drilling. It should be pointed out that all deformation operations, regardless of their type (bending or hydroforming) are grouped with the hydroforming ones.

The sequence of operations for each group is defined in Fig. 11. By definition, a single operation is assigned for the cutting and hydroforming operation groups. However, in the case of drilling, it may be useful to segregate operations using optimization methods such as diameter (i.e., tooling holes or attachment holes).

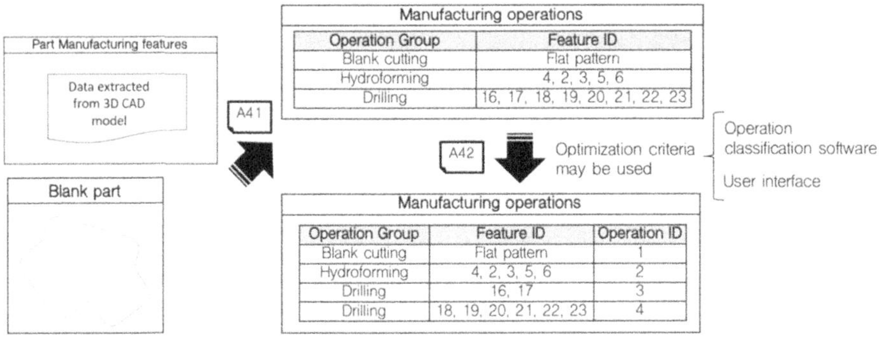

Operation Group	Feature ID
Blank cutting	Flat pattern
Hydroforming	4, 2, 3, 5, 6
Drilling	16, 17, 18, 19, 20, 21, 22, 23

Operation Group	Feature ID	Operation ID
Blank cutting	Flat pattern	1
Hydroforming	4, 2, 3, 5, 6	2
Drilling	16, 17	3
Drilling	18, 19, 20, 21, 22, 23	4

Fig. 11. Grouping and sorting of operations with activities A41 and A42.

Finally, the allocation of resources and the definition of process parameters. The user is responsible for the allocation of resources for each operation. The design of the hydroforming operation is much more complex than bending. In this case, it is necessary to design the shape of the die on which the sheet is deformed. In addition, it is usual when generating high flanges or bent geometries to use a blank holder to guide the deformation of the sheet metal. The hydroforming process is mainly controlled by pressure, pressure evolution in the hydroforming cycle, and temperature (Fig. 12).

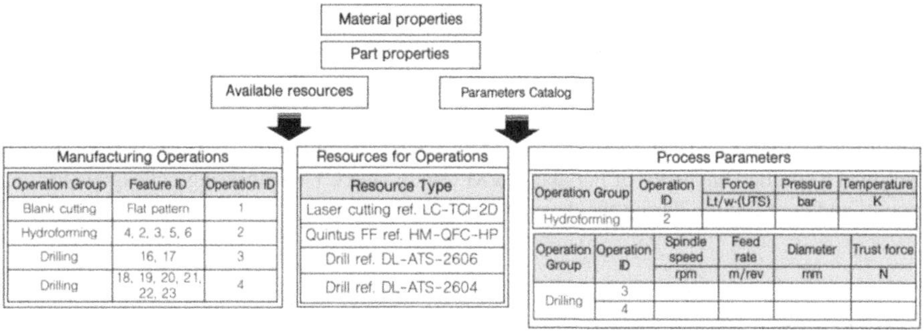

Fig. 12. Result of activity A43 "Defining operation".

5 Conclusions and Further Work

The use of semantic structures and ontological models is becoming more popular and has become an engineering discipline. In this regard, the MfM methodology seems to be very suitable for defining and reproducing data, functions, and behaviours of complex systems using graphical models. Its applicability for the intended purpose has been demonstrated in many projects in which it has already been applied.

The literature review brought up the state-of-the-art and relevant research on the addressed topic. However, it can be concluded that there is currently a gap in open source tools that can support manufacturing processes and still rely heavily on technical knowledge intrinsic to experienced professionals.

In this paper, a base model has been proposed for the characterization and definition of the sheet metal forming process, which is widely used in multiple sectors, including aerospace. Considering the needs of industries, where many new products are continuously being developed and, in many instances, external services are requested for carrying out the work, this model could be useful on many occasions.

A sample of 26 typical sheet metal parts was taken for this model and, first of all, the operations required to obtain them were analysed. In this research, it has been determined that bending and hydroforming are the most commonly used processes and, therefore, the need arose to create a model that could discern the parts according to the most appropriate technology to subsequently describe the manufacturing process.

Once the scope model was defined, the data and behavioural models were obtained. Note the iterative nature of the development process of these models, since to explain how the model behaves on each activity, it becomes necessary to have new inputs or information from other activities. Regarding the data and services layer, only structures have been proposed that are intended to be used as a guide for future projects.

For a better understanding of these concepts, the model was applied to a real part of the case study. This model explains in great detail the behaviour of each activity, the rules imposed, and the logic used to design them.

Finally, when a company needs to quote the cost and hours required to develop a work package, with a high volume of sheet metal parts, they might consider using this analysis to determine how many parts are hydroformed and how many are bent. In addition, an estimation of the total number of operations, cost, and time required could be provided.

In the same way, if there is also a model for the automated tool selection, the developed model can help to determine if all the resources are available, or if it is necessary to obtain or externalize any operation.

Acknowledgements. The authors express their sincere gratitude to the University of Sevilla, Pontifical Catholic University of Parana, M&M Group, and CT Engineering colleagues for their support and contribution during the development of this work.

References

1. Szejka, A.L., Mas, F., Junior, O.C.: Towards knowledge-based system to support smart manufacturing processes in aerospace industry based on models for manufacturing (MfM). In: Canciglieri Junior, O., Noël, F., Rivest, L., and Bouras, A. (eds.) Product Lifecycle Management. Green and Blue Technologies to Support Smart and Sustainable Organizations. pp. 425–437. Springer International Publishing, Cham (2022)
2. Mas, F., Rios, J., Menendez, J.L., Gomez, A.: A process-oriented approach to modeling the conceptual design of aircraft assembly lines. Int. J. Adv. Manuf. Technol. **67**(1–4), 771–784 (2013)
3. Morales-Palma, D., et al.: Metamodels approach supporting models for manufacturing (MfM) methodology. In: IFIP International Conference on Product Lifecycle Management. Cham: Springer International Publishing, pp. 398–409 (2021)
4. Holland, O.: Model-based systems engineering. In: Loper, M.L. (ed.) Modeling and Simulation in the Systems Engineering Life Cycle. SFMA, pp. 299–306. Springer, London (2015)
5. Mas, F., Racero, J., Oliva, M., Morales-Palma, D.: A preliminary methodological approach to models for manufacturing (MfM). In: Chiabert, P., Bouras, A., Noël, F., Ríos, J. (eds.) PLM 2018. IAICT, vol. 540, pp. 273–283. Springer, Cham (2018)
6. Ghaffarishahri, S., Rivest, L.: Feature recognition for structural aerospace sheet metal parts. Comput.-Aided Des. Appl. **17**(1), 16–43 (2020)
7. Thomé, A.M.T., Scavarda, L.F., Scavarda, A.J.: Conducting systematic literature review in operations management. Prod. Plan. Control **27**(5), 408–420 (2016)
8. Page, M.J., et al.: The PRISMA 2020 statement: an updated guideline for reporting systematic reviews. Int. J. Surg. **88**, 105906 (2021)
9. Van tooren, M., et al: Aircraft design support using knowledge engineering and optimisation techniques. In: 46th AIAA/ASME/ASCE/AHS/ASC Structures, Structural Dynamics and Materials Conference (2005)
10. Emberey, C.L., et al.: Application of knowledge engineering methodologies to support engineering design application development in aerospace. In: 7th AIAA ATIO Conf, 2nd CEIAT Int'l Conf on Innov and Integr in Aero Sciences, 17th LTA Systems Tech Conf; followed by 2nd TEOS Forum. p. 7708. (2007)
11. Mas, F., et al.: Knowledge-based application to define aircraft final assembly lines at the industrialisation conceptual design phase. Int. J. Comput. Integr. Manuf. **29**(6), 677–691 (2016)
12. Adamczyk, B.S., Szejka, A.L., Júnior, O.C.: Knowledge-based expert system to support the semantic interoperability in smart manufacturing. Comput. Ind. **115**, 103161 (2020)
13. Montáns, F.J., et al.: Data-driven modeling and learning in science and engineering. Comptes Rendus Mécanique **347**(11), 845–855 (2019)

14. Szejka, A.L., Junior, O.C.: The application of reference ontologies for semantic interoperability in an integrated product development process in smart factories. Procedia Manuf. **11**, 1375–1384 (2017)
15. Ríos, J., et al.: KBE application for the design and manufacture of HSM fixtures. Acta Polytechnica **45**(3) (2005)
16. Lee, J., Park, G.: Optimization of the structural and process parameters in the sheet metal forming process. J. Mech. Sci. Technol. **28**, 605–619 (2014)
17. Ghaffarishahri, S., Rivest, L.: A prototype of an automated feature recognition algorithm for aerospace sheet metal parts. Comput.-Aided Des. Appl. **19**(4), 624–661 (2022)

The Implementation of a PDM/PLM System: Case Study in an Industrial Process Cooling Company

Romeo Bandinelli[1(✉)], Virginia Fani[1], Ilaria Bucci[1,2], and Monica Rossi[2]

[1] University of Florence, Florence, Italy
romeo.bandinelli@unifi.it
[2] Politecnico di Milano, Milan, Italy

Abstract. The contemporary industrial scenario places significant emphasis on the efficient reduction of the time to market for new products, underscoring the necessity for companies to stay agile and competitive. Concurrently, the demand for product customization has become a defining characteristic, requiring companies to offer bespoke solutions that cater to diverse customer needs. Within this context, the implementation of a Product Data Management (PDM) and Product Lifecycle Management (PLM) system emerges as a strategic imperative. This research reports an extensive study in an industrial process cooling company in Tuscany, Italy. The study implemented a PDM/PLM system to improve and streamline the management of product data. The main goal was to optimize the processes for developing, producing, and delivering new products. This was achieved by integrating SolidRules Incentro software into the company's operations. This software improved research and archiving of product data, management of bills of materials and document retrieval. Consequently, the time to market for new products and customizations was significantly reduced. The initial phase of this project involved acquainting with the new system, gaining a comprehensive understanding, initiating its integration into the company, and providing training for future users. Simultaneously, meetings and seminars were conducted to thoroughly explore the capabilities of the PDM/PLM system. The benefits of implementing this were subsequently evaluated using key performance indicators, forming a comprehensive approach aimed at assessing and quantifying the positive impact on the company's overall operational efficiency and productivity.

Keywords: Product Lifecycle Management · PLM · Implementation

1 Introduction

Facing global competition and the rigorous economic context, product and process innovation has been considered a crucial weapon for engineering and manufacturing companies to maintain market position and achieve lasting competitive advantage [1]. To become effective and profitable, companies require more powerful information systems

Published by Springer Nature Switzerland AG 2025
P. Sureephong et al. (Eds.): PLM 2024, IFIP AICT 741, pp. 184–195, 2025.
https://doi.org/10.1007/978-3-031-93323-3_14

that help them reshape the workflow, improve operational efficiency, improve the cost structure, and reduce the time-to-market of new products. [2] Product data management or product lifecycle management systems (PDM/PLM), in particular, have become one of the most important investments for companies in the engineering and manufacturing sectors because high-quality product data management has been a fundamental capability to promote new product development and improve PLM effectiveness [3]. PDM, or Product Data Management, is a product data management system that systematically manages all product documentation, CAD files, technical drawings, and other data from start to finish. Building upon the foundational capabilities of PDM, a PLM (Product Lifecycle Management) system expands the scope to oversee the entire product lifecycle, integrating product data management with processes like supply chain management, manufacturing workflows, and customer feedback. This integration enables a comprehensive approach to product development, facilitating better collaboration, improved efficiency, and enhanced innovation across all stages of the product's life. [4].

In summary, this paper progresses as follows: Sect. 2 delves into PDM and PLM theoretical foundations, while Sect. 3 outlines our methodology within a real-world case study. Section 4 details the case study phases, addressing challenges and efficiencies during PDM to PDM/PLM integration. Section 5 presents results, analyzing temporal and qualitative aspects. Section 6 draws conclusions, reflecting on broader implications. This structured approach offers a comprehensive exploration of PDM/PLM integration in industrial contexts.

2 Theoretical Background

PLM, or Product Lifecycle Management, corresponds to the business activity that manages, in the most effective way, the products of a company throughout their entire life cycle; from the very first idea of a product to its withdrawal and disposal [5].

The main difference between the two is that PLM addresses a wider system in terms of scope and usually involves all business functions, while PDM is mainly a tool for product data management [6] and is a component of PLM [7].

It's crucial to emphasize that PLM aims to enhance efficiency, collaboration, and decision-making throughout the product development process [6].

The implementation of these systems proves crucial for the entire cycle of creating new products [7]. Integrating PDM and PLM allows for the coordination of product development processes and data management [8] and is crucial for efficient design, collaboration, and data consistency [9], therefore companies strive to create a unified system that not only manages product data but also optimizes the overall product lifecycle.

Elisa d'Avolio et al. (2016) [10] assess the performance of Product Lifecycle Management (PLM) and Product Development within the fashion industry. This evaluation is conducted through the utilization of a dashboard featuring indicators categorized into five key dimensions: Time, Costs, Quality, Flexibility, and Infrastructure.

This analysis focuses on assessing the advantages of implementing PLM software at Alcatel Alenia Space, utilizing a KPI dashboard to measure performance. Within this framework, the indicators are organized across five key dimensions: Quality, Time, Cost, Infrastructure, and Communication. These categories collectively help in evaluating the impact of PLM tools on enhancing operational efficiencies and overall effectiveness within the leading space company, as documented by Marco Alemanni et al. (2008) [11].

Another study was conducted by Vanessa Nappi and Henrique Rozenfeld (2013) [12], focusing on an investigation aimed at identifying and classifying performance indicators that address the sustainability perspective for a Product Lifecycle Management (PLM) system.

Key Performance Indicators (KPIs) are used in Product Lifecycle Management (PLM) to measure and improve the performance of business processes [13]. KPIs are used throughout the entire lifecycle of a product, from design to manufacturing to service [14]. They provide valuable insights into the efficiency and effectiveness of these processes, allowing organizations to identify areas for improvement and make datadriven decisions [15]. The use of KPIs in PLM can lead to increased agility and competitiveness for organizations, as they can identify bottlenecks, track progress, and make informed decisions to improve their processes [5].

3 Methodology

The research methodology adopted a comprehensive approach, beginning with a theoretical background, followed by a real-world case study within a company. The case study encompassed introducing the new system, implementing it, and evaluating performance at various stages. The implementation phase involved rigorous activities such as porting and coding, formulation, and management of bills of materials, and efficient documentation handling. The integration of the PDM/PLM system into the existing workflow required meticulous attention to detail and close collaboration with the company analyzed.

Following the implementation, the study transitioned into the performance measurement phase. Here, to measure the effectiveness and impact of the integrated PDM/PLM system, a set of KPIs were identified. The measurement phase aimed to provide a quantitative assessment of how the system influenced operational efficiency and productivity.

This approach allowed for a comprehensive assessment, enabling the research to quantify the positive impact on the company's overall operational efficiency and productivity. This methodology ensured a holistic understanding of the implications of PDM/PLM integration within the industrial domain.

4 Case Study

This case study investigates a leading company headquartered in Tuscany, Italy, specializing in the development, production, installation, sale, and support of cutting-edge process cooling systems tailored for industrial applications and temperature control. Initially focusing on designing integrated cooling systems for the plastics industry, it has expanded its portfolio over the years to cover various sectors such as rubber, food and beverages, energy, data centers, chemicals, and pharmaceuticals.

Examining the organizational structure, the company operates on a make-to-order basis, exclusively producing items based on incoming orders. Operating on a "maketo-order" basis involves creating a unique product of limited dimensions or executing a project tailored to a client's specific requirements. Each project or order is regarded as a collection of resources, both material and non-material, brought together for a defined period to achieve objectives, considering potential economic and temporal constraints.

Emphasizing the intricacies of company's operational process, effective communication between various departments is crucial. The use of a competitive PDM/PLM system is indispensable to ensure the entire product lifecycle aligns with the market's quality standards. The subsequent workflow among different departments is detailed in Fig. 1.

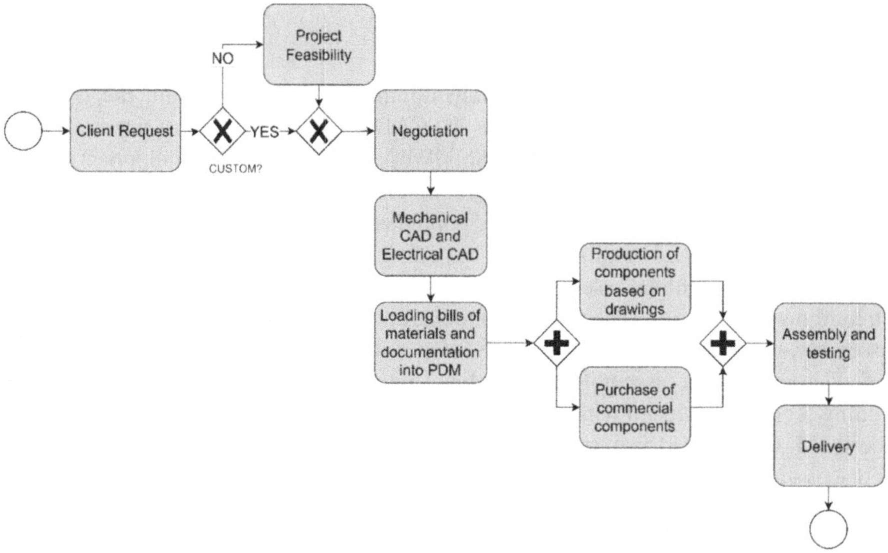

Fig. 1. Workflow of the company analyzed.

In the case study, the chosen PDM/PLM system is SolidRules Incentro, known for its capability to manage various formats, including 3D vector formats, 2D and 3D PDFs, CAD-derived technical bills of materials, and spare parts. The PDM portal, positioned at the heart of the company, seamlessly supports CAD operations. Key features encompass search, encoding, vault, automation, collaboration, and multi-document support.

4.1 Implementation Framework

The implementation of a PLM is a critical process that requires a structured approach [16]. Organizations often struggle with defining concrete goals and relating technology to business benefits during PLM implementation [17]. One of the challenges is organizing and migrating existing product data to the PLM system [18].

Fani et al. (2021) proposed a framework that aims to identify a PLM implementation roadmap in order to make the application steps of this software more effective within companies [19].

We borrow Fani's PLM implementation framework, originally designed for the fashion industry, to adapt and suit a different industrial context. While Fani's initial two phases (process and needs analysis) have been conducted, they are not the primary focus of our study. Instead, we align our 'Implementation Phase' with Fani's 'Detailed Implementation,' adapting it to our case study and subdividing it into three sub-phases. Furthermore, we expand upon Fani's 'Training and Testing' phase, transforming it into our 'Measurement Phase.' A significant innovation in our work is the incorporation of KPIs, which offer a novel approach to assessing the PLM implementation's impact and efficacy.

4.2 Implementation Phase

Porting and Coding Activities. In transitioning to the new system, the company replaced SolidWorks with SolidEdge for 3D CAD modeling and bill of materials management. The migration, facilitated by SolidWorks' tool, preserved component properties. The subsequent step involved coding components using SolidRules Incentro, with a 10-character system. Coded components populated the PDM, facilitating easy tracking, mass replacements, and modifications. The coding process included categorizing articles into four major functional assemblies (structural, hydraulic, refrigeration, and electrical) streamlining production. The production department was organized into zones for each assembly, resulting in time and cost advantages during machine assembly. The ability to code convenience assemblies streamlined mass coding, enhancing efficiency. The coding process concluded with field completion for an efficient and error-free search in the PDM. Customized features in SolidRules Incentro allowed quick, error-free mass replacement of uncoded components, ensuring complete coding in the new language.

Formulation and Management of Bills. SolidRules Incentro facilitates the customization of bills of materials (BOMs), allowing various types of BOMs to be analyzed on the PLM portal based on their intended use. Examples include valued BOMs containing price and production time information, production-oriented BOMs with technical details, and procurement-oriented BOMs containing information on commercially available components and suppliers. After completing the porting and coding phase, the machine BOM can be loaded onto the PLM directly from the 3D model or from the PDM by searching for the relevant item. These BOMs are subsequently checked against existing ones to prevent the accidental purchase or production of unnecessary components. In case of an error in the BOM, the system sends a message containing error details, typically related to the structure of the Excel file to be loaded. The resolution involves ensuring the correct structure of the Excel file. Common errors may include the lack of coding for some components. To resolve this error, coding for the listed components is required. Once errors are corrected or missing codes are added, the BOM is processed.

Documentation Management. As part of the implementation phase, all documents, drawings, and information were systematically renamed and stored in the dedicated section of the PDM. The new naming conventions followed a precise syntax to ensure compatibility with the PLM portal. Upon completion of these tasks, every department within the company will have comprehensive access to all necessary resources in the PLM portal. Leveraging the system's efficient search functionality, files, documents, and information of every kind will be readily available at any given moment. This meticulous approach to documentation management streamlines accessibility and ensures a cohesive organizational structure within the PLM framework.

4.3 Measurement Phase

Identification of KPIs for Company Dashboard. The selection of KPIs in this study was a deliberate and well-informed process, as they were not chosen randomly. Rather, they constitute a set drawn from existing literature [10–12], and their selection, guided by the expertise of the company's professionals, was a meticulous endeavor led by autohrs. This approach ensures that the chosen KPIs are not only rooted in established practices but also align with the insights and recommendations of industry experts within the organization.

In the measurement phase, the crucial step involves identifying KPIs for the company dashboard. This dashboard is a pivotal management tool that facilitates strategic coordination, detects deviations, and ensures swift corrective actions. It provides a concise and synthesized view of the company's overall health, aiding in the identification of correlations between causes and effects. The dashboard transforms raw data into graphical representations, enhancing accessibility compared to other communication methods. Its effectiveness lies in addressing key aspects beforehand, such as determining desired analyses, specifying information to be collected, and establishing efficient data sources. KPIs play a central role in constructing an effective dashboard, offering vital information to assess and explain how an organization progresses toward its business activity objectives.

Two primary hypotheses guided this process. The first hypothesis considered the timing of data collection, occurring during the case study when the company was actively implementing the new system. Due to this focus on implementation in preparation for the upcoming Go-Live in the current year, the estimation of future benefits from the new system was only possible.

The second hypothesis pertained to the selection of KPI categories, focusing on short-term evaluation. Long-term indicators, such as reduced time to market, increased revenue and customers, improved product quality, and cost reduction, were deferred for analysis due to the short-term evaluation scope.

The classification of KPIs in this study for assessing performance post the introduction of the new PDM/PLM system is based on two major categories: quality and time. Quality indicators aim to measure the product/service conformity with customer expectations. For the software's quality, specific indicators were chosen to evaluate results based on predefined standards, considering factors such as avoidable errors, released resources, details availability, system effectiveness and clarity, implemented functionalities, and identified non-conformities. In contrast, time-related KPIs facilitate the evaluation of processes from initiation to completion. Events and activities from the old and new systems were analyzed to estimate the differential time. Events represent occurrences independent of resource participation, while activities denote actions that resources must perform. To estimate the company's future performance post the introduction of the new system, several activities were conducted. Initial involvement of all company departments, which would work with the new software in the short and long term, provided insights into the current method of work and the existing PDM/PLM system. This preliminary involvement aimed to create an AS-IS overview of the company's processes and systems.

This detailed examination formed the basis for formulating a comprehensive KPI dashboard for performance measurement during the measurement phase of the case study (Table 1).

Performance Analysis. During the measurement phase, a comprehensive evaluation of both temporal and qualitative aspects was conducted for the AS-IS and TO-BE states of the PDM/PLM system. The identified challenges and proposed enhancements encompassed various facets crucial to the overall performance and quality of the system. The temporal challenges spanned activities related to PDM, including CAD model usage, BOM management, documentation retrieval, parts inventory search, and intersite project transfers.

Critical Temporal Issue 1 (KPI-T-01): Addressing the need for efficient modifications and mass replacements of released parts, the current system necessitates a timeconsuming machine-by-machine approach. The proposed system streamlines these operations by allowing instant mass replacements, significantly reducing time and potential errors.

Critical Temporal Issue 2 (KPI-T-02): The BOM management challenge highlighted inefficiencies in the current system's handling of BOM uploads, impacting both production and procurement. The new system offers real-time and error-free BOM uploads directly from the 3D machine model, enhancing efficiency and accuracy.

Critical Temporal Issue 3 (KPI-T-03): The challenge of time-consuming documentation search and retrieval across disparate systems is addressed by the new PLM portal. It consolidates information, files, and data, providing a unified and efficient search interface, eliminating the need to navigate multiple portals for information retrieval.

Critical Temporal Issue 4 (KPI-T-04): Efficient synchronization between the PLM system and the Material Requirements Planning (MRP) system is crucial. The new system ensures automatic and error-free synchronization, streamlining material requisition processes and reducing manual checks.

Critical Temporal Issue 5 (KPI-T-05): The lack of a unified global system posed significant challenges in knowledge exchange and project transitions. The new system, with its universal language and streamlined workflows, promises to enhance organizational efficiency, reducing lead times for projects and facilitating seamless global collaboration.

Qualitative challenges encompassed issues related to the current system's article and component coding, vaulting functionality, and overall user accessibility.

Qualitative Challenge 1 (KPI-Q-01) (KPI-Q-02): The current article and component coding system lacked clarity and consistency, impeding efficient inventory management. The proposed system introduces a clear and standardized coding system, enhancing readability and reducing the risk of errors. Additionally, it ensures the availability of technical data in all portals, eliminating the need for manual data retrieval.

Table 1. KPI dashboard chosen for analysis.

KPI ID	PLM KPIs	Description	Process
KPI-T-01	Average time for document change	Average time for a drawing or document change	Data management
KPI-T-02	Time to export the report of BOM items to ERP	Average time to run the scheduled export batch containing information about properties and the related materials	Import/Export configuration
KPI-T-03	Document searching time	Document searching cost	Data management
KPI-T-04	Part number delay time between product and process engineering	Average time before a part number moves to the engineer area to the production one and eventually all version created	Product develp-ment
KPI-T-05	Start up agerage time	Average time for a project start-up	Configuration management
KPI-Q-01	Data searching time	Refers to the capability of the user interface to be simple and friendly	Data management
KPI-Q-02	Product traceability	The introduction of configuration management facilities allows more accurate product information and traceability	Configuration management
KPI-Q-03	Improved documental management	Management function (revisioning, versioning and vaulting)	Configuration management
KPI-Q-04	PLM system scalability	Refers to the PLM capability to increase its performance to accommodate the resources growth	Data configuration
KPI-Q-05	Activated workflow number	Number of process involved into the system through the using of the workflow functionalities	Data management

Qualitative Challenge 2 (KPI-Q-03): The evaluation of the vaulting functionality revealed challenges in assigning and managing file states in the current PDM system. The proposed system introduces efficient state changes, automatic synchronization, and reservation features, minimizing errors and conflicts during the file management process. Qualitative Benefit 1 (KPI-Q-04): Among the qualitative benefits, the new system's ability to handle continuous updates externally reduces the internal workload for system management. The standardized and accessible CAD model information contributes to better decision-making, reducing the risk of costly errors.

Qualitative Benefit 2 (KPI-Q-05): The new system's impact on the user base is substantial, extending its usage to the testing phase with SolidRules Incentro. This ensures a more holistic and efficient approach to product testing, contributing to overall product quality.

In summary, the performance analysis captures the dual dimensions of temporal and qualitative challenges, demonstrating how the proposed PDM/PLM system effectively addresses each issue, leading to improved efficiency, reduced errors, and enhanced overall system quality.

5 Results

The examination of KPIs following the implementation of the PDM/PLM system offers a comprehensive view of the transformative impact on various facets of the company's operations.

Regarding Time-Related KPIs, one of the most significant improvements comes in the form of the "Average Time for Document Change" (KPI-T-01). The system has streamlined this process, addressing the limitations of mass replacements and the sluggishness in CAD file handling. With time savings ranging from 20 to 40 min per operation, particularly for hydraulic or refrigeration components, the newfound efficiency represents a noteworthy enhancement.

Similarly, the "Time to Export the Report of BOM Items to ERP" (KPI-T-02) has witnessed a marked improvement. The system's efficiency gains, saving 5 min per operation, contribute to a smoother workflow, especially when modifications to the 3D model occur or various departments request BOM-related information.

Document management has undergone a significant overhaul, as indicated by the "Document Searching Time" (KPI-T-03). The reduction in search time by 1 min per search operation translates into a substantial daily time saving of 1 h and 10 min, equating to approximately 300 h annually. This streamlined accessibility ensures that crucial information is readily available across all departments.

Addressing delays in part number transitions, the "Part Number Delay Time Between Product and Process Engineering" (KPI-T-04) demonstrates a commendable reduction of 1–2 h, occurring approximately twice a week. This not only signifies improved warehouse supervision but also leads to significant annual time savings ranging from 50 to 100 h.

The "Start-up Medium Time" (KPI-T-05) showcases a challenging-to-quantify yet noticeable reduction in the startup phase for projects. Although the frequency remains uncertain, the positive trend suggests enhanced efficiency in handling project transitions.

Turning our attention to quality-related KPIs, the "Data Searching Time" (KPI-Q-01) indicates a notable improvement in the user interface's capability for data searching. The clarity introduced by more optimal article coding enhances the interpretation of associated files, even without direct access.

Product traceability, assessed through "Product Traceability" (KPI-Q-02), has significantly benefited from the introduction of configuration management facilities. The avoidance of information gaps and the need to navigate multiple portals ensures more accurate product information and, crucially, prevents costly errors.

The "Improved Documental Management" (KPI-Q-03) underscores the pivotal role of the PDM/PLM system in bridging the gap between PDM and the management system. By doing so, it averts critical errors, presenting a more cohesive and error-resistant documentation structure.

The scalability of the PLM system, measured by "PLM System Scalability" (KPI-Q-04), has notably increased. This enhancement not only liberates high-know-how resources but also provides a detailed historical record of file operations, contributing to more informed decision-making.

Introducing one additional workflow process in "Activated Workflow Number" (KPI-Q-05) signifies a step towards more intricate process integration, aligning with the broader goals of improved efficiency and collaboration.

The comprehensive analysis of these results underscores the tangible and multi-faceted benefits derived from the adoption of the PDM/PLM system. The time-related improvements indicate streamlined processes, reduced operational bottlenecks, and enhanced overall efficiency. Simultaneously, the qualitative enhancements point towards a more robust and error-resistant data management structure, fostering better decision making and traceability. The successful integration of SolidRules Incentro has proven instrumental in achieving these positive outcomes, setting the stage for sustained operational excellence within the company.

6 Conclusions

In conclusion, this study lays the foundation for a performance evaluation of analyzed company, a Tuscany-based company, following the implementation of an upgraded PDM/PLM system. To achieve this objective, a comprehensive overview of the company's operations and product diversity was crucial, highlighting its multifaceted role in designing, producing, installing, selling, and supporting process cooling systems for various industries. The study delved into the theoretical background, offering insights into contemporary production systems, the history of PDM/PLM systems, and their integration with other platforms.

A detailed exploration of the chosen PDM/PLM system, SolidRules Incentro by Siemens, followed, with a focus on the internship's implementation of the PDM portal. Subsequently, the paper explored the realm of performance measurement systems, specifically defining KPIs for evaluating PDM/PLM system performance. Despite limited literature in this domain, validated KPIs were identified for short-term performance analysis, guided by the hypotheses that only short-term measurable improvements over the previous system would be considered.

The analysis concentrated on two major categories of KPIs: time and quality. Temporal KPIs revealed substantial improvements in various processes, including CAD modifications, BOM loading, document search efficiency, transition from engineering to production, and project initiation. Noteworthy time savings were estimated, ranging from minutes to months, demonstrating the system's efficiency gains across diverse operations.

Qualitative KPIs underscored enhancements in interface efficiency, information traceability, PDM portal vaulting functionalities, PLM system scalability, and the inclusion of additional business processes. The results showcased improved article coding, error prevention, streamlined PDM-PLM connectivity, resource liberation, detailed BOMs, and enhanced historical tracking.

However, it's crucial to acknowledge that these measurements are estimations awaiting validation post the anticipated Go-Live in the summer of the same year. This paper establishes the groundwork for ongoing performance measurement, emphasizing the dynamic nature of this process, necessitating continuous oversight and validation as the company moves forward with its upgraded PDM/PLM system.

References

1. Flores-Garcia, E., Bruch, J., Wiktorsson, M., Jackson, M.: Decision-making approaches in process innovations: an explorative case study. J. Manuf. Technol. Manag. **32**, 1–25 (2021). https://doi.org/10.1108/JMTM-03-2019-0087
2. Singh, S., Chandra Misra, S., Kumar, S.: Identification and ranking of the risk factors involved in PLM implementation. Int. J. Prod. Econ. **222**, 107496 (2020). https://doi.org/10.1016/j.ijpe.2019.09.017
3. Li, S.-H., Chen, J.-L., Yen, D.C., Lin, Y.-H.: Investigation on auditing principles and rules for PDM/PLM system implementation. Comput. Ind. **64**, 741–753 (2013). https://doi.org/10.1016/j.compind.2013.04.007
4. Eigner, M.: Forty years of product data management from PDM via PLM to SysLM. In: Eigner, M. (ed.) System Lifecycle Management: Engineering Digitalization (Engineering 4.0). Springer Fachmedien Wiesbaden, Wiesbaden, pp. 5–25 (2021)
5. Do, N.: Integration of design and manufacturing data to support personal manufacturing based on 3D printing services. Int. J. Adv. Manuf. Technol. **90**, 3761–3773 (2017). https://doi.org/10.1007/s00170-016-9688-8
6. Stolt, R., André, S.: Requirements Handling in Multidisciplinary Product Development - A Company Study. In: Transdisciplinarity and the Future of Engineering, pp. 115–124 (2022)
7. Lopota, A.V., Tsyrkov, A.V., Tsyrkov, G.A.: Methods and tools of project-operational management of a machine-building enterprise. In: International Conference "Quality Management, Transport and Information Security, Information Technologies" (IT&QM&IS), St. Petersburg, Russia, 2017, pp. 544-547 (2017). https://doi.org/10.1109/ITMQIS.2017.8085882
8. Kadam, O.B., Pirayesh, A., Fatahi Valilai, O.: Technological insights of interoperable models for integration of CAD/PLM/PDM and ERP modules in engineering change management. In: Kim, K.Y., Monplaisir, L., Rickli, J. (eds.) Flexible Automation and Intelligent Manufacturing: The Human-Data-Technology Nexus. FAIM 2022. LNME, pp. 556–564. Springer (2023)
9. Pérez Fernández, R.: A new approach for using cad and PLM integration. ICCAS 2022 (2022). https://doi.org/10.3940/rina.iccas.2022.30
10. d'Avolio, E., Bandinelli, R., Rinaldi, R.: Product Development and PLM Performance Measures: A Multiple-Case Study in the Fashion Industry (2016)
11. Alemanni, M., Alessia, G., Tornincasa, S., Vezzetti, E.: Key performance indicators for PLM benefits evaluation: the Alcatel Alenia space case study. Comput. Ind. **59**, 833–841 (2008). https://doi.org/10.1016/j.compind.2008.06.003
12. Nappi, V., Rozenfeld, H.: Sustainability performance indicators for product lifecycle management. In: Proceedings of the 22th International Confgress of Mechanical Engineering, pp. 4071–4082 (2013)

13. Wang, H., Peng, G.: Knowledge management through product lifecycle. In: Wang, H., Peng, G. (eds.) Collaborative Knowledge Management through Product Lifecycle: A Computational Perspective. Springer Nature, Singapore, pp. 1–37 (2023)
14. Wannes, A., Ghannouchi, S.A.: KPI-based approach for business process improvement. Procedia Comput. Sci. **164**, 265–270 (2019). https://doi.org/10.1016/j.procs.2019.12.182
15. Wang, L., Liu, Z., Liu, A., Tao, F.: Artificial intelligence in product lifecycle management. Int. J. Adv. Manuf. Technol. **114**, 771–796 (2021). https://doi.org/10.1007/s00170-021-06882-1
16. Gunjal, H.V., Belokar, R.M.: Systematic review: implementation of product lifecycle management in industries. In: Singh, R.P., Tyagi, M., Walia, R.S., Davim, J.P. (eds.) Advances in Modelling and Optimization of Manufacturing and Industrial Systems. Springer Nature, Singapore, pp. 263–279 (2023)
17. Koomen, B.: A Knowledge-Based Approach for PLM Implementation Using Modular Benefits Dependency Networks. In: Nyffenegger, F., Ríos, J., Rivest, L., Bouras, A. (eds.) Product Lifecycle Management Enabling Smart X, pp. 553–562. Springer International Publishing, Cham (2020)
18. Giddaluru, M.P., Gao, J.: A data preparation and migration framework for implementing modular product structures in PLM. In: Fortin, C., Rivest, L., Bernard, A., Bouras, A. (eds.) Product Lifecycle Management in the Digital Twin Era. Springer International Publishing, Cham, pp. 201–210 (2019)
19. Fani, V., Falchi, C., Bindi, B., Bandinelli, R.: Implementation framework for PLM: a case study in the fashion industry. Int. J. Adv. Manuf. Technol. **113**, 435–448 (2021). https://doi.org/10.1007/s00170-021-06623-4

Wasp Detection System Using AI Technology to Support Honey Bee Farming

Paweena Suebsombut[1](✉), Suttipong Raiphon[1], Sorawit Sritichai[1], Nontawat Thongkam[1], and Thongchai Yooyativong[2]

[1] School of Applied Digital Technology, Mae Fah Luang University, Chiang Rai, Thailand
paweena.sue@mfu.ac.th, {6431501123,6431501150, 6431501159}@lamduan.mfu.ac.th

[2] Center of Excellent in Artificial Intelligence and Emerging Technology, School of Applied Digital Technology, Mae Fah Luang University, Chiang Rai, Thailand
thongchai@mfu.ac.th

Abstract. This paper aims to propose wasp detection system using artificial intelligent (AI) technology to support beekeepers. The YOLOv8x was selected for wasp detection modelling. The proposed wasp detection model was trained, validated, and tested based on the dataset collected from Roboflow which is the open data source. There were 18,425 images collected meticulously categorized into training (88%), validation (8%), and test (4%) sets. An overall performance of this proposed model is around 85% of confidence score. According to the results, the proposed model can effectively identify and classify wasps with high precision and recall rates. Moreover, beekeepers can receive real-time notification as text and image via LINE chat application, which is crucial for timely intervention in protecting beehives from wasp attacks.

Keywords: Honey Bee · Wasp Detection · Artificial Intelligence (AI) · Beehive Protection

1 Introduction

Nowadays, number of beekeepers in Thailand is around 35,405 people according to an information from the Department of Agricultural Extension (DOAE), Thailand [1]. Bees are important in pollinating plants and are essential for agriculture and biodiversity. They enhance agricultural productivity and hold significant economic importance. Besides, bees produce honey, beeswax, royal jelly, bee pollen, and other commercial and health-beneficial products. Furthermore, honey is the main by-product. It represents a huge economic value. It is a revolving income for farmers of approximately 2,100 million baht per year excluding other types of produce and processed products [1]. Thailand has the second highest honey production volume in ASEAN and the 36th highest in the world.

Moreover, according to an information from the Food and Agriculture Organization of the United Nations (FAO) indicates that Thailand is the 20th largest exporter of honey

P. Sureephong et al. (Eds.): PLM 2024, IFIP AICT 741, pp. 196–206, 2025.
https://doi.org/10.1007/978-3-031-93323-3_15

in the world, shipping to Taiwan, Indonesia, the United States, Canada, and the People's Republic of China and Saudi Arabia [1].

Regarding beekeeping, beekeepers need to know the life-cycle and behavior of bees. The life-cycle of a bee consists of four main stages: egg, larva, pupa, and adult (see Fig. 1).

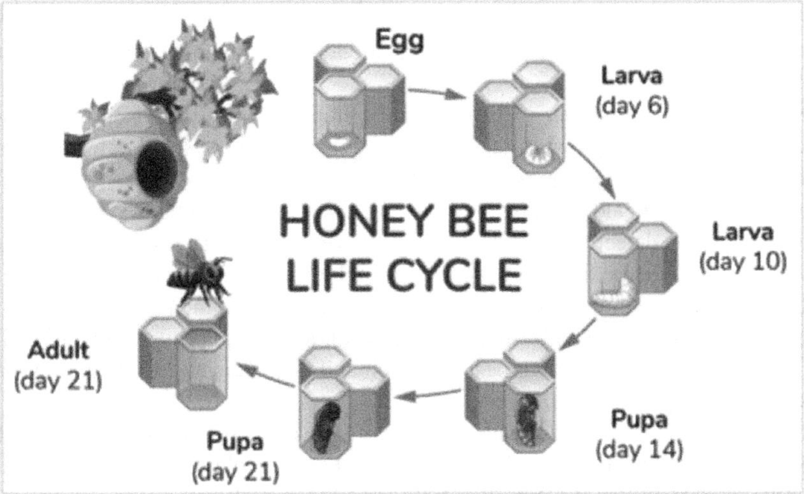

Fig. 1. Life cycle of a honey bee [2].

The queen bee lays eggs in the honeycomb's cells, which hatch into larvae after about three days. The larvae undergo several molts over five to six days before entering the pupal stage, which lasts approximately 12 days before transforming into adult bees. The bee society is intricately organized with distinct roles. The queen bee is responsible for worker bees, and laying eggs, which are non-reproductive females, operate various tasks such as cleaning the hive, caring for larvae, building honeycombs, and foraging for food. Male bees, known as drones, primarily mate with queens from other colonies. Each day, worker bees start forage in the morning, collecting nectar and pollen from flowers. Upon returning to the hive, some worker bees build and repair the honeycombs, care for the larvae, and clean the hive. They communicate about food sources through a dance language. In the evening, bees return to the hive to rest and regulate the hive's temperature to ensure the larvae's proper development. [2, 3].

Beside bee life-cycle knowledge, beekeeping involves preparing the necessary equipment, such as bee boxes, which serve as the bees' habitat, and frames or honeycombs for building and storing honey. Beekeepers need to locate their hives in areas with sufficient food sources. It is essential to understand the bees' food sources to ensure they can forage adequately and produce honey efficiently such as longan orchards, lychee orchards, cornfields, sunflower fields, wildflowers fields, etc. which helps bees gathering nectar and pollen throughout the year. During beekeeping, beekeepers always face of issues with wasps. They are natural predators of bees which the beekeepers will notice this issue quite late. Wasps will kill bees and steal larvae and honey to feed their colonies. In addition, the bees cannot be able to store enough food during the winter, weakening of the hive, or abandon the hive altogether which are the results of an attacking by wasps repeatedly. These impacts the honey production and the bees' role in crops' pollinating. There are three main wasp species in Thailand [4] comprising Lesser banded [5], Yellow-legged [6], and Greater banded [7]. The wasp will attack and weaken bee colonies. Wasps kill bees and steal larvae and honey to feed their colonies. Repeated attacks can weaken the hive, preventing bees from storing enough food to survive the winter or causing them to abandon the hive altogether. This impacts honey production and beekeepers' income.

Due to the large number of hives and the impracticality of constant monitoring, beekeepers need to adopt smart technology (i.e. monitoring system, detection system, notification system, etc.) helping to surveil for threats from wasps. Therefore, this paper proposes a detection and warning model using artificial intelligent (AI) to support beekeepers for wasp detection which can help the prevention of losses and mitigate the impacts. With the expeditious advancement of Internet of Things (IoT) and artificial intelligence (AI) technologies, several studies have explored their applications in agricultural and ecological monitoring. These technologies have proven to be effective in addressing challenges related to pest detection and management, which are critical for protecting vital species such as honeybees.

Recent researches have demonstrated the effectiveness of convolutional neural networks (CNNs) in detecting and classifying pests. For example, Fuentes et al. [8] utilized a combination of Faster R-CNN, R-CNN, and SSD models integrated with a residual network and visual geometry group network (VGG) to detect various pests and diseases on tomato plants. Similarly, Lin et al. [9] applied Fast R-CNN approach to develop an anchor-free regional convolutional network to wasp classes, achieving high accuracy metrics. In the realm of wasp detection for bee protection, several approaches have been implemented. A notable example is the use of machine learning algorithms to detect wasps, which pose a significant threat to bee colonies. These algorithms leverage large datasets of insect images to train models capable of distinguishing between bees and their predators. Furthermore, Sabanci et al. [10] proposed the pest-damaged wheat grain recognition using a convolutional recurrent hybrid network, achieving high accuracy rates. Similarly, Zhong et al. [11] developed a visual flying insect detection system using machine learning approach, demonstrating the feasibility for pest monitoring.

Despite these advancements, there remains a gap in research specifically targeting the protection of bees from wasps using Artificial Intelligent (AI) and Internet of Things (IoT) technologies. This paper aims to address this gap by developing a wasp detection model utilizing YOLOv8, coupled with automated countermeasures and notification mechanisms. By enhancing the accuracy and responsiveness of wasp detection, this system contributes to the conservation of bee populations, which are crucial for agricultural productivity and beekeepers' income.

2 Methodology

Figure 2 illustrates the proposed system diagram. The input data will be collected and analyzed to detect the wasp. The output data will be sent to Line application to inform bee farmers relevant to an attack of wasp to their honey bee. After data collection step, the dataset will be pre-processing before using for training the wasp detection model. The model will be trained, validated, and tested before implementation as shown steps of the proposed wasp detection modelling in Fig. 3.

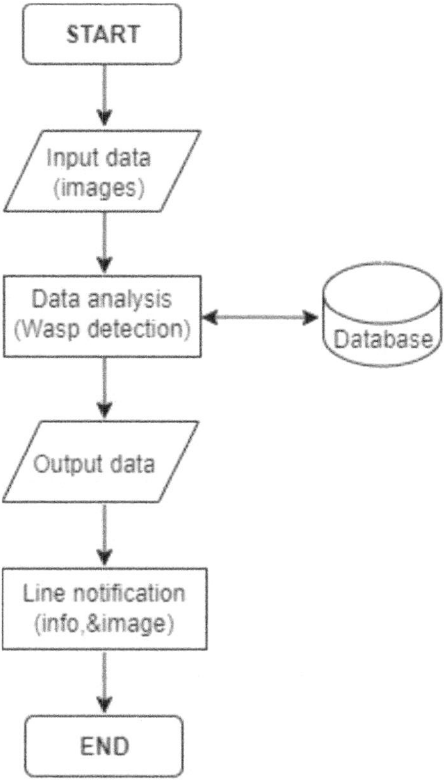

Fig. 2. Wasp detection System Diagram.

2.1 Dataset and Pre-processing

The study utilized the "Asian Hornet 2" dataset sourced from Roboflow [12], a comprehensive collection of 18,425 images meticulously categorized into training (88%), validation (8%), and test (4%) sets. This strategic partitioning ensured a robust evaluation of the model's performance on unseen data while intercept overfitting during training.

A series of pre-processing steps were meticulously applied to optimize the model's learning process and enhance its generalization capabilities. Firstly, all images underwent

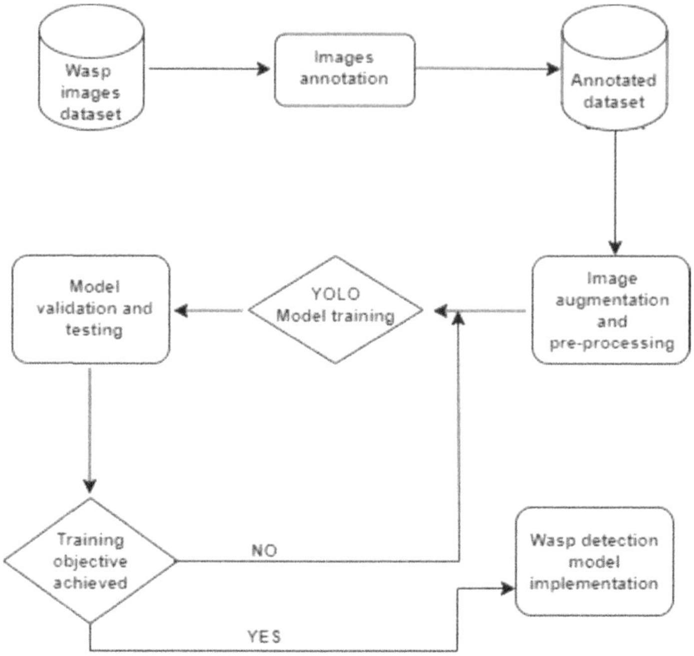

Fig. 3. The steps of the proposed wasp detection modelling.

auto-orientation to correct any rotational inconsistencies, ensuring standardized input for subsequent analysis. Subsequently, the images were resized to a uniform dimension of 640x640 pixels using a stretch-to-fit approach (see Fig. 4). This resizing not only facilitated efficient computation, but also ensured consistent input dimensions for the YOLOv8x model.

Furthermore, a range of data augmentation techniques were employed to further augment the dataset and introduce variations that would enhance the model's robustness against real-world scenarios. These augmentations included random rotations within ± 15 degrees, horizontal and vertical shear transformations of $\pm 15^{\circ}$, the range of brightness adjustment from -25 to $+25\%$, exposure adjustments within the same range, and the application of blur effects with a maximum radius of 2.5 pixels. Each original training image yielded three augmented variations, effectively tripling the size of the training set and enriching the diversity of input samples (see Fig. 5).

2.2 Model Selection and Training

Nowadays, there are various tools for data analytics. The You Only Look Once (YOLO) is one of extensive tools for object detection modelling. In this work, the YOLOv8 was selected because it offers a unified architecture that supports various tasks, including instance segmentation, object detection, and image classification. Notably, it introduces a novel anchor-free detection head, eliminating the need for predefined anchor boxes and stream lining the training process [13]. Additionally, YOLOv8 leverages a new backbone

Fig. 4. Example of dataset pre-processing (Bounding Box tools).

Fig. 5. Examples of augmentation image dataset.

design, the C2f module, which improves feature extraction and enhances the model's ability to capture intricate details in images [13]. Its architecture is modular and scalable, allowing for flexible customization and adaptation to different tasks and datasets.

However, there are several variants of LOYOv8. According to performance of each YOLOv8 variants (see Fig. 6 and Table 1), YOLOv8x is the largest model in the YOLOv8 family, offering the highest accuracy but at the cost of slightly increased computational requirements. The model's ability to discern subtle features and patterns was deemed crucial for achieving reliable detection results. While the inference speed of YOLOv8x is slightly slower than other YOLOv8 variants, it was still within acceptable limits for real-time or near-realtime applications. Consequently, YOLOv8x was selected using in this work due to its speed, accuracy, and ease of use. The need for high precision in Asian hornets' detection is significant, as misidentification could have significant consequences for bee colonies.

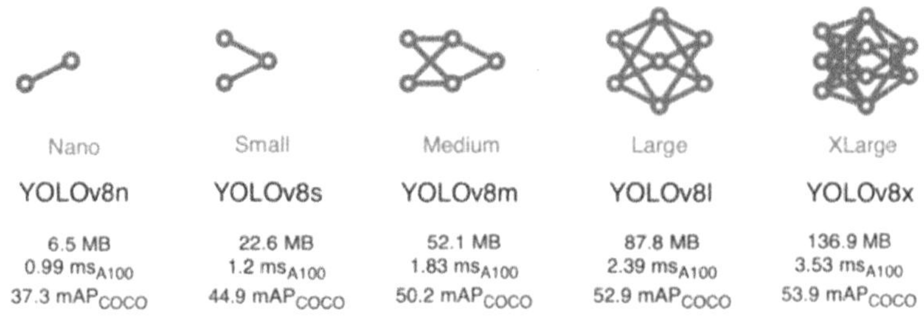

Fig. 6. YOLOv8 variants from YOLOv8 Nano to YOLOv8 X [13]

Table 1. Performance and Speed benchmarks of five YOLOv8 variants. [13]

Model	Size (pixels)	mAPval 50 95	Speed CPU ONNX (ms)	Speed A100 TensorRT	Params (M)	FLOPs (B)
YOLOv8n	640	37.30	80.40	0.99	3.20	8.70
YOLOv8s	640	44.90	128.40	1.20	11.20	28.60
YOLOv8m	640	50.20	234.70	1.83	25.90	78.90
YOLOv8l	640	52.90	375.20	2.39	43.70	165.20
YOLOv8x	640	53.90	479.10	3.53	68.20	257.80

After model selection process, the augmented dataset was utilized to train the YOLOv8x object detection model on the Ultralytics Hub platform. This cutting-edge architecture, known for its accuracy and speed in real-time object detection tasks, was chosen for its potential to effectively identify Asian hornets in diverse environments. The training process was conducted within the Google Colab environment with hyper-parameter tuning, leveraging its powerful GPU resources to expedite the computationally intensive training procedure.

Throughout the training process, the model's performance was continuously monitored and evaluated on the validation set, ensuring that the model was learning to generalize beyond the training data and avoid overfitting. The final model selection was based on a comprehensive assessment of its accuracy and inference speed. While the model achieved a commendable accuracy of 53.9%, its inference speed of 479.1 ms presented a slight trade-off between precision and real-time responsiveness. This trade-off was deemed acceptable, as the primary objective was to develop a system capable of reliably detecting Asian hornets in real-world scenarios, where timely alerts are crucial for effective pest control measures.

3 Results

After finished the training model, the model was validated and tested. Model performance was evaluated using metrics such as mean Average Precision (mAP), precision, F1-score, and recall by varying an mAP@0.5 of X percent and an mAP@0.5:0.95 of Y percent.

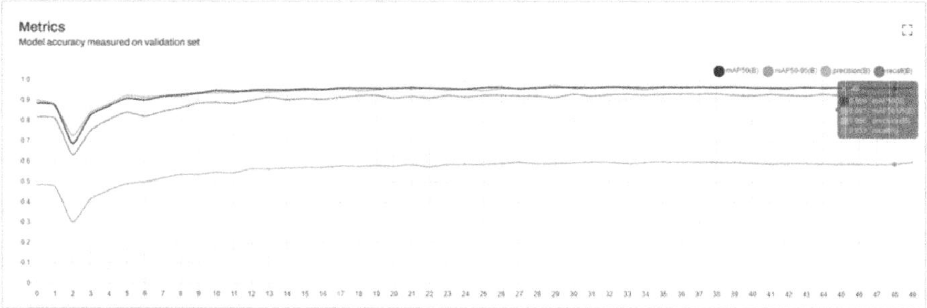

Fig. 7. Performance and Speed benchmarks of five YOLOv8 variants.

The model's performance metrics on the validation set, including mAP50(B), mAP50–95(B), Precision(B), and Recall(B), are illustrated in Fig. 7. The mAP50 (B), which indicates the mean average precision at an intersection over union (IoU) of 0.50, reaches approximately 0.969, showing high accuracy. The mAP5095(B), representing mean average precision across IoU thresholds from 0.50 to 0.95, is around 0.601, indicating more variability in precision at higher IoU values. Precision(B) stands at about 0.966, suggesting the model's predictions are highly accurate with few false positives. Recall(B) is approximately 0.933, reflecting the ability of model to detect most objects in the images, though some objects are still missed. However, in overall, the model demonstrates high precision and accuracy.

After validation process, the proposed wasp detection model was tested using the test dataset. The performance of detection was around 85% of confidence score (see Fig. 8).

Fig. 8. Example of proposed model testing result

In terms of implementation step, an initial field tests are conducted via a simulated environment and situation. The image and streaming video were captured by ESP32CAM. The live video feed from the camera can be viewed on computer, smartphone, or other device that is connected to the same network by setting ESP32-CAM as a web server allows video data to be sent over the network. The system's performance is evaluated under various conditions, including different lighting and weather scenarios. The results show consistent detection accuracy, validating the system's robustness.

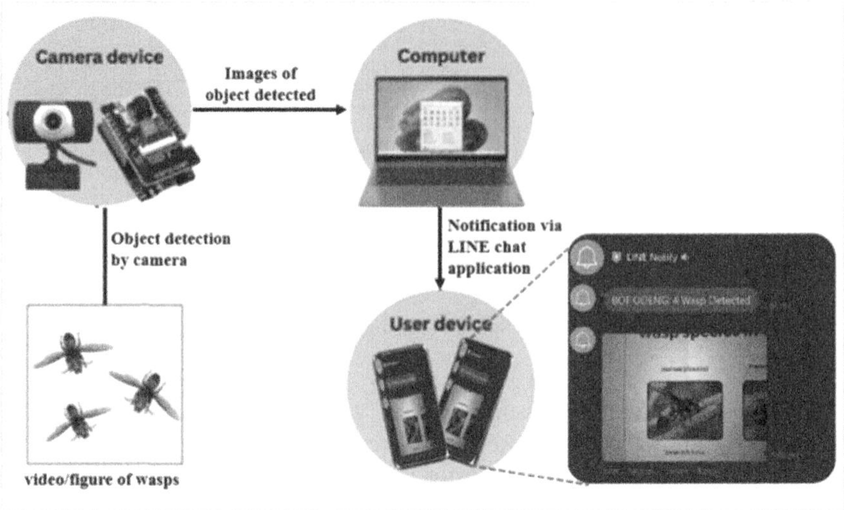

Fig. 9. System implementation diagram

Regarding display and notification information part, Line application is used as notification media because this application widely used in Thailand. The system will send notification to users via LINE application as illustrated in Fig. 9. The image frame containing the detection (along with the bounding box) will be sent to the LINE chat which provides visual confirmation and additional context for users assessing the situation. Consequently, users can notice the attack of wasp to their honey bees.

4 Conclusion and Discussion

This paper proposed the real-time wasp detection and classification to support honey bee farming using the YOLOv8x model. The proposed model was trained and validated on the "Asian Hornet 2" dataset from Roboflow and the supplementary IP-23 dataset, resulting in a robust and accurate system. By achieving an mAP@0.5 of X percent and an mAP@0.5:0.95 of Y percent, the YOLOv8x model showcased its potential to accurately identify wasp, even under challenging conditions. The implementation of this proposed model within a real-world context, such as a beehive monitoring system, could significantly benefit beekeepers and contribute to the preservation of bee populations. Beekeepers can receive a notification of an attack of wasp via LINE chat application. By providing timely alerts of wasp presence, beekeepers would enable to implement targeted pest control measures, mitigating the threat posed by these invasive predators.

Although, the YOLOv8x model achieved an impressive mAP@0.5 of X percent and mAP@0.5:0.95 of Y percent, a few challenges are observed such as varying environmental conditions, similarities of species, etc. Certain instances of misidentification occurred due to similarities in shape among different insect species, particularly those with subtle visual distinctions. Additionally, blurred images occasionally posed challenges for accurate detection, as the model struggled to discern fine details under such conditions. These findings are consistent with previous research on single-stage object detection architectures, which often exhibit slightly lower accuracy compared to two-stage models. To address these limitations, several potential strategies could be explored in future work. Increasing the diversity and volume of training images, particularly those depicting challenging scenarios with overlapping objects or varying lighting conditions, could further enhance the model's robustness. Additionally, incorporating image enhancement techniques, such as sharpening or de-blurring algorithms, could potentially improve detection accuracy for images with lower quality. Consequently, the future work will focus on expanding the dataset and refining the detection algorithms to enhance performance further. Furthermore, the integration of the model with a user-friendly interface and notification system would enhance its practicality for beekeepers and researchers alike.

Acknowledgments. The authors would like to acknowledge the supported of the Mae Fah Luang University, Thailand. We also would like to acknowledge all of our colleagues who worked together and provided encouragement to the authors. Furthermore, the authors also would like to acknowledge the supported by the INDIA ASEAN project funded by SERB (File No: CRD/2022/000504), Government of India.

Disclosure of Interests. The authors have no competing interests to declare that are relevant to the content of this article.

References

1. DOAE: Thailand. Accessed 12 Aug 2022. https://www.facebook.com/pr.doae
2. Geeksforgeeks: Honey Bee Life Cycle. Accessed 17 Jan 2024. https://www.geeksforgeeks.org/life-cycle-honey-bee/
3. Suwannapong G., Benbow ME., Nieh JC.: Biology of Thai honeybees: natural his and threats. In: Florio, R.M. (ed.) Bees: Biology. Threats and Colonies, pp. 1–98. Nova Science Publishers, Inc., New York (2011)
4. Flickr.: Wasp Species in Thailand. https://www.flickr.com/groups/1944588@N23/pool/
5. MarkEisingBirding.: lesser banded hornet (Vespa affinis). https://www.markeisingbirding.com/index/lesser-banded-hornet
6. Jack Rowe.: Yellow-Legged Hornets in the Southeast. Accessed 16 Aug 2023. https://www.aces.edu/blog/topics/bees-pollinators/yellow-legged-hornets-in-the-southeast/
7. Hawkeswood, T.J., Kabir, A.: Record of the greater banded hornet, Vespa tropica (L., 1758)(Hymenoptera: Vespidae) from Koh Samui. Surat Thani Province, Thailand feeding on fallen fruits of Longan. Dimocarpus longan Lour.(Sapindaceae). Int. J. Biol. Other Sci. Calodema **944**, 1–3 (2021)
8. Fuentes, A., Yoon, S., Kim, S.C. and Park, D.S.: A robust deep-learning-based detector for real-time tomato plant diseases and pests recognition. MDPI (Sensors) **17**(9), 2022 (2017). https://doi.org/10.3390/s17092022
9. Lin, Y., Feng, P., Guan, J., Wang, W., Chambers, J.: IENet: interacting embranchment one stage anchor free detector for orientation aerial object detection. arXiv preprint arXiv:1912.00969 (2019)
10. Sabanci, K., Aslan, M.F., Ropelewska, E., Unlersen, M.F., Durdu, A.: A novel convolutional-recurrent hybrid network for sunn pest–damaged wheat grain detec. Food Anal. Methods **15**(6), 1748–1760 (2022)

11. Zhong, Y., Gao, J., Lei, Q., Zhou, Y.: A vision-based counting and recognition system for flying insects in intelligent agriculture. MDPI (Sensors) **18**(5), 1489 (2018)
12. Roboflow.: Asian Hornet 2 Dataset (2023). https://universe.roboflow.com/cyppuhyr/asianhornet-2
13. Sharma, A.: Training the YOLOv8 Object Detector for OAK-D. Accessed 1 May 2023. https://pyimagesearch.com/2023/05/01/training-the-yolov8-object-detector-for-oak-d/

Comparing Different ChatGPT4.0 Generated Oriented Word Suggestions as Design Fixation Interferences for the Ideation Phase in Design Thinking

Fabrice Fischer[1(✉)] and Mickael Gardoni[1,2]

[1] Ecole de Technologie Supérieure, Montreal, Canada
fabrice.fischer.1@ens.etsmtl.ca, mickael.gardoni@etsmtl.ca
[2] INSA Strasbourg, Strasbourg, France

Abstract. The purpose of this article is to present a comparative case study (CCS) methodology in the generation of lists of words suggestions as interference to the phenomenon of fixation of the design in the ideation phase of a Design Thinking (DT) process. The base line is a random list of suggestions from the Large Language Model (LLM) ChatGPT 4.0. While the more advanced first lists of suggestions are related to knowledge from the discipline of the problem and the second list of suggestions is specifically not related to the discipline of the problem. This CCS is qualitative and use the Design Research Science Methodology (DSRM).

Keywords: Design thinking · Ideation · Fixation phenomenon · Word interference · Comparative case studies (CCS) · Design research science methodology (DSRM) · Large language model (LLM) · ChatGPT 4.0 · Knowledge limit · Generativity limit

1 Introduction

Innovation is at the heart of the discourses on value creation. Innovating could be slow or fast, incremental or disruptive, specific or generic, i.e., it is diverse. Contrary to be popular belief that most innovations result from the one-human genius of a Leonardo da Vinci, the innovative designs often result from a structured and systematic innovation methodology conducted by a design team.

There are a few well-established innovations methodologies such as CK, TRIZ, ASIT, which are a bit complex to understand. There is also an apparently easy to comprehend innovation methodology with Design Thinking (DT).

DT is often described as a non-linear five-phase innovation methodology: 1. Empathy, 2. Definition, 3. Ideation, 4. Prototyping, 5. Test. While most individuals have some views on each phase simply based on their names, in practice, DT is complex to implement with a high degree of freedom in the choice of the DT activities and tools, requiring a fair level of expertise and experience to navigate effectively.

Published by Springer Nature Switzerland AG 2025
P. Sureephong et al. (Eds.): PLM 2024, IFIP AICT 741, pp. 207–219, 2025.
https://doi.org/10.1007/978-3-031-93323-3_16

In DT, the ideation phase is the phase concerned with generating innovative ideas to the problem identified in phase "1. Empathy" and formalized in phase "2. Definition". DT presents a fair level of challenges and difficulties, ranging from the design fixation phenomenon (Purcell and Gero 1996), limitation of knowledge of the design team (Koh and De Lessio 2018), limitation in experience of the design team, to limitation in generativity (Thomas and Tee 2022), i.e., variations on the same themes but limited generation of truly new and diverse ideas, and more.

The design fixation phenomenon relates to the designers' mind fixating on a given solution, thereby limiting their ability to ideate. Interferences contribute to the mitigation of the fixation of the phenomenon.

With the focus on words as interferences, instead of preset words on cards, or a generated words manually or randomly with generative artificial intelligence, could we try to mitigate other limitations of the ideation phase such as the limitation of knowledge, and, the limitation in generativity? The limit of experience, while possibly not out of reach, seems different as it would require an injection of experience of some sort, and is not part of this study.

So, could we generate a list of words specific to DT projects, which might help to extend the knowledge effectively used by the design team, and, or, could these generated list of words contribute to increase the generativity of the design team while they ideate?

We could generate these lists of word manually or experiment with some digital automation tools. With the more recent advancements and generalization of generative artificial intelligence tools, could we generate a list of words of higher interest to the design team in their ideation effort? How could we generate such a list of words? Would it be of interest to the design team? How could we even qualify and evaluate the notion of "interest" or "value" to a design team, or more cautiously, to a single designer? Can we, as a minimal but sufficient approach, collect the practitioner's preference for a list over another with regards to limitation of knowledge, and with regards to the limitation in generativity?

2 The Problem

The problem we address is:

> "For a design team at the time of the ideation phase of a DT project, using generative artificial intelligence, could we generate lists of words, as design fixation interferences of higher interest with regards to the limitations of knowledge and the limitation of generativity."

Specifically, can we generate, with generative artificial intelligence, lists of words to use as design fixation interferences to a design team at the ideation phase of a DT project? Also, can we increase the list of words' perceivable interest or value with regards to the limitation of knowledge and the limitation of generativity?

This problem comes with many other questions. What Methodology should we use? Which artificial intelligence could we use to generate this list? What do we mean with an interference of "higher interest"? How can we evaluate this "higher interest"?

3 Research Objectives and Hypotheses

The primary research question is:

> Can we generate higher interest word suggestions as design fixation interferences for the ideation phase in Design Thinking using ChatGPT4.0?

The secondary research questions are:

> How can we generate the lists of word suggestions?
>
> How could we assess that some lists are more interesting than others with regards to the limitation of knowledge and the limitation in generativity?
>
> The specific objective of this study is to

- generate different lists of words as design fixation interferences in the ideation phase of a DT project.

  - Random words related to the DT project.
  - Words related to knowledge in the discipline of the DT project, with the intent to contribute to the mitigation of the knowledge limitation.
  - assess if some lists are of relative higher interest to a designer.

The hypothesis are:

- We can generate a random list of words related to a DT project using ChatGPT4.0
- We can generate a more specific list of words related to knowledge related to the discipline of the DT project using ChatGPT4.0
- We can generate a more specific list of words related to knowledge not related to the discipline of the DT project using ChatGPT4.0
- We can assess if a list of words presents a higher interest to a design team versus another list of words by asking practitioners about their preferences.

With this research we expect to contribute to the corpus of knowledge, methodologies and tools available to the designer of a DT project, and possibly other designer facing the design fixation phenomenon.

The scope of this CCS is limited to a few examples while applying the DSRM for the generation of the word lists as artifacts. Its findings will, therefore, be limited to these few examples. It does not intend to offer a generalization of its findings, which may or may not be the topics of other research work.

While respecting our time and resources constraints, this scope is sufficient to illustrate our concepts of the generation of higher interest list of words with generative artificial intelligence while contributing to the mitigation of the limit of knowledge and the limitation of generativity.

4 Theoretical Background

This research is at the crossroad the following theoretical topics:

- Design Thinking and its ideation phase.
- Word interferences as a design fixation mitigation prop
- Design Science Research Methodology (DSRM) (De Sordi 2021)
- Generative Artificial Intelligence (GAI) with its Large Language Models (LLM) to generate lists of words as artifacts of the DSRM
- Comparative Case Studies (CCS) (Kröger 2021)

4.1 Design Thinking

The terms "Design Thinking" are related to innovating and generally encompass different but related dimensions ranging from a concept, an approach, a methodology, activities, tools, up to the project of researching an innovative design solution to a human centric problem or opportunity leveraging all the other dimensions. The DT methodology DT is often described as a non-linear five-phase innovation methodology:

1. Empathy: where the designers empathize with the user to better understand the problem (or the opportunity) to be addressed with the design solution to innovate.
2. Definition: where the problem or opportunity is further clarified down to a clear formulation of the definition of the problem to address
3. Ideation: where the designers ideate potential design as solutions to the problem and then select the solutions of higher interest
4. Prototyping: where prototypes are being made in preparation for the tests on some characteristics of or the complete products/services
5. Test: the prototypes are tested on some characteristics of or the complete

These phases are non-linear and the approach includes iterating fast between the phases to learn fast and come up with a better design while remaining human centric.

4.2 Word Interferences to Mitigate the Design Fixation Phenomenon

As far as the design fixation is concerned, the phenomenon is that design teams' and individual designers' minds tend to rapidly fixate on a design solution, being a suggested solution or their own idea (Leahy et al. 2020). It is a problem because thisfixation phenomenon limits their ability to generate additional design solutions ideas. And previous research has shown that to get to the best final design, a DT team needs to generate many ideas, with the higher number of ideas being positively correlated with the best final design solution.

And, "... my phone just rang, and I lost track of my thoughts ... What? Oh, yes..." interferences have been proven to help mitigate the fixation phenomenon. All kinds of interferences help, from a pause in the ideation work, to the introduction of words, or the introduction of objects. Brian Eno (Eno and Schmidt 1975), with his oblique strategies, designed a box with a collection of cards with words to be used as design prompts, and are effective as fixation interferences.

Now, could some interferences appear to be more potent than others in relaunching the ideation flow of the design teams? A few-minute pause helps a design team that ran out of ideas, but less so compared to the introduction of words via the Brian Eno cards, or "How would an elephant or a ghost solve the problem, or an object. So, there should be a way to, if not maximize, but at least, enhance the potency of the interferences.

4.3 Design Science Research Methodology (DSRM)

The Design Science Research Methodology (DSRM) is concerned with the research of innovative solutions. It includes the development of artifacts with the intent to solve an identified problem.

The artifacts could be constructs, methods, models, and instantiations.

4.4 Generative Artificial Intelligence (GAI) with its Large Language Models (LLM) to Generate Lists of Words as Artifacts of the DSRM

Generative Artificial Intelligence (GAI) with its Large Language Models (LLM) is quickly becoming mainstream. Based on text or files entered in a prompt, these systems can generate text, images, videos and sounds. ChatGPT 4.0 is one of such LLM systems. It can be used to generate lists of words based on some relevant prompts. We could consider using ChatGPT as a method, and its answers as instantiations of the DSRM.

4.5 Comparative Case Studies (CCS)

Comparative Case Studies (CCS) come under various forms. One classic such form consists in establishing a case baseline to compare to, and one or many alternative cases to compare to the baseline.

5 Methodology

For this study of comparing potential list of suggestions, the Design Science Research Methodology (DSRM) is used for the creation of the list of suggestions, while the Comparative Case Study (Mello 2021) methodology is used to compare them.

The Design Science Research Methodology (DSRM) (Johannesson and Perjons 2021) is particularly pertinent in the context of the creation of artifacts to address a problem.

Also, DT projects generally come as singularities with each project having its own context and specificities. DT projects tend to be unique. While some institutions have portfolios of DT project, they are usually only a small number of projects in the low tens and almost never in the hundreds in any given one institution or accessible to a given research team. In that context, the research on DT projects tends to be more qualitative than quantitative. DSRM is well suited for qualitative research. This research is subjected to the same constraints as the other DT project-related research and will be qualitative in nature.

Furthermore, the interest of this research is to assess the "higher interest" nature of some suggestions. The relative adjective "higher" comes with a comparison. The Comparative Case Studies (CCS) methodology is meant to compare one or more alternatives versus a baseline.

The combination of DSRM with CCS offers the combination of the creation of artifacts to address a problem and the comparison of an alternative versus a baseline in the context of qualitative research, which suits well with the research objective and constraints of this research.

The design of the research is to generate three lists of suggestions, a random list as a baseline (A), a list of words related to knowledge in the discipline of the DT project (B), with the intent to contribute to the mitigation of the knowledge limitation, a list of words related to knowledge in a different discipline than the discipline of the DT project (C), and to compare then in terms of a designer interest or value at the time of the ideation phase of a DT project.

The research phases are as follows:

- Define the elements of the research and the context of the comparison: list of suggestions, generate, comparative interest, artifacts.
- Generate the lists of words using ChatGPT4.0 A, B, C.
- Compare the lists and draw learnings applicable to the samples compared.
- Share the findings (this paper).

6 The Creation of the Artifacts

6.1 Artifact Method

The method is to use ChatGPT to generate lists of ten suggestions composed of three-to-five-words.

6.1.1 Generating the List of Ten Random Three-to-Five-Words Related to the DT Project

ChatGPT4.0 prompt:

> Generate a list of 10 random 3 to 5 related words.

Answer:

> xxx

6.1.2 Generating a List of Ten Three-to-Five-Words of Knowledge Related to the Discipline of the DT Project

ChatGPT4.0 prompt:

> With the following definition of the problem: XXX
>
> Generate a list of 10 of 3 to 5 words presenting a knowledge related to the discipline of the problem.

With XXX to replace with the definition of the problem

6.1.3 Generating a List of 3 to 5 Words of Knowledge Not Related to the Discipline of the DT Project

ChatGPT4.0 prompt:

> With the following definition of the problem: XXX
>
> Generate a list of 10 of 3 to 5 words presenting a knowledge not related to the discipline of the problem.

With XXX to replace with the definition of the problem.

6.2 Artifact Instantiation: Identification of examples of DT problems

To identify examples DT problems, we search for DT papers and used identically or similarly worded but shortened versions of the identified problems.

Here are the five problems identified:

Problem #1: How can emergency departments redesign their processes to efficiently manage patient flow, reduce wait times, and improve satisfaction within existing resource constraints?

Problem #2: What innovative pedagogical strategies and technologies can higher education institutions implement to significantly increase student engagement and satisfaction?

Problem #3: How can affordable, sustainable lighting solutions be developed and made accessible to low-income families using local resources and renewable energy? (Brown and Wyatt 2009)

Problem #4: How can public libraries innovate their services and spaces to meet evolving community needs and increase user engagement in the digital age?

Problem #5: What innovative, economically and environmentally sustainable water management solutions can be tailored to meet the specific needs of rural communities?

6.3 Data and Sources

For this experiment we are using the following data:

1. Examples of DT problem definitions: a few examples of known DT problems.

   The sources for these examples are academic paper.

2. lists of word suggestions:
   a. a list of ten random three-to-five-words related to the DT project.
   b. a list of ten three-to-five-words of knowledge related to the discipline of the DT project.
   c. a list of ten three-to-five-words of knowledge did not relate to the discipline of the DT project.

   These lists are the output of respective prompts, one per list, using ChatGPT 4.0
3. Questionnaire to the practitioners

   The authors of this paper prepared the questionnaire.

4. Practitioners' answers to the Questionnaire
   The practitioners will fill the questionnaire with their answer and send the filled questionnaire to the author of this paper.
5. Synthesis of the practitioners' answers
   The filled questionnaires are the data source for this synthesis.

6.4 Data Preparation and Collection Methods

For this experiment we are using the following data:

- List of DT problem definition: we searched into DT academic research papers, for preferably well-cited papers with one DT problem example each. We then summarized and transposed the mentioned or illustrated DT problem in respective problem definition in our own words.
- Lists of word suggestions, we logged into our ChatGPT account, selected the version ChatGPT 4.0 and created a prompt for each list. We then removed the ChatGPT numbering and formatted the list to fit into our Questionnaire tables.

For the practitioners' preferences, we are using the following.

- The questionnaire is prepared by the authors of this research.
- The answers to the questionnaire will come from practitioners.

7 Evaluation

We can use the "forced ranking" evaluation method. A scale to measure or evaluate the solution would be great but would require a thorough clarification of the dimensions and a form a grading on each of these dimension. For this study, we will stop at a forced ranking by practitioners to get a directional evaluation of the effectiveness of the artifacts without getting into the evaluation of the intensity of that effectiveness.

8 Discussion

This paper presents a methodological approach to navigate qualitatively and directionally the use of GAI to generate higher interest word suggestions as design fixation interferences in a context of a collection of unique occurrences of DT projects.

8.1 Theoretical and Practical Contributions

The combination of DSRM with CCS in a GAI ADT is of interest to DT. While the formulation through this paper might feel a bit heavy to set up and explain, once the methodological framework is in place, its practical use is relatively fast and aligned with the DT bias and appetite for fast learning iteration cycles of ideation-prototype- test.

The DSRM methodology is strong with the contribution of the artifacts as key components to figure-out potentially solutions, which might lead to knowledge. The CCS, with its baseline and a forced ranking of alternatives by the practitioners, allows for a directional intuitive preference in a mostly qualitative context of unique occurrences of DT projects where the contribution to the mitigation of knowledge limitation and generative limitation are difficult to measure. It is practical by construct.

8.2 Limitations and Future Research

To start with, this study is only qualitative. While we would like to believe that more practitioners would answer the questionnaire in a similar way, it can't be confirmed until a quantitative study is conducted to deliver some statistically relevant findings. Then, the request on our questionnaire is a forced ranking. While it points directionally to a preference, it does not identify the underlying components of that preference, and it does not provide a clear evaluation grading in each of these sub-components.

9 Conclusion

This research contributes to the body of knowledge related the words as fixation phenomenon interferences of higher interest related to the mitigation of the knowledge limitation and the generativity limitations. The methodology, combining the DSRM with its method artifact, and the CCS to establish the DT practitioners' directional preferences are relevant to the qualitative nature DT projects singularities.

The research on using GAI tools to augment the DT process is in its infancy. Here are some research directions of interest. We expect that far more, to all, information from the earlier phases 1. Empathy and 2. Definition, will be consumed into building a better understanding of the problem to inform the GAI generating the list of word interference. Researching and better understand that contextual information within the DT process augmented with GAI could lead to recommendation of even higher interest. As far as the interferences format is concerned, with the fast progression of the sound, image, and video generation, research on these GAI interference formats comparative effectiveness might further enhance the effectiveness of the interferences. And while it might be possible to generate quality sound and visual interferences, would it be or not at the expense of a higher fixation anchoring or a lower generativity? As for the word interferences, the three-to-five-word suggestions length of this paper is restrictive to represent knowledge. What would be the most effective qualitative and quantitative GAI word interference? Is there such a thing as a specific length, or could it be adaptative to the problem concerned and specific to the knowledge introduced? At this point, we know so little about the potential uses and impacts on this fast-evolving Augmented Design Thinking (ADT).

Appendix - Questionnaire

To the creative processes practitioner:

In the ideation phase of Design Thinking (DT) projects, individual designers, as well as design teams, often end up running out of new ideas and are met with the design fixation phenomenon. This design fixation phenomenon, which can be represented as the designers' minds fixating on an early design idea, goes against the maximisation of the number of design ideas, that is positively correlated to a better final design. To mitigate this design fixation all kinds of interferences are being used by the practitioners ranging from taking a few minutes break, to introducing words, such as with Brian Eno's oblique strategy cards, or introducing objects.

In this experience, we present you with a few problems to be addressed by DT teams. For each problem, we present three lists of ten 3 to 5 related words to be used a word-interferences.

You are requested to rank these lists from the most interesting to the least interesting to a design team in mitigating respectively the limit of knowledge and the limit of generativity of the designers.

Here are clarifications on key terms:

- Limit of knowledge: While generating design ideas, individual designers tend to use the knowledge they frequently and recently use, and to a lesser extent knowledge they have but rarely use or have not used in a long time, and much less so new knowledge they seek specifically to address the DT project problem. This limit in the use of knowledge is referred to as the limit knowledge.
- Limit of generativity: While generating design ideas, individual designers find it difficult to generate truly new ideas that have not be found before and which are not similar to other existing ideas i.e. ideas that are further away from all the other ideas already existing or that have been generated earlier in the DT project. This limit to generate truly novel and different ideas is referred to as the limit of generativity.

Problem #1: How can emergency departments redesign their processes to efficiently manage patient flow, reduce wait times, and improve satisfaction within existing resource constraints?

A	B	C
Destinations, Adventure, Culture, Maps Stars, Planets, Moon Food, Cooking, Taste, Recipes Instruments, Rhythm, Harmony Waves, Seashells, Sand, Sun, Beach Stories, Reading, Literature, Books, Authors Animals, Leaves, Birds Sports, Win, Team, Games Computers, Internet, Software, Technology Winter, Snow, Cold, Ice, Sled	Patient triage systems Queue management theories Resource allocation models Lean healthcare principles Process optimization techniques Patient flow analysis Emergency department layout Staff scheduling strategies Patient satisfaction metrics Digital health technologies	Behavioral economics insights Customer service best practices Industrial design principles Software development cycles Agile project management Human-centered design Supply chain optimization Environmental sustainability strategies Data mining techniques Social media analytics

Problem #2: What innovative pedagogical strategies and technologies can higher education institutions implement to significantly increase student engagement and satisfaction?

A	B	C
Destinations, Adventure, Culture, Maps Stars, Planets, Moon Food, Cooking, Taste, Recipes Instruments, Rhythm, Harmony Waves, Seashells, Sand, Sun, Beach Stories, Reading, Literature, Books, Authors Animals, Leaves, Birds Sports, Win, Team, Games Computers, Internet, Software, Technology Winter, Snow, Cold, Ice, Sled	Active learning techniques Blended learning models Educational gamification methods Student-centered teaching Peer learning strategies Digital classroom tools Flipped classroom approach Formative assessment practices Online learning platforms Collaborative project work	Social media engagement Virtual reality experiences User interface design Behavioral psychology insights Corporate training programs Crowdsourcing ideas platforms Mobile app development Environmental sustainability projects Community service initiatives Digital storytelling techniques

Problem #3: How can affordable, sustainable lighting solutions be developed and made accessible to low-income families using local resources and renewable energy?

A	B	C
Destinations, Adventure, Culture, Maps Stars, Planets, Moon Food, Cooking, Taste, Recipes Instruments, Rhythm, Harmony Waves, Seashells, Sand, Sun, Beach Stories, Reading, Literature, Books, Authors Animals, Leaves, Birds Sports, Win, Team, Games Computers, Internet, Software, Technology Winter, Snow, Cold, Ice, Sled	Renewable energy sources Solar lighting technologies Energy-efficient design Low-cost manufacturing Sustainable materials use Off-grid power systems Local resource utilization Waste-to-energy solutions Community-based projects Lifecycle cost analysis	Behavioral change strategies Microfinance models Social entrepreneurship principles Community engagement tactics Crowdfunding platforms Digital literacy programs User experience design Mobile payment systems Supply chain logistics Policy advocacy methods

Problem #4: How can public libraries innovate their services and spaces to meet evolving community needs and increase user engagement in the digital age?

A	B	C
Destinations, Adventure, Culture, Maps Stars, Planets, Moon Food, Cooking, Taste, Recipes Instruments, Rhythm, Harmony Waves, Seashells, Sand, Sun, Beach Stories, Reading, Literature, Books, Authors Animals, Leaves, Birds Sports, Win, Team, Games Computers, Internet, Software, Technology Winter, Snow, Cold, Ice, Sled	Digital lending technologies Community programming strategies User experience design Interactive learning spaces Digital literacy workshops E-resource management Collaborative workspace design Virtual event hosting Library mobile apps Maker spaces integration	Social media marketing Augmented reality experiences E-commerce platforms Urban planning insights Customer relationship management Sustainable building practices Data analytics techniques Community art projects Wellness program development Smart city technologies

Problem #5: What innovative, economically and environmentally sustainable water management solutions can be tailored to meet the specific needs of rural communities?

A	B	C
Destinations, Adventure, Culture, Maps Stars, Planets, Moon Food, Cooking, Taste, Recipes Instruments, Rhythm, Harmony Waves, Seashells, Sand, Sun, Beach Stories, Reading, Literature, Books, Authors Animals, Leaves, Birds Sports, Win, Team, Games Computers, Internet, Software, Technology Winter, Snow, Cold, Ice, Sled	Rainwater harvesting systems Solar-powered desalination Gravity-fed water supplies Aquifer recharge techniques Drip irrigation methods Greywater recycling practices Bio-sand filtration units Water-efficient crop selection Community water management Watershed conservation strategies	Microfinance funding models Community organizing principles Behavioral change campaigns Mobile technology applications Social entrepreneurship strategies Participatory design processes Digital mapping tools Educational outreach programs Policy advocacy techniques Remote sensing technologies

For each problem, please rank the lists - A, B, C - in decreasing interest in helping to mitigate the presented limit:

The responses to the questionnaire

Table 1. Forced ranking of the lists A, B, or C from the most interesting to the least interesting to mitigate the knowledge limitation and the generativity limitation

Problem	Knowledge Limitation	Generativity Limitation
#1		
#2		
#3		
#4		
#5		

References

Brown, T., Wyatt, J.: Design thinking for social innovation. Stanford Soc. Innov. Rev. **8**(1), 31–35 (2009). https://doi.org/10.48558/58Z7-3J85

De Sordi, J.O.: Design Science Research Methodology: Theory Development from Artifacts. Springer, Heidelberg (2021). https://doi.org/10.1007/978-3-030-82156-2

Eno, B., Schmidt, P.: Oblique strategies: Over one hundred worthwhile dilemmas, 5 edn. (1975)

Johannesson, P., Perjons, E.: An Introduction to Design Science. Springer, Heidelberg (2021). https://doi.org/10.1007/978-3-030-78132-3

Koh, E.C.Y., De Lessio, M.P.: Fixation and distraction in creative design: the repercussions of reviewing patent documents to avoid infringement. Res. Eng. Des. **29**(3), 351–366 (2018). https://doi.org/10.1007/s00163-018-0290-y

Leahy, K., Daly, S.R., McKilligan, S., Seifert, C.M.: Design fixation from initial examples: provided versus self-generated ideas. J. Mech. Des. **142**(10), 101402 (2020). https://doi.org/10.1115/1.4046446

Mello, P.A.: Qualitative Comparative Analysis: An Introduction to Research Design and Application. Georgetown University Press, Washington, D.C. (2021)

Mello—2021—Qualitative comparative analysis an introduction .pdf. (n.d.)

Purcell, A.T., Gero, J.S.: Design and other types of fixation. Des. Stud. **17**(4), 363–383 (1996). https://doi.org/10.1016/S0142-694X(96)00023-3

Thomas, L.D.W., Tee, R.: Generativity: a systematic review and conceptual framework. Int. J. Manag. Rev. **24**(2), 255–278 (2022). https://doi.org/10.1111/ijmr.12277

Lifecycle Approaches Based BIM: Challenges and Opportunities

Assam Hammi(✉), Djamel Ouahrani, Abdelaziz Bouras, and Khalid Naji

College of Engineering, Qatar University, P.O. Box: 2713, Doha, Qatar
{assam.hammi,djamel,abdelaziz.bouras,knaji}@qu.edu.qa

Abstract. Building Information Modeling (BIM) is a smart process that uses 3D models and information management to assist Architecture/ Engineering/Construction and Facility Management (AEC-FM) practitioners in design, construct and operate buildings and infrastructure more efficiently and more economically. BIM integrates new technologies and tools to facilitate the exchange of information between the different stakeholders involved in a project, from the initial conception to the final demolition. However, one of the challenges BIM is facing is how to manage all those information generated throughout the entire lifecycle of any built asset, which can span several decades and involve multiple changes and updates. Unlike the manufacturing industry (such as aeronautics or automobiles), which has established a set of procedures and standards to document and control the product development and delivery across the whole lifecycle, the AEC-FM sector is still in the process of adopting and implementing lifecycle management, assessment concepts and practices. In this paper, we review the literature on how life-cycle approaches can be integrated with BIM enabled projects in the AEC-FM sector. We also discuss some of the benefits that lifecycle approaches can bring along with the digitalization of the sector, in terms of mitigating the environmental impacts and increasing the performance of buildings.

Keywords: BIM · Lifecycle Approaches · Integration · Architecture · Engineering · Construction and Facility Management

1 Introduction

The AEC-FM industry sector is known for its slow adoption of innovations [1]. This is due to the use of those innovations by multiple disciplines separately within the sector which gives the struggle aspect during their implementation. Moreover, the construction industry involves a long chain of actors, such as clients, architects, contractors, subcontractors, suppliers, regulators, and end-users. Each of these actors may have different interests, expectations, and requirements for the projects. This may make it difficult to align their goals and ensure their satisfaction with the outcomes. The use of BIM sets it apart from traditional computer-aided design technologies (CAD) due to the level and type of information involved. BIM processes depend on the collaborative creation and sharing of information-rich 3D models. [2] Therefore, BIM targets the integration

Published by Springer Nature Switzerland AG 2025
P. Sureephong et al. (Eds.): PLM 2024, IFIP AICT 741, pp. 220–231, 2025.
https://doi.org/10.1007/978-3-031-93323-3_17

between disciplines in the AEC- FM industry sector where separation of processes is obvious. [3] As a result, BIM is often considered as a disruptive innovation that has a profound impact on various professional bodies, the technologies they employ, the processes they implement, and more broadly, their collaborative practices. [4] The adoption of BIM is promoted in numerous countries; nevertheless, its implementation poses difficulties and gives rise to multiple challenges at both micro and macro levels [5]. Amongst those challenges, there are limitations to managerial, technological, and collaborative capabilities throughout the project and operational phases of a facility's lifecycle. Hence, continued evolution of BIM depends on the development and adoption of new business models that are based on an integrated lifecycle approaches [6]. Buildings are becoming complex products with multiple kind of materials and components that interact with each other and with the environment. This justifies to look at their lifecycle and implies that buildings need to be designed, constructed, operated, maintained, renovated, and demolished with consideration of their whole lifecycle impacts and benefits. For instance, the National Museum of Qatar is indeed a remarkable architectural feat made from thousands of what it called Ultra-High-Performance Fiber-Reinforced Concrete (UHPFRC) disks to create a stunning desert rose-inspired structure. (Fig. 1). It consists of fiber-reinforced concrete with stainless steel embeds. UHPFRC was used to create the 76,000 panels that form the skin of the museum's disks [7].

Fig. 1. The National Museum of Qatar[1]

The museum's design and construction required sophisticated tools to manage the lifecycle of these complex components, from design to maintenance. One of the most commonly used processes in the manufacturing industry, supporting this idea is "The Product Lifecycle Management". PLM is a business strategy that creates and sustains a product-centric knowledge environment. It is rooted not only in design tools and

[1] https://new.qatargreenleaders.com/project/national-museum-of-qatar-nmoq/.

data warehouse systems but also in product maintenance, repair, and dismissal support systems [8]. The application of PLM concepts, methods, and technologies along with the recent maturity in the deployment of BIM in the AEC-FM sector will drive changes in assumptions about data, information, and knowledge management across the building lifecycle. Potential applications of Lifecycle Approaches in BIM enabled projects from related work will be highlighted in the next section, while Sect. 3 will shed light on the comparison of different materials for the iconic National Museum of Qatar, based on international standards, that involves evaluating various aspects such as sustainability, durability, and aesthetic value. Section 4 will raise discussion on the importance to integrate those approaches with BIM for the benefits of the AEC-FM industry sector, and finally Sect. 5 will conclude the article with broader range of recommendations for further research in the field.

2 Literature Review

The AEC-FM industry sector is often seen as unique in terms of interoperability, but it has similarities with other industries [9]. This is because there are many stakeholders from different trades and disciplines involved. Different BIM tools that use different software types are needed for different fields of application to perform specific analyses on the built asset throughout its whole lifecycle [10]. Therefore, BIM can serve as a basis for lifecycle approaches by being used for various purposes.

In this context, *Barholli* looked for ways to combine building information modeling (BIM) and life cycle assessment (LCA) to improve the information management in the engineering project development process and to make the buildings more environmentally friendly [11]. She classified the integration methodology into 3 categories (Fig. 2) a) the conventional method, b) the static method with BIM tools and c) the dynamic BIM integrated method. Many construction companies currently use the conventional method. It is also the procedure for the generic LCA tool, such as Spreadsheet-based and Online component catalogues, which are mentioned above. The process begins with manually extracting the bill of quantities from a 2D, 3D projects or automatically from a BIM model. Then, the tool multiplies the values with the environmental impact indicators to produce the LCA results. The indicators are either manually entered in the tool or linked with the LCA database. The process depends on the building phase of the project, as it requires sufficient information to perform the LCA. The process is generally time-consuming and prone to errors, as the calculations are done manually. Another drawback is that the information flow is unidirectional, as the LCA results are not connected to the model. This can make the decision-making process more difficult and complicated. Moreover, an expert with specialized knowledge on the field is needed to perform the LCA procedures and calculations with conventional tools. In the static method LCA calculations are done by LCA tools third party from the information in the BIM model. The BIM model exports the information in a standard exchange format that enables the tools to perform the calculations. This method uses a standard exchange format (IFC or gbXML) to transfer the information from the BIM model to the LCA tool.

This avoids the manual entry of data and improves the accuracy of the LCA results. However, this method has some limitations, such as flow is one-way, interoperability issues…etc. The dynamic method requires a BIM Model that can automatically extract the Bill of Quantities. Then, a parametric tool can link the BOQ with an LCA database and do the LCA analysis. The BOQ and the LCA parameters have to be mapped and related for the procedure to work. The building phase is also very important in this process. The dynamic method allows the LCA values to be assigned back to the BIM model and complete the information. The decision making is still not easy, but in this case, it is possible to get direct feedback during the design process.

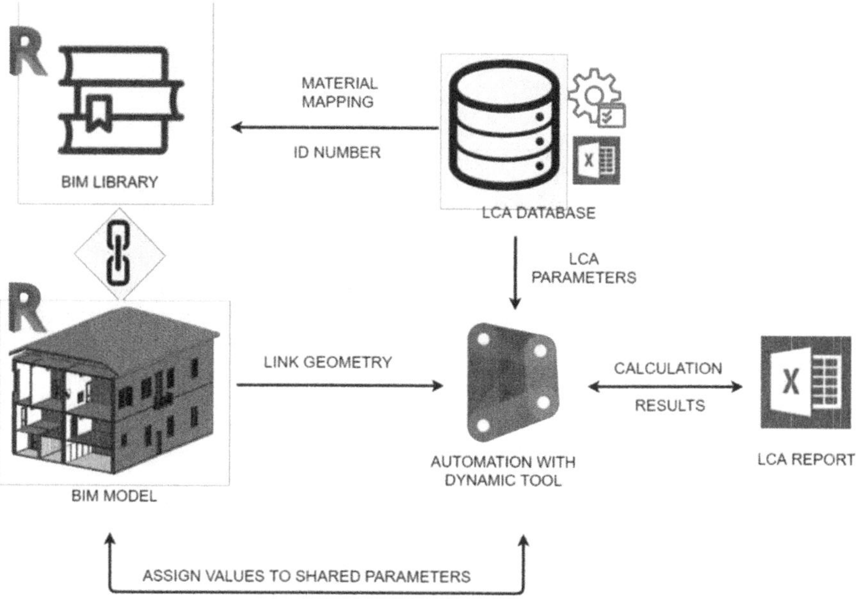

Fig. 2. BIM-LCA integration with dynamic approach [11].

Tchana.et al. addressed the problem of managing linear infrastructures, such as roads, railways, pipelines, etc., from the idea to the in-use phases [12]. He proposed a Linear Infrastructure Management System (LIMS) as a way to combine the benefits of both PLM and BIM processes. (Fig. 3) showed that PLM could provide functions such as storing, securing, editing and tracking data. However, BIM could provide functions such as creating and automating models. Both BIM and PLM rely on a digital representation of the project. They use a common digital model that can be accessed, modified and viewed by different users.

They also enable data exchange between various software and tools through standard formats, such as IFC for BIM and STEP for PLM. However, they differ in their functionalities and objectives. BIM is mainly concerned with visualizing, calculating and checking the properties and interactions of the elements in the model. It does not manage the documents or their versions and states. PLM, on the other hand, handles the processes of validation, modification and communication of the documents. It allows tracking the changes and history of the project. Therefore, BIM is a passive tool that shows the current state of the project, while PLM is an active tool that shows the evolution of the project. Many researchers believe that combining BIM and PLM could result in a complete management system for the infrastructure.

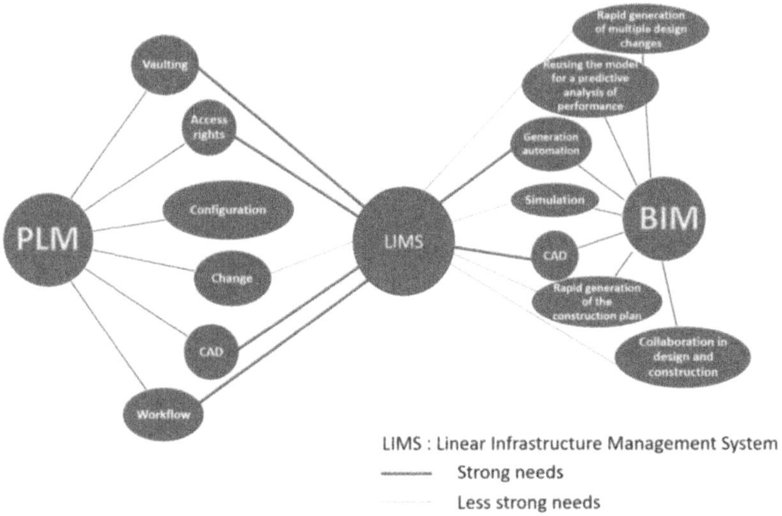

Fig. 3. LIMS built on both BIM and PLM features. [12]

Ghnaya compared the functionalities of the document management tools adopted in the construction and manufacturing companies to manage and control documents during a Design/Engineering Change Management (D/ECM) process [13]. These tools namely *SmartUse, BIM360* for BIM enabled project, and *Smarteam, 3DExperience* for Project based PLM. The results of the comparison revealed that the PLM tools have more advanced features than the ones used by the construction (SmartUse and BIM 360 Team). For example, Smarteam and 3DExperience can automate the D/ECM process by digitizing phases such as engineering change request and engineering change orders. The construction partners, on the other hand, rely more on e-mail and phone calls to communicate and track changes. Moreover, the revision management feature is more comprehensive in Smarteam and 3DExperience, as it allows users to differentiate between minor and major revisions and to create branches from previous revisions. Additionally, the check-in/check-out functionality in these tools helps prevent overwriting revisions after modifications. *Kumar et al.* proposed a framework for designing and simulating smart buildings that are energy-neutral and have embedded sensors, using semantic web technologies. This framework integrates ontologies of Lifecycle Assessment (LCA), Building Information Modelling (BIM), and Product Lifecycle Management (PLM). [14].

This framework connected LCA, BIM and PLM at different levels, from L1 to L6 namely L1-data integration layer, L2-data exchange layer, L3- software implementation layer, L4-design implementation layer, L5- stakeholder integration layer, and L6-domain integration layer. These levels represent a unified framework for the collaborative design

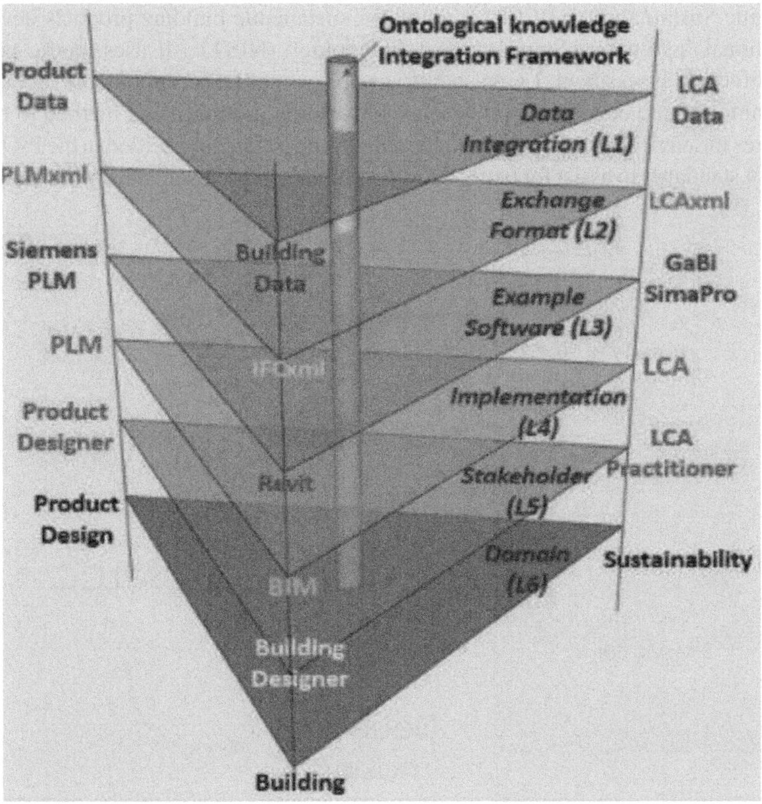

Fig. 4. A model framework for the ontological integration of BIM, PLM and LCA for the design of energy-neutral smart buildings. [14]

of smart buildings that are energy-neutral. The three corners of the triangular levels are LCA (sustainability domain), BIM (building domain), and PLM (product design domain) (Fig. 4). At L1, product and building data are combined with LCA data. At L2, a common data layer is created by merging the exchange formats of LCA, BIM and PLM. At L3, a software environment that integrates the functions of LCA, BIM and PLM is used to process the data. At L4, a design implementation layer that has data processing and knowledge integration capabilities is envisioned. This layer leads to a collaborative design platform (L5) that involves experts from different fields, such as LCA practitioners, architects and product designers. At L6, the final integration of the domains of sustainability, building design, and product design is achieved, resulting in an interdisciplinary domain for smart building design.

3 Towards a Sustainable Lifecycle Assessment Approach

To achieve a building's sustainable lifecycle assessment, it's essential to examine every phase of its existence, from obtaining raw materials and production to distribution, setup, usage, as well as recycling and disposal processes. Building for Environmental and

Economic Sustainability (BEES) is a tool for sustainable building products developed by National Institute of Standards and Technology (NIST)[2]. It uses metric process-based lifecycle assessment (LCA) and life-cycle costing (LCC) approaches to assess the environmental and economic performance of building products. The method to be used measures the environmental performance of building products specified in the ISO 14040 series of standards to assist for better decision making in the sustainable product/material choice. (Fig. 5).

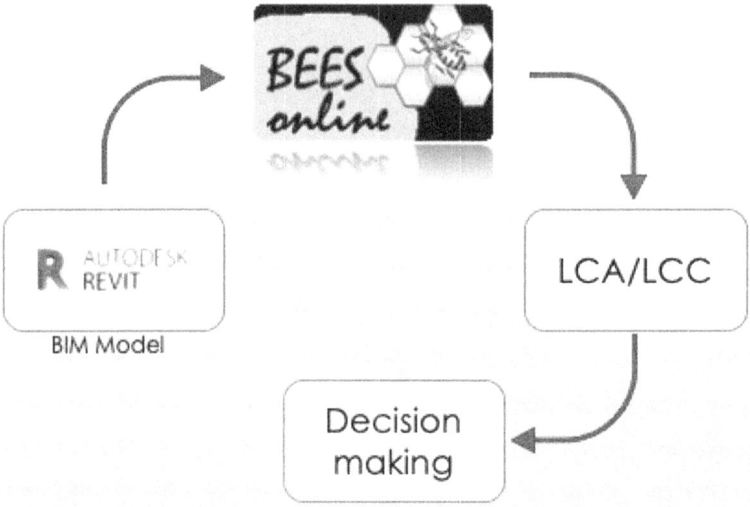

Fig. 5. BEES workflow-based LCA/LCC

For our case study, we will examine the National Museum of Qatar mentioned earlier in the paper where they used Gehry Technologies system to manage the BIM model during Design & Built phases. The museum's structure incorporates 130 disks to form its roofs, floors, and walls. There are two distinct types of disks: those for the roof and those for vertical applications. Additionally, this disk system has been adapted to function as sun shades and to form the underside of the floors with interlinking beams. Across all these typologies, the concrete cladding maintains uniformity in shape, material, color, and texture, covering an area of 120,000 m^2. (See Red Box in Fig. 6).

For the example, we consider the roof sheathing, also called roof decking. Once we extract the quantity of the product from the BIM model, we shall set the parameters to start the analysis, based on current Product Category Rules (PCRs) defined in UNIFORMATII (Fig. 7).

The products listed below (Fig. 8) include the two products that meet the selected criteria namely OSB (Oriented Strand Board) and Plywood Sheathings. The stacked bar

[2] https://www.nist.gov/services-resources/software/bees.

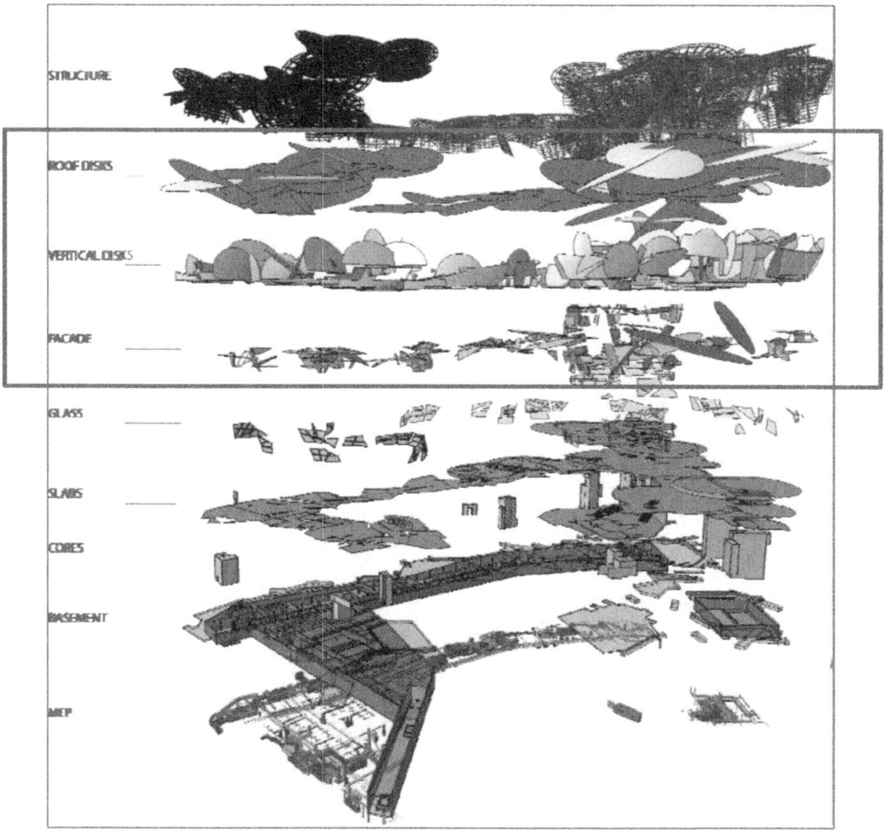

Fig. 6. Topology of the QNM taken from the BIM model (Ateliers Jean Nouvel) [7]

graph compares the results for the selected products for a given impact category by lifecycle stage from Raw materials to EOL (End of Life). Those impact categories contain **Ozone depletion** (gradual reduction of the ozone layer in Earth's atmosphere), **Global Warming Potential** (GWP) that provides a common unit of measure, aiding policymakers in addressing climate challenges effectively. **Smog** which is a specific type of air pollution, **Acidification** (environments transition toward acidic conditions), **Eutrophication** which is the gradual increase in the concentration of phosphorus, nitrogen, and other plant nutrients in an aging aquatic ecosystem, **Water Use and Primary energy demand - non-renewable and renewable.** As sample, we have taken the results of GWP where we noticed that the highest value of this impact category would happen during the manufacturing phase (0.171 & 0.154). Such results will be exported to the PLM system to manage the whole lifecycle of the chosen product and would allow performing the necessary interpretation all along the building lifecycle.

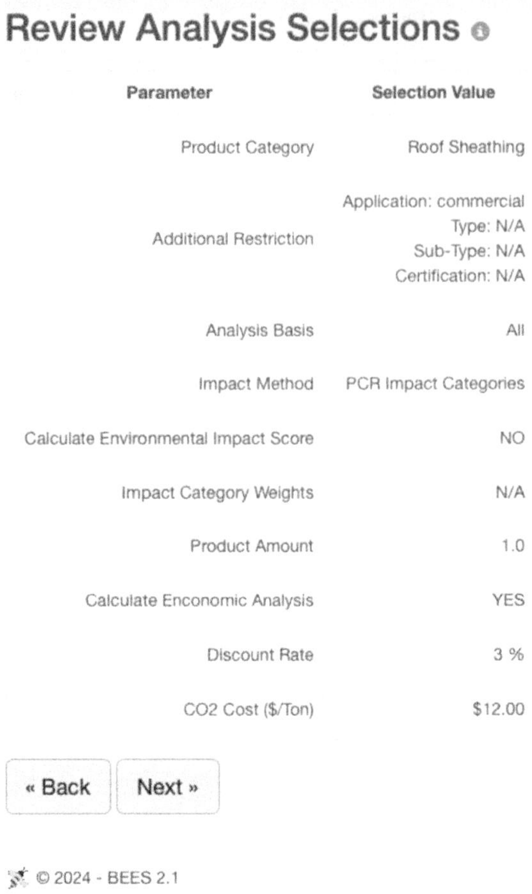

Fig. 7. Parameters Setup on BEES platform

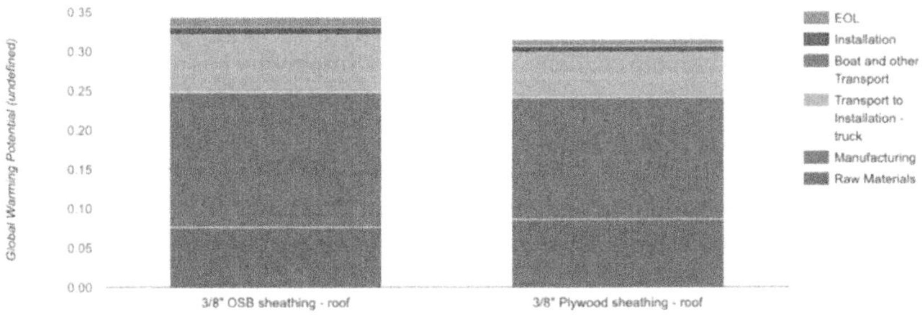

Fig. 8. Roof sheathings Comparison based on Global Warming Potential Impact Category

4 Discussion

Information technology has brought many changes to various industrial sectors, including the AEC-FM industry sector. However, the built environment is still relatively slow to be adapted to the digital transformation that affects how buildings and infrastructure are designed, constructed and operated. Some parties in the AEC-FM industry sector have already adopted a distributed way of working, where they collaborate to develop services and offer them in the digital market. This shows some degree of integration between the built environment and information system in a lifecycle vision, but it is not yet mature [15]. This is due to multiple factors among other the shortage of people who know how to integrate the lifecycle approaches in the built environment. For example, when an organization implements lifecycle-based BIM process, the people involved need to be assessed and trained. Since manufacturing industries produce many units of the same product, they can benefit from the increased efficiency and productivity that PLM provides. Therefore, the return of investment in PLM is very short for them. However, in the construction industry, each building project is unique and complex, requiring different design, engineering, and management processes. Therefore, PLM in construction should not focus on the whole building as a form, but rather on the elements that are industrially produced and standardized, such as building material quality including certified performance, cooling components system, lighting components. In another hand there is a lack of clarity about the situation of BIM based lifecycle approaches mandates, legislations must require the use of standards such as ISO 19650 standard for example, which is essential for managing BIM asset information throughout the life cycle. [16] Another aspect that is missing is the use of strategies to dynamically link the lifecycle technical data with the BIM models. These strategies include specific tools, to enable the integration of BIM-Lifecycle approaches, defined processes to update the database according to reality, planning methods for the engineering project execution process that take into account the special features of a multidisciplinary project development flow. IT systems are a common term for BIM and PLM, suggesting that they rely heavily on information technology for their execution, but they are actually more than that, they are methods and processes that incorporate IT aspects. The comparison of both industries shows that the AEC-FM industry is undergoing a transformation and facing many challenges in terms of culture, economy, technology and society. The Manufacturing industry also has similar issues, but with more differences. Both industries need new skills, standards, interoperability, training and an IT system that covers the entire lifecycle of their products or buildings [17]. Moreover, the AEC-FM industry is increasingly adopting innovative technologies and methodologies, as BIM has gained more awareness. However, BIM alone cannot meet all its needs and fulfil all its requirements, as it lacks some functionalities and features that PLM has for complex products. BIM should either adopt PLM elements to become more complete and effective, or evolve into BLM (Building Lifecycle Management). However, there is no consensus on how PLM and BLM are related [18]. Some argue that they are the same approach for different products, while others suggest that they can learn from each other and foster innovation. One of the main benefits that AEC-FM industry can get from PLM is the extension of BIM process and protocols throughout the lifecycle, especially in the facility management phase, by using a "unifying platform that captures, integrates and shares the object-based information generated by BIM- based authoring, analysis, and simulation applications" [19]. In summary, there is a need for a collaborative lifecycle management system for constructions that BIM cannot fulfill. PLM can provide some solutions that are already mature in the complex product manufacturing, but they need to be customized for the construction industry context and integrated with BIM models.

5 Conclusion and Future Work

The AEC-FM industry must adopt standardization, new technologies and automation to enhance its activities. These three terms are essential for improving the quality and productivity of projects, which are often affected by incompatibilities, software programs interoperability, information dispersion and communication barriers. BIM associated with lifecycle approaches is regarded by the leading enterprises as a solution to most of these problems. In this paper, we have explored the current state-of-the-art of integrating lifecycle approaches with BIM in the AEC-FM industry sector, which is concerned by the management of technical data throughout the building lifecycle. This initiative is mainly driven by the need to increase the productivity of enterprises, by managing the information better and covering the entire lifecycle of the project. We also investigated NMoQ for environmental performance assessment goals despite the limitation of the used website application in terms of available building products. We plan to delve deeper into research on the BIM-based lifecycle approach, as it represents an emerging and promising area of study.

Acknowledgment. This publication was made possible by internal grant # QUCG-CENG- 23/24–191 from Qatar University. The statements made herein are solely the responsibility of the author.

References

1. Harty, C.: Innovation in construction: a sociology of technology approach. Build. Res. Inf. **33**(6), 512–522 (2005). https://doi.org/10.1080/09613210500288605
2. Eastman, C.M.: BIM Handbook: A Guide to Building Information Modeling for Owners, Managers, Designers, Engineers and Contractors. John Wiley & Sons, Hoboken (2011)
3. Hammi, A., et al.: Integration potentiality of Blockchain technology within BIM enabled projects. Value Manag. J. **1**(01) (2022)
4. Hochscheid, E., et al.: Practitioners' concerns about their liability toward BIM collaborative digital mockups: case study in civil engineering. J. Constr. Eng. Manag. **149**(9), 05023010 (2023)
5. Kassem, M., Succar, B.: Macro BIM adoption: comparative market analysis. Autom. Constr. **81,** 286–2992017. https://doi.org/10.1016/j.autcon.2017.04.005
6. Jupp, J.R., Singh, V.: A PLM perspective of BIM research initiatives. Int. J. Prod. Lifecycle Manag **9**, 180–197 (2016)
7. Menétrey, P.: UHPFRC cladding for the Qatar national museum. In: Proceedings of the International Symposium on Ultra-High-Performance Fiber-Reinforced Concrete, Marseille, France (2013)

8. Terzi, S., et al.: Product lifecycle management-from its history to its new role. Int. J. Product Lifecycle Manag. **4**(4), 360–389 (2010)
9. Turk, Z.: Interoperability in construction-Mission impossible? Dev. Built Environ. **4**, 100018 (2020)
10. Utkucu, D., Sözer, H.: Interoperability and data exchange within BIM platform to evaluate building energy performance and indoor comfort. Autom. Constr. **116**, 103225 (2020)
11. Braholli, O.: Automated Procedures for LCA Analysis on a BIM Project, Master in Building Information Modelling, Politecnico Di Milano (2020)
12. de Tchana, Y.T., et al.: Project management for linear infrastructure: studying PLM and BIM for an efficient set of basic functionalities. In: 12 th International Conference on Modelling, Optimization and SIMulation-MOSIM'18 (2018)
13. Ghnaya, O.: Comparaison des outils informatiques pour la gestion des modifications de conception/d'ingenierie entre les industries soutenues par le BIM et le PLM. Master Thesis, Ecole de technologie superieure (2023)
14. Kumar, T., Mani, M.: Design of energy-neutral smart buildings: an ontological framework to integrate LCA, BIM and PLM. In: BSA Conference 2022, vol. 5. IBPSA-Italy (2022)
15. Guignone, G., et al.: BIM and LCA integration methodologies: a critical analysis and proposed guidelines. J. Build. Eng. **73,** 106780 (2023)
16. Standardization, I.O.F.: ISO 19650-1: 2018-Organization and digitization of information about buildings and civil engineering works, including building information modelling (BIM)-Information management using building information modelling-Part 1: Concepts and principles. International Organization for Standardization, Geneva (2018)
17. Mangialardi, G., et al.: BIM and PLM associations in current literature. In:Product Lifecycle Management and the Industry of the Future: 14th IFIP WG 5.1 International Conference, PLM 2017, Seville, Spain, 10–12 July 2017, Revised Selected Papers 14. Springer, Heidelberg (2017)
18. Kubler, S., et al.: Building lifecycle management system for enhanced closed loop collaboration. In: Product Lifecycle Management for Digital Transformation of Industries: 13th IFIP WG 5.1 International Conference, PLM 2016, Columbia, SC, USA, 11–13 July 2016, Revised Selected Papers 13. Springer, Heidelberg (2016)
19. Aram, S., Eastman, C.: Integration of PLM solutions and BIM systems for the AEC industry. In: Proceedings of the 30th ISARC, Montreal, Canada, p. 1046-1 (2013)

Effect of Internal & External Financing on Labor Productivity – (2009–2014) Panel Data Study of Ecuadorian Companies

Luis Simbaña-Taipe[1(✉)], Gonzalo Rodas[1], Bryan Mena[1], Ximena Morales[2], and Alfredo Salazar-Baño[1]

[1] Department of Economic Administrative and Commerce Sciences, University of the Armed Forces ESPE, Sangolquí, Ecuador
lesimbania@espe.edu.ec

[2] Business, Society and Technology Research Center (ESTec), Department of Administrative and Economic Sciences, Technological University Indoamerica, Quito, Ecuador

Abstract. This research analyzes the effect of internal and external financing on the labor productivity of Ecuadorian companies from 2009 to 2014. For this purpose, the database of the National Institute of Statistics and Censuses (INEC) is used. In addition, ordinary least squares (MCO) regression for panel data is applied as an estimation strategy. The study focuses on explanatory variables with financial characteristics (external and internal funding) and variables with commercial characteristics (number of employees, source of foreign capital, innovation of new services, and purchasing of machinery and equipment). The results demonstrate that external financing, number of employees, source of foreign capital and purchase of machinery and equipment, all have a positive impact on labor productivity. Internal financing, on the other hand, has no effect on the dependent variable. Meanwhile, the introduction of a new service reduces labor productivity. The Ecuadorian economy presents several barriers that prevent the development of national companies. These obstacles could be a lack of social and scientific awareness, an absence of innovative initiative and technological development, and deficient management of internal and external financing. Therefore, identifying the factors that affect the labor productivity of Ecuadorian companies is a relevant issue for the country's business economic development. Due to this information, the companies could develop capabilities and strategies to stay in the market and improve their competitiveness.

Keywords: labor productivity · quantile regression · external & internal financing · innovation

JEL Classification: D24 · L

Published by Springer Nature Switzerland AG 2025
P. Sureephong et al. (Eds.): PLM 2024, IFIP AICT 741, pp. 232–252, 2025.
https://doi.org/10.1007/978-3-031-93323-3_18

1 Introduction

The current business environment has been characterized by a constant development of new technologies, which generates a dynamic and competitive market. Therefore, organizations must focus their efforts on the acquisition of new knowledge, accompanied by innovation and technological advancement; Tavani et al. [1].

In this sense, Dess et al. [2] states that organizations require financial resources, which allow them to keep the current business context, through a process of constant innovation. This argument is in line with what Zhang [3] said, who defines the current market as a changing environment. Within these conditions, companies are forced to implement competitiveness strategies, which allows them to make their market participation relevant and subsequently generate attractive economic benefits accompanied by business development [3].

Currently, according to Münch Galindo [4], most companies do not carry out an adequate decision-making process inherent in their operation financing. However, to stay competitive in the market, it is essential that companies base their management on appropriate financing strategies [5].

Similarly, Van Horne et al. [6] interpret funding sources as important strategic tools that pursue the financial sustainability of the organization. This means that an adequate financing policy translates into an increase in entrepreneurial capacity to generate sufficient funds to boost their permanence in the market through knowledge, innovation, and technology [7].

Currently, there are numerous types of financing as an influential factor in productivity. Pilay-Villacís et al. [8], analyze the main financing forms for SMEs in Santa Elena-Ecuador. They determine that most companies opt for external financing for their growth projections. However, the success of these operations will depend on decision-making practice to acquire more timely credits [7].

On the other hand, Mora-Torres [9] studying the sources of short-term financing as a strategy for increasing business productivity in SMEs, establishes external financing in the form of bank loans, as one of the main short-term financing resources of SMEs. Overall, the min consensus in the literature is that access to finance has the potential to explain the link between financial development and business productivity [10].

The main objective of this research is to determine the influence of the type of financing on the labor productivity of Ecuadorian companies. For this study, microenterprises are discarded, and the National Survey of Science, Technology, and Innovation Activities is taken as a source of information, from 2009–2014.

The obtained results provide relevant contributions because they help to understand in a theoretical explanation the effects of financing type on labor productivity. Additionally, policies and strategies could be generated, to facilitate the procurement of better labor productivity indices, for the decision-making of managers, business owners and government agencies.

On the other hand, the results will help to formulate a theoretical debate on the main methods of calculating productivity, considering key advantages of each outcome. It also allows us to take advantage of the information available from the National Survey of Science, Technology, and Innovation Activities Censos [11].

The structure of this study is distributed as follows. In the first section, the literature review is presented. In the second section the methodology, sample and estimation strategy are detailed. The third section considers the univariate and multivariate results of the estimates. Finally, in the fourth section, conclusions, main contributions, limitations of research, and future lines of research are proposed.

2 Literature Review

One of the seminal studies to analyze the impact of different variables on labor productivity is the one proposed through the theoretical model of Griliches et al. [12]. It states that innovation is a function of productivity. Additionally, the explanatory variables represent determining business factors in the innovation process Wadho et al. [2]. This model, also known as Griliches's [12] theory of technological capital, modifies the aggregate productivity function, adding a new coefficient (R&D Expenses) to estimate R&D results.

$$Y_{it} = A_u F(L_{it}, K_{it}, M_{it}, G_{it} \quad (1)$$

where i and t denote, the company and the period. Y represents productivity. L, K and M, G and A respectively are, the amount of labor, capital, intermediate inputs, capital of technological knowledge & innovation and the degree of efficiency achieved by the organization.

Bravo [13] analyzes the effect of cooperation and innovation on productivity in 3,249 Chilean manufacturing companies, during 2008–2012. The author uses the CDM model of Crépon et al. [14] and highlights the theoretical contributions of Griliches. In the research, he models the relationship between innovation and its determinants, within a function of productivity Wadho et al. [2]. In addition, it determines that cooperation increases investment in innovation by 118%, which stimulates the introduction of technologies to the production process. This effect motivates an increase in productivity of 43%.

On the other hand, Oyarzún et al. [15] use the theory of Griliches to analyze the impact of R&D tax incentive laws on productivity. A sample of 5,620 Chilean companies was used during 2009–2014, they determined that the average effect of benefiting from the tax incentive law is positive and significant on labor productivity.

An emblematic study of the application of Griliches theory is that of Hashi et al. [16]. They apply a multi-stage approach of innovation to the mature market economies of Western Europe and the advanced economies in the transition of Central and Eastern Europe. The authors use the fourth Community Innovation Survey (CIS4), which covers 90,000 companies. The findings reveal a positive relationship between innovation activities and productivity. Competition encourages organizations to participate in innovation and fosters them to invest in it, to impact productivity.

Finally, Lordén et al. [17], in their study of the effect of internal R&D expenditure on the technological efficiency of 1,930 Spanish companies during 2008-2012, refer to the theoretical contribution established by Griliches, to provide their main research hypothesis. The authors were able to demonstrate that companies that invest in R&D (internal costs) can adapt better in times of crisis and have resilience reactions García et al [18]. Table 1 presents the main studies that use Griliches' theory of technological capital.

Table 1. Main studies using Griliches' Theory of Technological Capital

Authors	Count ry	Period	Sample (companies)	Dependent Variable	Independent Variable
Bravo [13]	Chile	2008–2012	3,249	Productivity	Cooperation Innovation
Oyarzún Acuña et a.l [15]	Chile	2009–2014	5,620	Productivity	Tax incentive laws
Hashi et al. [16]	Europe	2002–2004	90,000	Productivity	Innovation R&D expenditure
Mate Lordén et al. [17]	Spain	2008–2012	1,930	Productivity	Internal R&D expenditure

Source: Own collection of Griliches' main studies

2.1 Labor Productivity

There could be several barriers that prevent the growth and development of an economy, the obstacles could be a lack of social and scientific awareness, innovative initiative, technological development, deficient management for internal and external financing [19, 20]. Under this scenario, it is important to identify the factors that affect the labor productivity of companies with this information, actions can be taken to keep in the market and improve competitiveness [21].

Labor productivity can be defined as the company value added per employee, and it is an important dynamic measure of both organizational performance and competitiveness. Hence, labor productivity is a key driver of economic prosperity, as it is strictly related to economic growth, and it is a key determinant of wealth creation [5]. Moreover, labor productivity is often combined with the company's innovation performance [2–5]. This statement supports innovative processes generating added value and transforming it into higher economic returns [22].

Although there is research on productivity measures, there is no absolute consensus on their most appropriate calculation [23]. Most studies focus on two methods for calculating productivity: labor productivity (PL) and total factor productivity (PTF) [24]. Labor productivity (PL) is calculated as the ratio of sales to the number of employees in the organization [25].

According to Carvalho et al. [26] the main advantage of this calculation is the abundant availability of data and the simplicity of the formula. However, a criticism arises, since this measure shows much more productivity improvements due to efficiencies in the acquisition of materials, than the gains from a more efficient use of labor and energy.

2.2 Determinants of Labor Productivity

2.2.1 External Financing

Differences in the companies' financial development affect the labor productivity growth regarding the operating sectors. Thus, for example, external financing, which depends on the particular environment in which the country is immersed, gives a specific productivity result.

Avarmaa et al. [27] studied the impact of financial leverage on labor productivity of Baltic companies during 2001–2008. A positive relationship was established between external financing and labor productivity. In addition, it was determined that, for local companies, an increase in external financing generates an improvement in the dependent variable. The result differs in multinational companies.

Likewise, Osei-Assibey [28] analyzed 5,057 Ghanaian companies and found that, compared to domestic financing, external financing sources are positively associated with labor productivity growth. Additionally, this type of financing is complemented by technical and management advice. This allows to generate a greater probability of growth in labor productivity.

Meanwhile, Hsu et al. [29] studied 32 developed and developing countries during 1976–2006. They suggest that innovative activities are positively linked to more developed stock markets, but not to credit markets. The results consistently show that sectors that rely on external financing promote greater innovation and labor productivity, due to the state of progress of their markets.

Furthermore, Cole et al. [30] modeled the relationship between technology acquisition with external financing as a critical determinant. This study, conducted in India, Mexico, and the United States, determines like a system with access to external finance where the adoption of advanced technologies is vital. This process drives greater labor productivity. In addition, the use of advanced production technologies involves high initial investment costs. However, these generate significant profits at the end of the company's development cycle. Considering the results obtained by Avarmaa et al. [26]; Cole et al. [29]; Hsu et al. [28]; Osei-Assibey [27], that present a positive relationship between external financing on labor productivity, the following hypothesis is proposed:

> H1: The external financing percentage has a positive relationship with labor productivity.

2.2.2 Internal Financing

Self-financing or own financing is conceptualized as the financial resources generated by the company without resorting to the financial market. This type of financing is considered an important source of resources, which motivates the sustainable development of the company. In such a way these resources increase their probability of survival in the market, Rodríguez-Pose et al. [31].

Carpenter et al. [13] examined the assumption that the growth rate of small businesses depends on the availability of internal financing. A sample of more than 1,600 companies was analyzed and it was determined that the growth of most small companies is limited by internal financing, along with a small leverage effect. In contrast, it also notices in a few numbers of companies that opt for a constant issuance of new shares exhibit growth rates well above those that only support by internal financing.

In contrast, Adegboye et al. [32] used a dataset from the World Bank's survey of companies. They discussed the role of access to finance and focused on improving innovation and productivity among Nigerian SMEs. This study classifies the financing of the company, into external and internal; and determines that the difficulty in accessing bank credit is the main reason for opting for domestic financing. Therefore, internal sources become the main engine of expansion and development.

Likewise, Li et al. [33] explored a sample of more than 600,000 Chinese companies (1998–2009). They focused on the effects of credit constraints on firm productivity. Their findings established that internal financing through the company's cash flow and the supply of external credit significantly promotes the company's productivity and growth rates. In addition, they determined that there is a substitution effect between domestic financing and the supply of external credit. Additionally, they recognized that the effect of domestic financing on firm productivity is weaker when companies have sufficient external credit. However, this effect is noticeably less significant for state-owned enterprises.

Considering the positive impact of the results obtained by Adegboye et al. [13]; Carpenter et al. [31]; and Li et al. [32], the following hypothesis is formulated:

H2: The domestic financing percentage has a positive relationship with labor productivity.

2.2.3 Number of Employees

The company's human talent, represented by the number of employees, is an important factor that causes the accumulation of knowledge and improvement of productivity [34]. Therefore, many researchers consider it as a determinant.

Salis et al. [35] analyzed if the adoption of human resource management practices improves employee performance, which promotes labor productivity growth. The authors studied more than 500 British business establishments, drawn from the 2004 Workplace Labor Relations Survey. They showed a positive association between the number of employees and productivity if human resource management practices are adopted on a continuous improvement basis. The findings also suggested that it is possible to increase labor productivity by implementing practices focused on knowledge sharing through personal interactions between employees.

Moreover, Hagsten [36] considered the relationship between the level of information and communication technologies (ICT) access of employees and labor productivity, among 14 European countries during 2001–2010. It was established that in most countries, there is a significant and positive relationship between the number of employees with access to ICT and labor productivity. However, the strength of the relationship varies between countries and industries. In addition, the research determined that this relationship has a greater impact on manufacturing than service companies since the industrial sector has a greater number of employees.

Another study by Heshmati et al. [37] mentions that labor productivity can be defined simply as total production or sales per employee at the company level. For this research, they constructed a multisectoral dataset from several countries in Europe between 1999 and 2005. The results showed that the number of workers has a positive effect on labor productivity. Its effect varied between the range of 0.32 and 0.42 and it was statistically significant to less than 1 percent. One possible explanation is that job training and education programs generally increase job skills, which produce higher labor productivity.

Considering the results obtained by Hagsten [35]; Heshmati et al. [36]; and Salis et al. [34] and considering their recurrent positive relationship, the following hypothesis is suggested:

H3: The number of employees has a positive relationship on labor production.

2.2.4 Origin of Foreign Capital

Liu et al. [38] argue that foreign direct investment (FDI) can have a positive impact on industries' labor productivity, through the direct introduction of capital, technology, and management skills. For this purpose, 41 subsectors of the Chinese electronics industry during 1996–1997 were analyzed, with different levels of FDI. The results suggest that the presence of foreign industry investment is associated with higher labor productivity.

In addition, Vahter [39] studied the effects of foreign direct investment on labor productivity in the manufacturing industries of two countries in transition, Estonia and Slovenia. The author determined that different types of foreign direct investment may have different effects on the host country. Furthermore, the existence of positive side effects may depend on the level of economic development of the host country.

Likewise, Boghean et al. [40] evaluated the relationship between foreign direct investment and labor productivity in countries of the European Union, using data extracted from the Eurostat website from 2000 to 2012.

Foreign direct investment (FDI) in the economies of host countries found to be mainly related to increased labor productivity through technology transfer, management, and trade competition. This process enables technological progress and long-term economic growth.

In addition, Hoang et al. [41] considered whether foreign direct investment (FDI) had a positive effect on labor productivity in Vietnam's companies from 1986 to 2014. The study provides strong statistical evidence that FDI has a positive impact on labor productivity on a long-term basis. These findings imply that workers are expected to further improve their knowledge and skills. Additionally, the researchers conclude that policymakers should propose concrete plans to establish more efficient production units.

When considering the results from the studies of Boghean et al. [39]; Hoang et al. [40]; Liu et al. [37]; and Vahter [38], who found a positive relationship between foreign capital on labor productivity, the following hypothesis is considered:

H4: Foreign capital has a positive relationship with labor production.

2.2.5 Innovation of a New Service

Griffith et al. [42] compared the role that innovation plays in productivity in four European countries, France, Germany, Spain, and the United Kingdom. They used internationally harmonized company-level data based on the Community Innovation Survey

(CIS3). Overall, the results suggest that the systems that drive innovation (R&D intensity, gross R&D expenditure, labor productivity growth) are remarkably similar in the four studied countries. However, they also found interesting differences, particularly in productivity variation, which is associated with the level of innovative activities.

Process innovation appears to be the main driver of labor productivity, while the influence of product or service innovation is not significant. This argument is established by Mairesse et al. [43] who investigated the effect of innovation on labor productivity for manufacturing and service French companies from 1998 to 2004. The impact of each type of innovation on labor productivity is recognized.

Considering the positive relationship obtained between the innovation of a new service and labor productivity shown by Griffith et al. [42]; and Mairesse et al. [43]; the following hypothesis is presented:

H5: New service innovation has a positive relationship with labor productivity.

2.2.6 Machinery and Equipment Acquisition

The study conducted by Goodrum et al. [44] examines changes in equipment technology as a factor that explains the increase in labor productivity. A sample of 200 activities over 22 years is used.

Through ANOVA and regression analysis, the results show that activities that experienced a significant technology change had a positive impact on labor productivity. While those that had no technological influence did not experience any change.

Goodrum et al. [44] observed 200 construction activities to determine the effect of equipment technology on labor productivity from 1976 to 1998. It was determined that activities that experienced significant changes in equipment technology have gained substantially greater long-term improvements in labor productivity than those that have not acquired a change in equipment technology.

Likewise, Goodrum et al. [45] assessed changes in materials technology and the effect they have on productivity. The studied sample included 100 construction activities between 1977 and 2004. Through ANOVA analysis of variance and regression analysis, the findings show that changes in technology had a significant positive relationship with labor productivity.

Considering the studies carried out by Goodrum et al. [44], Goodrum et al. [45], Goodrum et al. [46] and the positive relationship of machinery & equipment acquisition on labor productivity; the following hypothesis is formulated:

H6: Machinery & equipment acquisition has a positive relationship with labor productivity.

3 Methodology

3.1 Data and Sample

The samples were selected from the first "National Survey of Science, Technology and Innovation Activities" (ACTI), 2009–2011. This survey was conducted in 2013. A second and final survey was developed during the years 2012–2014 and it was accomplished in

2015. The latest survey was intended to update information on the science, technology, and innovation activities of Ecuadorian companies. Both surveys were coordinated by Ecuador's INEC; Censos [11].

The initial population of the surveys reached a total of 6,328 Ecuadorian companies, representing 19,686 observations. Some criteria were established to filter the data, to obtain a homogeneous and comparable sample. Like this, 263 companies were eliminated because they had zero sales values or nothing in the number of employees. Also, 4,517 companies were eliminated because their information was not reflected in both surveys. On the other hand, 483 companies that were created since 2012 were excluded, due to the lack of comparing information to be shown in the studied years. Similarly, 178 companies were discarded because they showed atypical data.

The last criterion for the depuration of the database was the removal of Ecuadorian microenterprises; this was to avoid biases in the estimation. To eliminate this type of bias, Article 3 of Decision 702 of the CAN was considered. It determines within this category of microenterprises those that have from 1 to 9 employees and a gross value in annual sales equal to or less than 100,000 USD, CAN [47].

3.2 Definition and Measurement of Variables

To carry out the comparison, two variables with financial characteristics were used: external financing and internal financing (fire, irp); and four variables with business characteristics: number of employees, origin of foreign capital, innovation of a new service, and acquisition of machinery & equipment.

Barbosa et al. [48] mentioned that labor productivity relates to the labor costs used to produce several products. In addition, labor productivity is closely linked to innovation outcomes and knowledge of products and processes. On the other hand, labor productivity is calculated through the net sales ratio per employee. Labor Productivity: (Net Sales)/(Number of employees) [2].

External financing (fire) presents a way to execute the company's operations through external funds [17]. The information is taken from the value recorded by each company in its financial reports, and it represents the percentage of external financing of each company regarding the total liabilities and equity.

Internal funding (firp) is a key to better operational performance and company profitability [48]. As competition intensifies due to changes in the industry structure and new technologies emerge, organizations are determined to reduce their operating costs. At the same time, through the search for new strategies to improve their productivity by accessing internal financing [49].

Refereeing to the number of employees (emp) is considered a significant resource in producing efficient goods and services. Therefore, improving productivity in the organization is based on human resources [17]. In addition, all companies that have an employee amount equal to or greater than ten are considered for the study.

Foreign capital financing (cex) for many developing countries has become a crucial strategy for economic growth. According to a Zhang-led study [50], when host nations liberalize trade, enhance education, and maximize human capital conditions, foreign capital investment boosts worker productivity. The indicator is obtained from information recorded in financial reports, such as the total foreign capital used for each company [5].

Table 2 shows the description of the variables with their respective calculation formulas.

Table 2. Variables Description

Type of variable	Group	Variable	Code	Description	Stata Formula
Dependent variable		Labor Productivity	Prd	Sales/Employees	prd_{it}
Independent variables	Financial Charact.	External Financing	Fire	Percentage of external financing per company	$fire_{it}$
		Internal Financing	Firp	Percentage of internal financing per company	$firp_{it}$
		Number of employees	Emp	Number of employees per company >= 10 employees	emp_{it}
		Origin of foreign capital	Cex	Percentage of foreign capital used by each company	cex_{it}
	Business Charact.	New Service Innovation	Isn	0 – the company did not introduce a new or significantly improved service. 1 – the company introduce a new or significantly improved service.	isn_{it}
		Acquisition of machinery & equipment	Ame	0 – the company did not acquire machinery & equipment 1 – the company acquire machinery & equipment	ame_{it}

Source: Own analysis of variables and their stata formulas

On the other hand, new service innovation (isn) is crucial in the research of labor productivity. Huergo et al. [51] stated that companies that develop technological activities give rise to process and product innovations, which increase global productivity, superior to the rest of the companies. Calvo [52] argued that Spanish manufacturing companies show that 30% of their turnover is due to the introduction of new products that originated from innovation. This variable is the kind of dichotomous variable, determined by 0 (the company that did not introduce a new or significantly improved service) and 1 (the company that introduced a new or significantly improved service).

Finally, obtaining machinery & equipment (ame) is a determining business factor for a strategic plan. Moreover, it contributes to generating value for the entity, ensuring its stability, competitiveness, and market growth [53]. The variable is dichotomous, with 0 (the company that did not acquire machinery & equipment) and 1 (the company that acquired machinery & equipment).

3.3 Estimation Strategy and Model Specification

Numerous investigations have focused on analyzing internal and external financing, innovation, and its influence on labor productivity [3]. Coad et al. [54] mentioned in their study that a company experiences modest growth, which may or may not be related to innovative activity. For example, they relate innovation to sales growth for high-tech sectors. The methodology used was a quantile regression approach to panel data.

On the other hand, Baronio et al. [55] mentioned that the main advantages of using panel data are the management of many observations and, in the same way, the removal of collinearity by improving the estimates of econometric studies. Wang et al. [56] analyzed the role played by eight factors with business and financial characteristics within China's industrial sector productivity. They also used a quantile regression for panel data as the principal estimation model, supported by the ordinary least squares model. Furthermore, they reached a complete conclusion about the mechanism of the influence of independent variables on productivity. These methodologies are typical in the impact evaluation literature because they allow studies to control heterogeneity. Consequently, they help to understand the treatment's causal relationship based on impact variables [57].

The present research uses a quantile regression estimation strategy for panel data where ordinary least squares is applied to strengthen the results. Based on the research of Koenker et al. [31], the representation of the quantile regression model is as follows:

$$Y_{it} = \beta 0 X_{it} + \mu t + \varepsilon_{it} \tag{2}$$

where $\mathbf{Y_{it}}$ is the dependent variable, $\boldsymbol{\beta_0}$ is the parameter to be estimated (corresponding to the quantile), $\mathbf{X_{it}}$ is the independent variable, $\boldsymbol{\mu_1}$ corresponds to the random perturbation (corresponding to the quantile), and $\boldsymbol{\varepsilon_{it}}$ is the random error. many writers [25, 53–55], who investigate the impact of variables with financial and commercial characteristics on productivity, employ the same research suggested methodology. The expression of the model used is the following:

$$prd_{it} = \beta 0 + \beta 1 firp_{it} + \beta 2 fire_{it} + \beta 3 emp_{it} + \beta 4 ame_{it} + \beta 5 isn_{it} + \mu l + \varepsilon_{it} \tag{3}$$

prdit: is labor productivity growth in the company. B0: Coefficient of explanatory variables. $\boldsymbol{\mu}$**1**: Corresponds to the invariant effects of the company over time. $\boldsymbol{\varepsilon_{it}}$: is the error of observation i in the time t.

4 Empirical Results

4.1 Univariate Analysis

Figure 1 presents the distribution of the growth rate of labor productivity. The grain density function is tent-shaped (Laplace). It presents heteroscedasticity and a positive kurtosis of 4.3973. Furthermore, the observations have the distribution as the main point, and thicker tails are exhibited at the extremes [68, 69]. Due to the existence of atypical data in the right tail of the distribution, a positive asymmetry is notable.

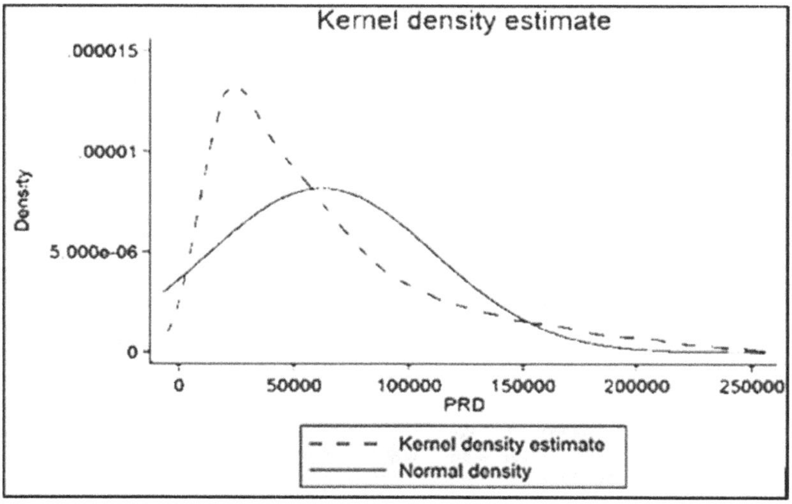

Fig. 1. Kernel estimation of Labor Productivity variable. Source: (Heterocedasticidad - Rafael de Arce y Ramn | Course Hero, 2016)

Table 3 shows the descriptive statistics. The average value of the labor productivity variable is $62,324.10 USD.

While the percentage of external financing is 0.85%, internal financing is 48.43%, and the number of employees is 336. For its part, foreign capital represents 8.69%. Furthermore, companies that present the innovation of a new service reach an average of 9.69%. Finally, the acquisition of machinery and equipment reaches 44.85%.

Table 4 shows the correlation matrix of the studied variables. Most of the coefficients of the variables do not exceed 0.3 in significance. Thus, there are no problems with collinearity.

4.2 Multivariate Analysis

Table 5 presents the results obtained using ordinary least squares, regression, and quantile regression for panel data. The variable percentage of external financing (fire) has a significant positive relationship with labor productivity only at the 0.25 quantile. For the rest of the quantiles, it has no significance. Therefore, hypothesis 1 is partially

Table 3. Descriptive Statistics

Variable	Obs	Mean	Std. Dev	Min	Max
prd	4.428	62.324,10	48.860,93	57	248.772
vts	4.428	23.900.000	64.100.000	100.500	943.000.000
fire	4.428	0,8550136	7,66289	0	100
firp	4.428	48,43089	45,56899	0	100
emp	4.428	336,033	724,8023	11	11.205
cex	4.428	8,696477	25,63787	0	100
isn	4.428	0,0968835	0,2958324	0	1
ame	4.428	0,4485095	0,4973978	0	1

Source: Own analysis of descriptive statistics
Note: Obs= Number of observations, Mean= Media, Std. Dev= Standard deviation, Min= Minimum value, Max= Maximum value.

Table 4. Correlation Matrix

	prd	fire	firp	emp	cex	isn	ame
prd							
	1.0000						
fire	0.0378	1.0000					
firp	0.0360	−0.0663	1.0000				
emp	0.0823	0.0425	0.1073	1.0000			
cex	0.1190	0.1204	0.0390	0.1927	1.0000		
isn	−0.0113	0.0128	0.2418	0.0557	−0.0043	1.0000	
ame	0.0581	0.0740	0.4639	0.0345	−0.0043	0.1375	1.0000

Source: Own

accepted. The explanation of this association applies to businesses with lower labor productivity possess the want for more funding to promote growth. These findings are in line with those obtained by Avarmaa et al. [26]; Cole et al. [29]; Hsu et al. [28]; and Osei-Assibey [27]. According to these studies, raising internal financing boosts worker productivity, promotes the innovation and the adoption of advanced technologies, since this type of financing has technical and management advice. Regarding the percentage of internal financing (firp) variable, it is observed that it has no impact on labor productivity. Therefore, hypothesis 2 is rejected. This relationship can be explained since financing through shareholders generates double returns, one through the shares sales and the other through the dividend's remuneration. This has a big impact on the company's profit margin. This result is adjusted with the studies of Adegboye et al. [13]; Carpenter et al. [31]; Li et al. [32]. This research contends that domestic funding is a limiting factor in corporate growth. There is difficulty in accessing external financing, which causes SMEs to use internal financing as an engine of development. Investments with domestic financing tend to have diminishing returns on labor productivity because there is a substitution effect of internal financing due to sufficient access to external credits.

Regarding the number of employees (emp) variable, there is evidence of a significant positive relationship with labor productivity. If this result is in all quartiles except quartile 0.10, then hypothesis 3 is accepted. The impact of this variable on labor productivity is due to the proper management of human talent in the company. This strategy focuses on taking advantage of growth opportunities. These results are consistent with those obtained by Hagsten [35]; Heshmati et al. [36]; Salis et al. [34]. All these studies mention that the human adoption resource management practices, the access level to ICTs, and employee control procedures have a significant positive relationship with labor productivity. These benefits continued the growth of companies, reinforced by human resource management policies. These policies, along with job training and education programs, increase employees' job skills, which results in higher labor productivity. The variable percentage of foreign capital (cex) showed a significant positive impact on all quartiles and the MCO regression. Therefore, hypothesis 4 was rejected. The effect of this variable on labor productivity is that the presence of FDI enhances labor productivity. Since it optimizes, simplifies, and automates processes. These results agree with those obtained by Boghean et al. [39], Hoang et al. [40], Liu et al. [37], and Vahter, [38]. The studies argue that companies are associated with higher labor of productivity through the direct introduction of capital and technology, management skills, and transfer of business competence. Moreover, technological progress is allowed, depending on the economic sector of the country where it is developed. Thus, foreign capital brings knowledge and technical advice, which improves the competencies and skills of company workers.

On the other hand, the variable innovation in new services (isn) has a significant negative relationship in the 0.10 and 0.25 quartiles and in the ordinary least squares regression. As a result, there is no impact on productivity. Hypothesis 5 is, therefore, rejected. This effect is due to the fact that investments and innovation of new products and services do not always have an acceptable return margin. This result increases when organizations lack investment policies and strategies. This scenario leads to an increase in unnecessary costs. This scenario leads to unnecessary cost increases. These findings agree with those obtained by Mairesse et al. [42], who determined that investment in new innovative products and services does not show a significant relationship with labor productivity.

Finally, at the quantiles of 0.10, 0.25, and 0.50, the acquisition of machinery and equipment (ame) variable significantly increases labor productivity. Therefore, hypothesis 6 is accepted. This variable's effect on worker productivity stems from businesses' ongoing use of technology to automate processes, which lowers expenses and boosts productivity. These results agree with those obtained by Goodrum et al. [44], and Goodrum et al. [42]. These authors mention that activities with technology and equipment changes substantially improve labor productivity on a long-term basis. Business managers should promote the acquisition of more efficient machinery and equipment in order to improve lower company productivity levels. With this change of strategy, knowledge increases new skills and best practices among workers. Which finally ends in a positive impact on their labor productivity. The participation of highly innovative companies is positively related to productivity. Furthermore, organizations benefit if they are involved in an environment of constant innovation [58].

Table 5. Relationship between productivity and internal and external financing

Variable	MCO	Q10	Q25	Q50	Q75	Q90
fire	148.9111*	−30.21489	296.1742***	150.1186	214.27	324.5535
	(0.029)	(0.508)	(0.000)	(0.154)	(0.257)	(0.286)
firp	−3.703623	.2308277	7.019764	23.7555	13.1673	−25.77936
	(0.782)	(0.979)	(0.564)	(0.242)	(0.718)	(0.661)
emp	17.06171***	.3287775	2.347297***	6.066069***	7.175676***	11.23021***
	(0.000)	(0.499)	(0.000)	(0.000)	(0.000)	(0.001)
cex	100.4269***	62.9702***	88.92598***	208.9239***	373.0173***	346.6296***
	(0.000)	(0.000)	(0.000)	(0.000)	(0.000)	(0.000)
isn	2765.478	−2909.492*	−5035.736**	−2703.823	1667.486	−13167.84
	(0.134)	(0.015)	(0.002)	(0.328)	(0.737)	(0.099)
ame	296.687	2917.78***	4228.883***	5344.433**	4176	8596.649
	(0.817)	(0.000)	(0.000)	(0.003)	(0.200)	(0.101)
Number of groups	635	635	635	635	635	635
Number of observations	3810	3810	3810	3810	3810	3810
R°2/Pseudo R°2	0.0019	0.0119	0.0091	0.0152	0.0257	0.0466

Source: Own analysis

Note: The significance of the values in Table 5 is represented by. * $p < 0.05$; ** $p < 0.01$ and *** $p < 0.001$. Robust standard errors are in arentheses.

5 Conclusion

His research analyzed, the effects of the type of financing on the labor productivity of 635 Ecuadorian companies from 2009 to 2014. For this purpose, It was necessary a quantile regression model for panel data. On the other hand, to eliminate heteroscedasticity in the sample, it opted for excluding the microenterprises, eliminating the atypical data, and only considering companies that present data greater than zero in the six years of study were considered. From the multivariate analysis, external financing has a positive impact on the labor productivity of companies. This effect is possibly due to the need to expand the market. Meanwhile, the variable percentage of internal financing does not include an impact on labor productivity. The cause could be that internal funding tends to generate diminishing returns due to various causes, such as excessive tax burdens and R&D investments.

Regarding the employee variable, there is a significant positive relationship with labor productivity. Companies manage human resources control strategies and policies, which exert a positive influence on the dependent variable. Concerning the origin of the foreign capital variable, it shows a significant positive impact on all quartiles. This result is probably attributable to the presence of foreign capital, which gives companies a competitive advantage by allowing them to import superior technologies.

Regarding innovation of new services, this variable has a significant negative relationship in quartiles 0.10 and 0.25. This effect turns into economic losses and cost increases derived from failed investment policies. In addition, the impact of company strategies with low productivity rates and no alignment to a dynamic current market.

Finally, the acquisition of machinery & equipment variable showed a significant positive impact on labor productivity in the quantiles 0.10, 0.25, and 0.50. This result focused on new technologies that improve the execution of activities, optimize labor practices, and increase labor productivity.

The obtained findings allow us to conclude that accessing external financing, having foreign capital, providing support from human resources, and acquiring machinery & equipment contribute to the labor productivity growth of Ecuadorian companies. In contrast, the results show that access to domestic financing and innovation in new services are not determinants of labor productivity.

The main contributions of this research study are based on multivariate analysis. It uses an econometric technique of quantile regression, reinforced with an ordinary least squares (MCO) regression. Likewise, within the financial variables, it was possible to determine external financing as the main influencing factor on labor productivity in Ecuadorian companies. These results contribute to adequate business decision-making and consequent formulation of more efficient investment strategies, which will become determinants for productivity growth.

Like any other research, the present study has some limitations. Firstly, the finding obtained cannot extended to all Ecuadorian organizations since the sample did not consider microenterprises. They were excluded due to their different types of growth compared to small, medium, and large companies. Secondly, a limited number of financial variables was used, considering that more financial indicators can be used for a better determination of productivity growth.

References

1. Tavani, D.Y., Zamparelli, L.: Growth, income distribution, and the 'entrepreneurial state.' J. Evol. Econ. **30**(1), 117–141 (2020). https://doi.org/10.1007/s00191-018-0555-7
2. Dess, G.G., Sánchez Soto, G.A., Lumpkin, T.G., Eisner, A.B., Mejía Estañol, R., Vargas Pérez, M.: Administración estratégica: textos y casos. McGraw-Hill (2011)

3. Zhang, K.H.: Does foreign direct investment promote economic growth? Evidence from East Asia and Latin America. Contemp. Econ. Policy **19**(2), 175–185 (2001). https://doi.org/10.1111/j.1465-7287.2001.tb00059.x
4. Van Horne, J.C., Wachowicz, J.M.: Fundamentals of Financial Management. Pearson Education, Boston (2001)
5. Motta, V.: Lack of access to external finance and SME labor productivity: does project quality matter? Small Bus. Econ. **54**(1), 119–134 (2018). https://doi.org/10.1007/s11187-018-0082-9
6. Boermans, M.A., Willebrands, D.: Financial constraints matter: empirical evidence on borrowing behavior, microfinance and firm productivity. J. Dev. Entrepre. JDE **23**(02), 1850008 (2018). https://doi.org/10.1142/s1084946718500085
7. Mora-Torres, C.J.: Las fuentes de financiamiento a corto plazo como estrategia para el incremento de la productividad empresarial en las PYMEs. Dominio de las Ciencias **3**, 338–351 (in Spanish). https://doi.org/10.23857/dc.v3i1.401
8. Zhang, D.: Capacity utilization under credit constraints: a firm-level study of Latin American manufacturing. Int. J. Finan. Econ. **27**(1), 1367–1386 (2022). https://doi.org/10.1002/ijfe.2220
9. Seven, U. Y., Tumen, S.: Agricultural credits and agricultural productivity: cross- country evidence. Singapore Econ. Rev. **65**(supp01), 161–183 (2020). https://doi.org/10.1142/s0217590820440014
10. Griliches, Z.: Issues in assessing the contribution of research and development to productivity growth. Bell J. Econ. **10**(1), 92 (1979). https://doi.org/10.2307/3003321
11. Wadho, W.Y., Chaudhry, A.: Innovation and firm performance in developing countries: the case of Pakistani textile and apparel manufacturers. Res. Policy **47**(7), 1283–1294 (2018). https://doi.org/10.1016/j.respol.2018.04.007
12. Bravo, S.: Cooperación, innovación, y productividad: estimación de un modelo CDM para Chile (2016). https://repositorio.uchile.cl/handle/2250/144218. Consultado: el 9 de junio de 2021. (in Spanish)
13. Oyarzún Acuña, T., Peña León, K., Sánchez Castillo, M.: Impacto de la Ley de Incentivo Tributario en I+D sobre la productividad de las empresas chilenas (2018). http://repositorio.udec.cl/jspui/handle/11594/2507. Consultado: el 9 de junio de 2021. (in Spanish)
14. Almendarez Hernández, M.A.: Determinantes de las patentes y otras formas de propiedad intelectual de los estados mexicanos (2018). Economía Sociedad y Territorio **18**(58), 657–695. https://doi.org/10.22136/est20181223. (in Spanish)
15. Mairesse, J., Robin, S.: Innovation and productivity: a firm-level analysis for French Manufacturing and Services using CIS 3 and CIS 4 data (1998–2000 and 2002–2004) Preliminary version, March 2008 (2008). https://www.semanticscholar.org/paper/a0ee408917f5cc492136720de8b4e1d4eb9ef6d1
16. Lordén, M.M.Y., Molero, J.: Efecto del gasto en I+D interno en la eficiencia tecnológica de empresas españolas. Análisis comparativo durante el periodo de crisis de 2008–2012. CTS **15**(44) (2020). Disponible en: http://ojs.revistacts.net/index.php/CTS/article/view/159. Consultado: el 9 de julio de 2021. (in Spanish)
17. García, J.J.M., Fernández, J.M.R.: Crecimiento de la productividad e inversión en I+D: un análisis empírico de las empresas manufactureras españolas. Economía Industrial (347), 99–110 (2002). https://dialnet.unirioja.es/servlet/articulo?codigo=751339. Consultado: el 26 de junio de 2023. (in Spanish)
18. Towo, N., Mori, N.Y., Ishengoma, E.: Financial leverage and labor productivity in microfinance co-operatives in Tanzania. Cogent Bus. Manag. **6**(1), 1635334 (2019). https://doi.org/10.1080/23311975.2019.1635334
19. Brito, J.: Produtividade, Competitividade e Inovação na Indústria Brasileira (2009). https://www.eco.unicamp.br/neit/images/stories/arquivos/ie_ufrj_et05_produtividade.pdf

20. Hsu, P.-H., Tian, X., Yan, X.: Financial development and innovation: Cross-country evidence. J. Finan. Econ. **112**(1), 116–135 (2014). https://doi.org/10.1016/j.jfineco.2013.12.002
21. Cole, H.L., Greenwood, J., Sanchez, J.M.: Why doesn't technology flow from rich to poor countries? Econometrica **84**(4), 1477–1521 (2016). https://doi.org/10.3982/ECTA11150
22. Rodríguez-Pose, A., Ganau, R., Maslauskaite, K., Brezzi, M.: Credit constraints, labor productivity, and the role of regional institutions: Evidence from manufacturing firms in Europe. J. Reg. Sci. **61**(2), 299–328 (2021). https://doi.org/10.1111/jors.12514
23. Carpenter, R.E., Petersen, B.C.: Is the growth of small firms constrained by internal finance? Rev. Econ. Stat. **84**(2), 298–309 (2002). https://doi.org/10.1162/003465302317411541
24. Li, Y.A., Liao, W., Zhao, C.C.: Credit constraints and firm productivity: microeconomic evidence from China. Res. Int. Bus. Finan. **45**, 134–149 (2018). https://doi.org/10.1016/j.ribaf.2017.07.142
25. Salis, S., Williams, A.M.: Knowledge sharing through face-to-face communication and labor productivity: evidence from British workplaces: Face-to-face communication and productivity in Britain. Br. J. Ind. Relat. **48**(2), 436–459 (2009). https://doi.org/10.1111/j.1467-8543.2009.00762.x
26. Hagsten, E.: Broadband connected employees and labour productivity: a comparative analysis of 14 European countries based on distributed Microdata access. Econ. Innov. New Technol. **25**(6), 613–629 (2016). https://doi.org/10.1080/10438599.2015.1105547
27. Brzozowski, M.L.: Impact of credit market development and stability on productivity: New evidence from the industry level, Aeconf.com (2020). http://aeconf.com/Articles/May2020/aef210105.pdf. Accessed 26 June 2023
28. Adegboye, A.C., Iweriebor, S.: Does access to finance enhance SME innovation and productivity in Nigeria? Evidence from the world bank enterprise survey. Revue Africaine de Developpement **30**(4), 449–461 (2018). https://doi.org/10.1111/1467-8268.12351
29. Heshmati, A., Rashidghalam, M.:Labour productivity in Kenyan manufacturing and service industries. SSRN Electron. J. (2016). https://doi.org/10.2139/ssrn.2776928
30. Liu, X., et al.: The impact of foreign direct investment on labor productivity in the Chinese electronics industry. Int. Bus. Rev. (Oxford, England) **10**(4), 421–439 (2001). https://doi.org/10.1016/s0969-5931(01)00024-5
31. de E. y Censos, N.: Encuesta Nacional de Actividades de Ciencia, Tecnología e Innovación-ACTI, Instituto Nacional de Estadística y Censos 2015 (2021). https://www.ecuadorencifras.gob.ec/encuesta-nacional-de-actividades-de-ciencia-tecnologia-e-innovacion-acti/
32. Altomonte, C., Gamba, S., Mancusi, M., Vezzulli, A.: R&D investments, financing constraints, exporting and productivity. Econ. Innov. New Technol. **25**(3), 283–303 (2016)
33. Andina, C.: Decisión No 702 - Sistema Andino de Estadística de la PYME (2008). https://www.ecolex.org/details/legislation/decision-no-702-sistema-andino-de-estadistica-de-la-pyme-lex-faoc115929/. Consultado: 07 sep 2021. (in Spanish)
34. Asada, H.: Effects of foreign direct investment and trade on labor productivity growth in Vietnam. J. Risk Finan. Manag. **13**(9), 204 (2020). https://doi.org/10.3390/jrfm13090204
35. Barbosa, F., et al.: Reinventing construction: a route to higher productivity (2017). https://www.designingbuildings.co.uk/wiki/Reinventing_construction:_a_route_to_higher_productivity. Consultado 06 sep 2021
36. Baronio, A., Vianco, A.: Datos de Panel Guía para el uso de Eviews, pp. 1–24 (2018). (in Spanish)
37. Belsley, D.A.: A guide to using the collinearity diagnostics. Comput. Sci. Econ. Manag. **4**(1), 33–50 (1991). https://doi.org/10.1007/BF00426854
38. Bernstein, B., Singh, P.J.: Innovation generation process: applying the adopter categorization model and concept of "chasm" to better understand social and behavioral issues. Eur. J. Innov. Manag. **11**(3), 366–388 (2008). https://doi.org/10.1108/14601060810889017

39. Boghean, C., State, M.: The relation between foreign direct investments (FDI) and labour productivity in the European union countries. Procedia Econ. Finan. **32**, 278–285 (2015). https://doi.org/10.1016/S2212-5671(15)01392-1
40. Calvo González, J.L.: Una caracterización de la innovación tecnológica en los sectores manufactureros españoles: algunos datos. Econ. Ind. (331), 139–150 (2000). (in Spanish)
41. Carvalho, L., Avellar, A.P.M.: Innovation and productivity: empirical evidence for Brazilian industrial enterprises. Revista de Administração **52**(2), 134–147 (2017). https://doi.org/10.1016/j.rausp.2016.12.009
42. Chudnovsky, D., López, A., Pupato, G.: Innovation and productivity in developing countries: a study of Argentine manufacturing firms' behavior (1992–2001). Res. Policy **35**(2), 266–288 (2006). https://doi.org/10.1016/j.respol.2005.10.002
43. Coad, A., Rao, R.: Innovation and firm growth in high-tech sectors: a quantile regression approach. Res. Policy **37**(4), 633–648 (2008). https://doi.org/10.1016/j.respol.2008.01.003
44. Crepon, B., Duguet, E., Mairessec, J.: Research, innovation and productivity: an econometric analysis at the firm level. Econ. Innov. New Technol. **7**(2), 115–158 (1998). https://doi.org/10.1080/10438599800000031
45. Crespi, G., Zuniga, P.: Innovation and productivity: evidence from six latin American countries. World Dev. **40**(2), 273–290 (2012). https://doi.org/10.1016/j.worlddev.2011.07.010
46. de Arce, R., Mahía, R.: Conceptos básicos sobre la heterocedasticidad en el modelo básico de regresión lineal tratamiento con E-VIEWS (2009). https://www.coursehero.com/file/17996707/heterocedasticidad/. Consultado sep. 06, 2021. (in Spanish)
47. Evert, P.-V., et al.: Las fuentes de financiamiento como estrategias de competitividad en las Pymes en la provincia de Santa Elena, Redalyc.org (2019). https://www.redalyc.org/journal/5732/573263325011/573263325011.pdf. Consultado: el 26 de junio de 2023. (in Spanish)
48. García Jiménez, L.A., Narváez Zurita, C.I., Erazo Álvarez, J.C.: Gestión integral de activos fijos con enfoque en el control contable administrative. Visionario Digital **3**(2), 305–328 (2019). https://doi.org/10.33262/visionariodigital.v3i2.479. (in Spanish)
49. Goedhuys, M.: The impact of innovation activities on productivity and firm growth: evidence from Brazil, EconBiz (2007). https://www.econbiz.de/Record/the-impact-of-innovation-activities-on-productivity-and-firm-growth-evidence-from-brazil-goedhuys- micheline/10010712090
50. Gogokhia, T., Berulava, G.: Business environment reforms, innovation and firm productivity in transition economies. Eurasian Econ. Rev. **11**(2), 221–245 (2021). https://doi.org/10.1007/s40821-020-00167-5
51. González-Díaz, R.R., Becerra-Pérez, L.A.: PYMES en América Latina: clasificación, productividad laboral, retos y perspectivas. CIID J. **2**(1), 1 (2021). https://doi.org/10.46785/ciidj.v1i1.100. (in Spanish)
52. Goodrum, Paul M., Haas, Carl T.: Partial factor productivity and equipment technology change at activity level in U.S. construction industry. J. Constr. Eng. Manag. **128**(6), 463–472 (2002). https://doi.org/10.1061/(ASCE)0733-9364(2002)128:6(463)
53. Goodrum, P.M., Haas, C.T.: Long-term impact of equipment technology on labor productivity in the U.S. construction industry at the activity level. J. Constr. Eng. Manag. **130**(1), 124–133 (2004). https://doi.org/10.1061/(asce)0733-9364(2004)130:1(124)
54. Goodrum, P.M., Zhai, D., Yasin, M.F.: Relationship between changes in material technology and construction productivity. J. Constr. Eng. Manag. **135**(4), 278–287 (2009). https://doi.org/10.1061/(ASCE)0733-9364(2009)135:4(278)
55. Griffith, R., Huergo, E., Mairesse, J., Peters, B.: Innovation and productivity across four European countries. Oxford Rev. Econ. Policy **22**(4), 483–498 (2006). https://doi.org/10.1093/oxrep/grj028

56. Hashi, I., Stojčić, N.: The impact of innovation activities on firm performance using a multi-stage model: evidence from the Community Innovation Survey 4. Res. Policy **42**(2), 353–366 (2013). https://doi.org/10.1016/j.respol.2012.09.011
57. Huergo Orejas, E., Moreno Martín, L.: La innovación y el crecimiento de la productividad en España. Ekon. Rev. Vasca Econ. **56**, 208–231 (2004)
58. Kimari, F.N.: Effect of credit risk management on financial performance of deposit taking savings and credit cooperative societies in Kenya. Univ. Nairobi (2013). Consultado: sep. 06, 2021. https://www.semanticscholar.org/paper/Effect-of-credit-risk-management-on-financial-of-in-Kimari/d385f308c08deed9d1f88c9e612c2611ac3083c3
59. Koenker, R., Bassett, G.: Regression quantiles. Econometrica **46**(1), 33–50 (1978). https://doi.org/10.2307/1913643
60. Krkoska, L.: Foreign direct investment financing of capital formation in Central and Eastern Europe. In: Presentado en Round Table "Industrial Restructuring in European Transition Economies: Experience to Date and Prospects" (2002 : Geneva) (2002). Consultado: sep. 06, 2021. https://digitallibrary.un.org/record/458350
61. Lööf, H., Mairesse, J., Mohnen, P.: CDM 20 years after. Econ. Innov. New Technol. **26**(1–2), 1–5 (2017). https://doi.org/10.1080/10438599.2016.1202522
62. Miró-Pérez, A.-P.Y., Torrent-Sellens, J.: Transformación digital y productividad total de los factores (PTF) en las empresas españolas del sector oleícola: Una aproximación digital. Revista de Estudios Regionales **2**, 77–113 (2020). https://ideas.repec.org/a/rer/articu/v2y2020p77-113.html
63. Naqeeb, U.R., et al.: Barriers to growth of SMEs in Western Balkan countries. J. Manag. Dev. **38**(1), 2–24 (2019). https://doi.org/10.1108/JMD-09-2018-0273
64. Olaya, E.: Efecto del gasto en investigación y desarrollo en el ingreso de los establecimientos de Ecuador. Revista Económica **3**(1), 7–18 (2017). https://ideas.repec.org/a/ris/lojare/0021.html. (in Spanish)
65. Oliveira, G.R., Resende, G.M., Costa, D.F., da Silva, C., Gonçalves, N.: Micro-impacts of the brazilian regional development funds: does lending size matter? Rev. Dev. Econ. **23**(1), 293–313 (2019). https://doi.org/10.1111/rode.12539
66. Osei-Assibey, E.: Source of finance and small enterprise's productivity growth in Ghana. Afr. J. Econ. Manag. Stud. **4**(3), 372–386 (2013). https://doi.org/10.1108/AJEMS-03-2012-0017
67. Pakes, A., Griliches, Z.: Patents and R&D at the firm level: a first report. Econ. Lett. **5**(4), 377–381 (1980). https://doi.org/10.1016/0165-1765(80)90136-6
68. Palmero Urquiza, D.E., Benítez Narváez, R.M.: Las estadísticas como herramienta para el estudio de parámetros medioambientales y ecológicos. Rev. Conrado, 101–104 (2017). (in Spanish)
69. Romano, L.: Explaining growth differences across firms: the interplay between innovation and management practices. Struct. Change Econ. Dyn. **49**, 130–145 (2019). https://doi.org/10.1016/j.strueco.2019.01.004
70. Simbaña, L., David, R.-P., Fernández-López, S., Morales, X., Diana, M.: Análisis descriptivo de las características empresariales y financieras de las pymes en Ecuador, vol. 13, pp. 87–102 (2018). (in Spanish)
71. Simbaña Taipe, L.E.: Determinantes de crecimiento de las empresas ecuatorianas (2017). http://purl.org/dc/dcmitype/Text. Universidade de Santiago de Compostela. Consultado: sep. 07, 2021. https://dialnet.unirioja.es/servlet/tesis?codigo=147327
72. Tello, M.D.: Effects of obstacles to innovation: are they complementary? lessons of the case of Peru. J. Innov. Econ. (2), 187–217 (2021). https://doi.org/10.3917/jie.pr1.0094
73. Vahter, P.: The effect of foreign direct investment on labor productivity: evidence from Estonia and Slovenia. SSRN Electron. J. (2004). https://doi.org/10.2139/ssrn.623184

74. Vergeer, R., Kraan, K., Dhondt, S., Preenen, P.: Labour productivity and innovation performance: the importance of internal labor flexibility practices. Econ. Ind. Democr. **38**(2), 271 (2017). https://doi.org/10.1177/0143831X15572836
75. Wang, K.-L., et al.: Combining the biennial Malmquist-Luenberger index and panel quantile regression to analyze the green total factor productivity of the industrial sector in China. The Sci. Total Environment **739**, 140280 (2020). https://doi.org/10.1016/j.scitotenv.2020.140280
76. Woo, C., et al.: Impact of green innovation on labor productivity and its determinants: an analysis of the Korean manufacturing industry: impact of green innovation on labor productivity and its determinants. Bus. Strategy Environ. **23**(8), 567–576 (2014). https://doi.org/10.1002/bse

Digital Transformation for Non-communicable Disease Prevention and Control in Thailand by Using a Mixed Methods Approach

Pradorn Sureephong[1], Anothai Laothiang[2], Kannikar Inthawong[3], Saengrung Satprasoet[1](✉), and Noppanan Auttajinda[1]

[1] College of Arts, Media and Technology, Chiang Mai University, Chiang Mai, Thailand
[2] Chiang Mai Provincial Public Health Office, Chiang Mai, Thailand
[3] Faculty of Public Health, Chiang Mai University, Chiang Mai, Thailand

Abstract. Thailand is experiencing a rapid increase in its aging population, leading to significant healthcare challenges, particularly in managing non-communicable diseases (NCDs) among the elderly. By 2035, over 20% of the population will be aged 60 or older, increasing the burden on an already strained healthcare system. This study investigates the implementation of various NCD monitoring programs in community hospitals in Chiang Mai Province, focusing on their features and effectiveness. The research framework includes four steps: Review and Observation, Content Analysis to Features, Recheck and Edit, and Discussion and Conclusion. The analysis identified that the NCD-x program and the Diabetes Mellitus Remission Program (DMRP) are the only programs incorporating all essential monitoring features. The NCD-x program, supported by the Thai Health Promotion Foundation, aligns with WHO recommendations and has been effectively implemented with extensive training. The DMRP, developed locally, focuses on personalized patient monitoring through direct consultations and the Line application but poses a significant workload on healthcare staff. In contrast, the Health Charter program, while inclusive and community-driven, has the fewest features, limiting its effectiveness in specialized NCD management. Key features like Behavioral modification, Daily Health monitoring, Symptom and Complication tracking, Appointment Tracking, and Health Team Communication are critical across most programs. However, features like Medication management and Alerts and Health Information are less common due to the need for physician oversight and timely health monitoring. Effective NCD management requires adapting programs to local contexts and patient needs, ensuring comprehensive and efficient healthcare delivery.

Keywords: Non-Communicable Diseases · Long Term Care · Community Hospitals · Monitoring programs

Published by Springer Nature Switzerland AG 2025
P. Sureephong et al. (Eds.): PLM 2024, IFIP AICT 741, pp. 253–264, 2025.
https://doi.org/10.1007/978-3-031-93323-3_19

1 Introduction

Thailand is facing a rapidly aging population, with the proportion of elderly people continuously increasing. It is projected that by 2035, [1] Thailand will become a completely aged society, with over 20% of the population being 60 years or older. This increase in the elderly population poses inevitable challenges in healthcare, as the elderly are at a higher risk of developing non-communicable diseases(NCDs), which can lead to severe complications and increased mortality rates if not properly managed. The Thai public health system has been struggling to provide adequate care for elderly patients with chronic diseases due to a shortage of medical personnel. The current ratio of doctors to population in Thailand is 1:1,843 [2], which is below international standards, and the distribution of medical personnel is uneven. Some areas, such as Bangkok, have a high doctor-to-population ratio of 1 to 630, while other areas face severe doctor shortages. Moreover, treating elderly patients with chronic diseases is an increasing financial burden for the Ministry of Public Health, and this budget is expected to rise further in line with the growing proportion of the elderly population.

Chiang Mai is one of the provinces in Thailand with a high proportion of elderly people, accounting for 15.4% of the total population in 2019 [3]. The province also has a high prevalence of NCDs among its population, especially diabetes and hypertension, which are the leading causes of death and disability [4]. Therefore, researching the use of monitoring programs for patients with NCDs in Chiang Mai can provide valuable insights into the challenges and opportunities of managing chronic diseases in an aging society. Furthermore, Chiang Mai has a well-established network of community hospitals that serve as the primary healthcare providers for the rural and urban population. These hospitals have implemented various monitoring programs for patients with NCDs, ranging from self-care education to telemedicine, depending on the availability of resources and the needs of the patients. Studying these programs can help identify the best practices and the areas for improvement in delivering effective and efficient care for elderly patients with chronic diseases.

The purpose of this research is to investigate how programs that monitor the outcomes of patients with chronic diseases are used and to contrast the various functionalities of each program at community hospitals in Chiang Mai province. To achieve the research objective, this study will use a mixed-methods approach, combining quantitative and qualitative data collection and analysis. The quantitative data will be obtained from secondary sources. The quantitative data will be used to describe the characteristics of the monitoring programs, such as the number of patients enrolled, the frequency and duration of the monitoring, the types of devices and platforms used, and the outcomes measured. The qualitative data will be collected through semi-structured interviews with key stakeholders, such as nurses and caregivers. The qualitative data will be used to explore the experiences, perceptions, and challenges of using the monitoring programs, as well as the benefits and limitations of each program. The mixed-methods approach will allow for a comprehensive and in-depth understanding of the monitoring programs for patients with non-communicable diseases at community hospitals in Chiang Mai province.

2 Literature Review

Thailand has experienced rapid social, cultural, and environmental changes in recent decades, leading to an increased burden of non-communicable diseases (NCDs) [3]. Urbanization, globalization, and an aging population are key factors driving the rise in NCDs [4]. Evidence indicates that urbanization is associated with higher mortality from NCDs [5]. Social and environmental changes in Thailand have led to behavioral changes, such as increased consumption of high-fat foods and lack of exercise [6]. These behaviors contribute to physiological risk factors like obesity and high blood pressure [6]. Economic, social, and cultural shifts can increase the risk of developing NCDs, but higher income levels associated with these changes may improve access to healthcare and healthier lifestyle choices, potentially reducing NCD risk [7]. Chronic NCDs are now among the top 20 causes of morbidity with high complications in developed countries [6]. The prevalence of NCDs increases with age: 40% of those over 44, 50% of those aged 65–74, and 70% of those over 85 experience chronic conditions [9]. NCDs present significant challenges globally, especially in managing and monitoring chronic conditions. Effective surveillance systems are crucial for managing NCDs, particularly in aging populations. This literature review examines various features of monitoring systems designed for NCDs, including daily health monitoring, medication management, behavioral tracking, symptom and complication tracking, appointment tracking, health team communication, health data analysis, and alerts.

Digital technology has enhanced the monitoring of NCDs patients, reducing illness incidence, and research of Murthy et al. (2023) [10] found that digital health (DH) technologies increased the efficiency of NCD monitoring by 40%, especially during the COVID-19 pandemic. Digital tools have improved health systems and patient care for NCDs, reducing complications and form the research of Geldsetzer et al. (2022) [12] explain to explored mobile health (mHealth) applications in Low- and Middle-Income Countries (LMICs), which support healthcare and manage disease conditions in communities. These applications have been effective for screening, diagnosing, and monitoring NCDs, demonstrating significant potential for improving health systems and patient care for the elderly.

Research highlights the global burden of NCDs, particularly in LMICs. In Chiang Mai, Thailand, various surveillance systems have been implemented to manage NCDs effectively, focusing on their impact and effectiveness in improving NCDs management, for example:

1. Monitoring Program in Thailand
   1) NCD-x Program

      The NCD-x program at Hospital to the increasing trend of noncommunicable diseases driven by lifestyle behaviors and improved access to healthcare services. The program operates by preparing medical staff and targeting patients, organizing activities, and evaluating outcomes to assess effectiveness. Results indicate that the program has been successful, but challenges such as insufficient staff and budget constraints persist. Plans are in place to expand the program into community areas to enhance efficiency and reach in the future.
   2) Diabetes School

The Diabetes Education School noncommunicable diseases by educating and managing patients with high HbA1C levels. The program, which has expanded due to increased patient referrals from larger hospitals to alleviate overcrowding, involves planning appropriate diets, exercise, and emotional management. It also includes monitoring by medical professionals and community health workers (CHWs). The outcomes of this program include reduced HbA1C levels and decreased cardiovascular risk. Despite challenges with using IT equipment, which can be a barrier for the elderly, this program has successfully educated patients and their families, leading to behavioral changes. There are plans to continue and expand the program further.

3) Diabetes self management education program

The Diabetes Self-Management Education (DSME) program is designed to combat the increasing prevalence of chronic noncommunicable diseases, especially diabetes, due to lifestyle choices affecting new patients within specific regions. The program focuses on monitoring blood sugar levels and promoting self-care across seven aspects including diet, exercise, medication, stress management, and complications of diabetes. It involves data collection, targeted screenings, DSME training sessions, biweekly tracking of behaviors and glucose levels, and reassessment using Hb1 C tests. Although definitive results of patient improvement are yet pending, there is a positive trend. The program faces challenges with its diverse instructional models, which complicate behavioral modifications and self-monitoring of glucose levels. An evaluation suggests that the program should adapt and possibly integrate other models to better suit its context.

4) Health Charter

The Health Charter program addresses the rise in noncommunicable diseases by improving community involvement in health issues. This escalation in diseases is attributed to better screening and planning, increased convenience in lifestyles, and a lack of health awareness and behavioral change among the public. The program operates through a process where community members actively participate in identifying health problems and drafting a health charter. This includes analyzing local health issues, providing feedback to the community, drafting solutions, and formalizing these commitments through MOUs. The outcome has been positive with the community adhering to the agreed health practices. However, the program faces challenges in implementation due to limitations in its scope and execution. The strengths of the program lie in its community-driven approach, which ensures active participation and sustainable outcomes. It is recommended to continue and expand the program, adapting it to different contexts as needed.

5) Diabetes Mellitus Motivation Interview program

The Motivational Interviewing (MI) program targets diabetes management through behavior modification, aiming to lower HbA1 c levels by engaging patients in selfmotivated change. Despite challenges such as operational delays, inconsistent patient attendance, communication barriers, and staffing shortages, the program has successfully empowered patients to make informed health decisions, achieving significant reductions in HbA1c levels for 80% of participants. This approach has proven effective, particularly for educated patients who understand their health needs but struggle with lifestyle adjustments, demonstrating the program's adaptability and cost-effectiveness.

6) Hypertension Control program

The HT Control program focuses on improving hypertension management through community screenings and systematic patient follow-ups. By setting criteria for immediate treatment for BP levels above 140/90, the program facilitates early detection and continuous monitoring, including lab check-ups and home blood pressure monitoring (HBPM) to track patients who miss follow-ups. Despite challenges such as delayed screening initiation, insufficient staff for screenings, and data management issues, the program has increased patient access to treatment and enhanced control over hypertension. Overall, it has proven to be beneficial by making healthcare access easier and treatment protocols more manageable, supporting the continuation and expansion of the program to cover more areas.

7) Diabetes Mellitus remission program

The Diabetes Mellitus remission program aims to address the rising incidence of this chronic disease by implementing a comprehensive management and education strategy focused on dietary changes, such as Low Carb and Intermittent Fasting (IF) diets. The program includes educating participants, selecting appropriate dietary methods, and monitoring results through body weight, body measurements, and fasting blood sugar levels. It also involves multidisciplinary follow-ups every three months, group food ordering via online platforms, and regular telephone check-ups. Although the program has a structured and flexible approach that allows for clear outcome measurement and adaptation to local contexts, it requires significant discipline and commitment from participants, making it unsuitable for all patients. This limitation highlights the need for stringent participation rules, which may not align with every patient's capacity or lifestyle.

2. Guideline NCDs monitoring from WHO

1) Noncommunicable Disease Facility-Based Monitoring Guidance

This document focuses on managing noncommunicable diseases (NCDs) within healthcare facilities by establishing indicators and measures for effectively monitoring and evaluating the treatment of NCD patients. Features include daily health monitoring, medication management, behavioral health tracking, symptom and complication assessment, appointment tracking, health team communication, health data analysis, and alerts and health information. These elements enhance the health system's capacity to care for and prevent NCDs.[12]

2) Global NCDs Surveillance Monitoring Framework

The "Global NCDs Surveillance Monitoring Framework" document by the World Health Organization (WHO) provides a comprehensive framework for monitoring and managing noncommunicable diseases (NCDs). It sets clear targets and indicators for evaluating national progress in managing NCDs. These indicators cover daily health monitoring, medication management, behavioral monitoring, symptom and complication assessment, appointment tracking, health team communication, health data analysis, and alerts and health information dissemination. The framework emphasizes the use of data to support decision-making and improve patient care on a global scale.[13]

3) Noncommunicable Diseases Progress Monitor 2022

The "Noncommunicable Diseases Progress Monitor 2022" report by WHO focuses on tracking the progress of countries in managing NCDs using defined indicators. It evaluates national efforts to reduce NCD-related mortality and morbidity. The report highlights advancements in many countries but also identifies ongoing challenges, particularly in low- and middle-income countries. The indicators used for assessment include national target setting, policy and strategic plan development, media campaigns, drug therapy management, and mortality data reporting. The report underscores the need for continued efforts to meet global health goals.[14]

4) Assessing National Capacity for the Prevention and Control of Noncommunicable Diseases

The "Assessing National Capacity for the Prevention and Control of Noncommunicable Diseases" document by WHO assesses the capabilities of countries in preventing and controlling NCDs. It surveys the presence of policies and strategic plans, budget allocations, risk factor monitoring, and essential health services. The document indicates progress in developing national capacity but highlights disparities in access to healthcare services and medication management. It also assesses the impact of the COVID-19 pandemic on NCD efforts, emphasizing the need to strengthen health systems to better respond to future pandemics.[15].

From the study of the guidelines for monitoring noncommunicable diseases (NCDs) by the World Health Organization (WHO) from the four documents, the key monitoring features can be summarized as follows.

1. Daily Health Monitoring: Regular collection of patient health data, such as blood pressure measurements and blood glucose levels, to assess health status and improve treatment.
2. Medication Management: Monitoring and provision of essential medications for treating NCDs, such as drugs for hypertension, diabetes, and cardiovascular diseases.
3. Behavioral Monitoring: Tracking and promoting healthy behaviors, such as smoking cessation, reducing alcohol consumption, encouraging physical activity, and maintaining a healthy diet.
4. Symptom and Complication Assessment: Monitoring and evaluating patient symptoms, as well as checking for complications that may arise, such as chronic kidney disease in diabetic patients.
5. Appointment Tracking: Managing and tracking patient appointments with healthcare providers to ensure continuous and effective care.
6. Health Team Communication: Promoting communication among healthcare team members, including doctors, nurses, and other health professionals, to coordinate and manage patient care effectively.
7. Health Data Analysis: Collecting and analyzing patient health data to inform treatment planning and improve decision-making processes.
8. Alerts and Health Information: Providing alerts regarding medication, appointments, and other health information to ensure that patients and caregivers can follow the treatment plan appropriately.

These features are crucial components that enhance the effectiveness of monitoring and managing noncommunicable diseases, providing a framework for strengthening health systems in various countries to sustainably manage NCDs.

3 Research Methodology

The aim of this study is to explore the features and benefits of integrated digital health solutions in managing non-communicable diseases (NCDs) in Chiang Mai, Thailand. The study adopts a mixed-methods approach, combining qualitative and quantitative data from various sources, such as literature review, observation, interviews, surveys, and health records. The study follows a three-step research framework, as shown in Fig. 1. Each step involves different data collection and analysis methods, which are described in detail in the following sections.

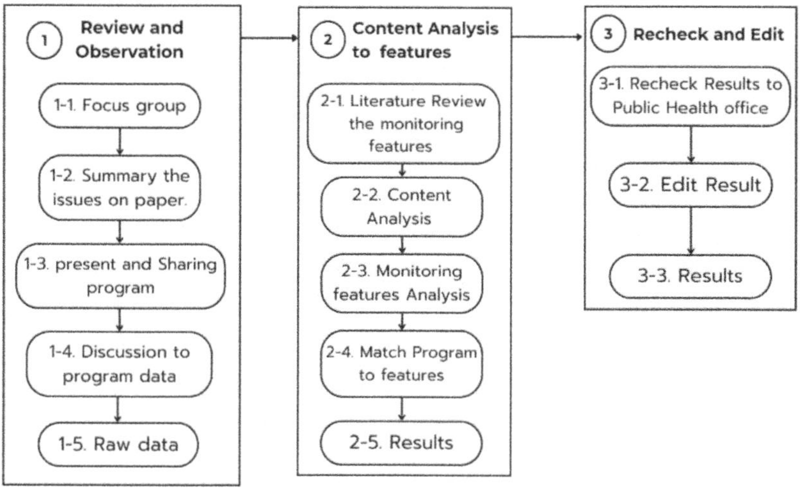

Fig. 1. Three-step research framework

Step 1: Review and Observation
1–1. Focus Group: Organize 7 hospital (Table 1) focus group discussions with healthcare providers and other stakeholders to gather qualitative insights on the current use and effectiveness of monitoring features.

1–2. Summary the Issues on Paper: Summarize the issues focus group discussions.
1–3. Present and Sharing Program.

1–4. Discussion to Program Data: Facilitate a discussion session to validate and refine the initial findings and gather additional feedback.

1–5. Raw Data Collection: Gather raw data from focus Group to monitoring features.

Step 2: Content Analysis to Features 2–1. Literature Review the Monitoring Features: Conduct a comprehensive review of existing literature on monitoring features relevant

Table 1. Programs chosen by hospitals for monitoring Program of NCDs.

No	Program	Hospital
1	NCD-x program	Hospital 1
2	Diabetes School	Hospital 2
3	Diabetes self – management education program	Hospital 3
4	Health Charter	Hospital 4
5	Diabetes Mellitus Motivation Interview program	Hospital 5
6	Hypertension Control program	Hospital 6
7	Diabetes Mellitus remission program	Hospital 7

to NCDs and Identify key components and best practices in monitoring NCDs in elderly populations.

2–2. Content Analysis: Perform content analysis on the raw data to identify common themes and patterns, and then utilize qualitative data analysis to monitoring features. 2–3. Monitoring Features Analysis: Analyze the specific features of different monitoring systems used for NCDs between raw data and research review shown in Table 2. 2–4. Match Program to features: Match the monitoring features from hospital.

2–5. Results: Analysis and comparison of features for monitoring features NCDs patients according to the implementation methods of each hospital.

Step 3: Recheck and Edit

1. Recheck Results to Public Health Office: Present the preliminary results to the Public Health Office for review and gather feedback and make necessary revisions to the results.
2. Edit Result: Edit the results based on the feedback received to ensure accuracy.
3. Result: Summary to results

4 Results

In this section, we present the results of our study on the features and methods used by different hospitals for monitoring NCD patients. We conducted a survey with 10 hospitals in Thailand and asked them about their current practices and challenges in managing NCD patients. We also collected data from their electronic health records (EHRs) and analyzed the types and frequencies of features that they use for monitoring NCD patients. We compared the results across the hospitals and identified the common and unique features and methods for each hospital. The main findings are summarized in Table 3.

From Table 2, in the section on monitoring features, it can be observed that the Behavioral feature is present in every program. Following this, the features Daily Health, Symptom and Complication, Appointment Tracking, and Health Team Communication

Table 2. Content Analysis to features from research review.

No	Reference	Monitoring Feature							
		Daily Health	Medication	Behavioral	Symptom and Complication	Appointment Tracking	Health Team Communication	Health Data Analysis	Alerts and Health Information
1	Ferguson et al. (2022)[16]	✓		✓	✓			✓	✓
2	Tuangratananon et al. (2021)[17]		✓		✓		✓		
3	Potempa et al. (2022)[18]		✓		✓			✓	
4	Tsai et al. (2020)[19]		✓		✓		✓	✓	
5	Peyroteo et al. (2021)[20]	✓	✓	✓	✓	✓	✓	✓	✓
6	Kostova et al. (2021)[21]							✓	
7	Ãgh et al. (2021)[22]		✓		✓		✓	✓	
8	Kabir et al. (2022)[23]		✓		✓		✓	✓	
9	Sheehan et al. (2021)[24]					✓	✓		
10	World Health Organization [12–15]	✓	✓	✓	✓	✓	✓	✓	✓

are included in 7 programs. Health Data Analysis is present in 5 programs. The features with the fewest programs implementing them are Medication and Alerts and Health Information.

To organize the monitoring features for patients with noncommunicable diseases (NCDs), the features are divided into 8 eatures. It can be seen that the NCD-x program and the Diabetes Mellitus Remission Program include all the features. Following this, the Diabetes School includes 7 features. The Hypertension Control Program has 6 features. The Diabetes Self-Management Education Program and the Diabetes Mellitus Motivation Interview Program each include 5 features. The program with the fewest features is the Health Charter.

From the analysis of the implementation of each program across all features, it is evident that the NCD-x program and the Diabetes Mellitus Remission Program (DMRP) cover all aspects comprehensively when compared to other programs. The NCD-x program, when applied to 280 patients, showed positive results in 190 cases, translating to a success rate of 67.86%. In contrast, the DMRP is still in the testing phase, with its effectiveness limited by the number of users.

5 Discussion

The analysis revealed that the NCD-x program and the Diabetes Mellitus Remission Program (DMRP) are the only programs that incorporate all essential monitoring features. The NCD-x program, supported by the Thai Health Promotion Foundation, serves

Table 3: Analysis and comparison of features for monitoring features NCDs patients

No	Reference	Monitoring Feature							
		Daily Health	Medication	Behavioral	Symptom and Complication	Appointment Tracking	Health Team Communication	Health Data Analysis	Alerts and Health Information
1	NCD-x program NCD-x program NCD-x program	✓		✓	✓			✓	✓
2	Diabetes School		✓		✓		✓		
3	Diabetes self – management education program		✓		✓			✓	
4	Health Charter		✓		✓		✓	✓	
5	Diabetes Mellitus Motivation Interview program	✓	✓	✓	✓	✓	✓	✓	✓
6	Hypertension Control program program							✓	
7	Diabetes Mellitus remission program		✓		✓		✓	✓	

as a pilot model for hospitals and aligns well with WHO recommendations and findings from Peyroteo et al. On the other hand, the DMRP, independently developed by local hospitals, focuses on individualized patient monitoring through direct consultations and the Line application. This personalized approach enhances the program's effectiveness but imposes a significant workload on healthcare staff. Conversely, the Health Charter program, a community-driven initiative, includes the fewest features, limiting its effectiveness in specialized NCD management despite its inclusive approach.

In analyzing the features, the Behavioral feature is present in all programs, as modifying patient behavior is crucial for disease control and symptom management. The Daily Health feature, which involves recording daily health data, is also commonly used, along with Symptom and Complication monitoring. These features are essential for detecting and managing complications. Appointment Tracking and Health Team Communication are critical for ensuring patients adhere to medical schedules and promoting patient education. However, features like Medication management and Alerts and Health Information are less commonly used due to the need for physician oversight and dependence on daily health monitoring.

In conclusion, while all monitoring features are essential for effective NCD management, the choice of features varies based on the program's local context, societal factors, and patient acceptance. Future program development should focus on integrating key features and addressing current limitations to enhance practical usability. Continuous patient care, facilitated through effective appointment tracking and follow-up mechanisms, will be crucial in ensuring the long-term success and sustainability of these NCD management programs.

6 Conclusion

In conclusion, the comprehensive analysis of various NCD management programs highlights the importance of integrating essential monitoring features to enhance their effectiveness. The NCD-x program and the Diabetes Mellitus Remission Program (DMRP) exemplify the benefits of a holistic approach, supported by extensive training and tailored patient monitoring, respectively. However, the limited features in community-driven programs like the Health Charter suggest a need for balance between inclu-sivity and specialized care. Behavioral modification emerges as a universally critical feature, underscoring the role of self-care in long-term disease management. Ulti-mately, the successful implementation of these features depends on their adaptability to local contexts and the specific needs of the patient population, ensuring that NCD management is both comprehensive and effective.

Acknowledgements. This research was supported by the DigitalHealth-Asia Project (https://digihealth-asia.eu/), under Grant Agreement No. 619193-EPP-1–2020-1-BE-EPPKA2-CBHEJP. We would like to extend our appreciation to the Chiang Mai Public Health Office for their assistance in facilitating the research site and data collection process. Furthermore, we are deeply grateful to the repre-sentatives from the seven participating hospitals for their cooperation, which was instrumental in the successful completion of this study.

References

1. Foundation of Thai Gerontology Research and Development Institute (TGRI): Thai elderly population ratio (2020). https://thaitgri.org/?p=38866. Accessed 15 Mar 2024
2. National Statistical Office and the Office of the Permanent Secretary, Ministry of Public Health (2020). https://www.thecoverage.info/news/content/2807. Accessed 15 Mar 2024
3. Dans, A., Ng, N., Varghese, C., Tai, E.S., Firestone, R., Bonita, R.: The rise of chronic non-communicable diseases in Southeast Asia: time for action. The Lancet **377**(9766), 680–689 (2011)
4. World Health Organization: Global Status Report on Noncommunicable Disease. World Health Organization, Geneva (2010)
5. Angkurawaranon, C., Wattanatchariya, N., Doyle, P., Nitsch, D.: Urbanization and non-communicable disease mortality in Thailand: an ecological correlation study. Trop. Med. Int. Health **18**(2), 130–140 (2013)
6. Yiengprugsawan, V., Caldwell, B.K., Lim, L.L.-Y., Seubsman, S.-A., Sleigh, A.C.: Lifecourse urbanization, social demography, and health outcomes among a national cohort of 71,516 adults in Thailand. Int. J. Popul. Res. **2011**, 464275 (2011)
7. Dye, C.: Health and urban living. Science **319**(5864), 766–769 (2008)

8. Dennis, S.: Secondary prevention of chronic health conditions in patients with multimorbidity: what can physiotherapists do? J. Comorb. (2016). https://doi.org/10.15256/joc.2016.6.82
9. Australian Institute of Health and Welfare: Chronic diseases (2015).http://www.aihw.gov.au/chronic-diseases/. Accessed 5 Apr 2016
10. Murthy, S., Kamath, P., Godinho, M.A., Gudi, N., Jacob, A., John, O.: Digital health innovations for non-communicable disease management during the COVID-19 pandemic: a rapid scoping review. BMJ Innov. **9**(1), 3 (2023). https://doi.org/10.1136/bmjinnov-2021-000903
11. Geldsetzer, P., Flores, S., Wang, G., et al.: A systematic review of healthcare provider-targeted mobile applications for non-communicable diseases in low-and middle-income countries. NPJ Dig. Med. **5**(1), 99 (2022). https://doi.org/10.1038/s41746-022-00644-3
12. World Health Organization: Noncommunicable Disease Facility-Based Monitoring Guidance Framework, Indicator and Application. WHO
13. World Health Organization: NCD Global Monitoring Framework. https://www.who.int/publications/i/item/ncd-surveillance-global-monitoringframework. Accessed 25 Oct 2023
14. World Health Organization: Noncommunicable Diseases Progress Monitor 2022. WHO (2022)
15. World Health Organization: Assessing National Capacity for the Prevention and Control of Noncommunicable Diseases: Report of the 2021 Global Survey. WHO (2021)
16. Ferguson, T., Olds, T., Curtis, R., et al.: Effectiveness of wearable activity trackers to increase physical activity and improve health: a systematic review of systematic reviews and meta-analyses. Lancet Dig. Health **4**(8), e615–e626 (2022). https://doi.org/10.1016/S2589-7500(22)00111-X
17. Tuangratananon, T., Julchoo, S., Phaiyarom, M., et al.: Healthcare providers' perspectives on integrating NCDs into primary healthcare in Thailand: a mixed method study. Health Res. Policy Syst. **19**(1), 139 (2021). https://doi.org/10.1186/s12961-021-00791-1
18. Potempa, K., Rajataramya, B., Singha-Dong, N., et al.: Thailand's challenges of achieving health equity in the era of non-communicable disease. Pac. Rim Int. J. Nurs. Res. **26**(2), 187–197 (2022)
19. Tsai, C.H., Eghdam, A., Davoody, N., et al.: Effects of electronic health record implementation and barriers to adoption and use: a scoping review and qualitative analysis of the content. Life (Basel) **10**(12), 327 (2020). https://doi.org/10.3390/life10120327
20. Peyroteo, M., Ferreira, I.A., Elvas, L.B., et al.: Remote monitoring systems for patients with chronic diseases in primary health care: systematic review. JMIR mHealth uHealth **9**(12), e28285 (2021). https://doi.org/10.2196/28285
21. Kostova, D., Richter, P., Van Vliet, G., et al.: The role of noncommunicable diseases in the pursuit of global health security. Health Secur. **19**(3), 288–301 (2021). https://doi.org/10.1089/hs.2020.0121
22. Ágh, T., van Boven, J.F., Wettermark, B., et al.: A cross-sectional survey on medication management practices for noncommunicable diseases in Europe during the second wave of the COVID-19 pandemic. Front. Pharmacol. **12**, 685696 (2021). https://doi.org/10.3389/fphar.2021.685696
23. Kabir, A., Karim, M.N., Islam, R.M., et al.: Health system readiness for non-communicable diseases at the primary care level: a systematic review. BMJ Open **12**(2), e060387 (2022). https://doi.org/10.1136/bmjopen-2021-060387
24. Sheehan, J., Laver, K., Bhopti, A., et al.: Methods and effectiveness of communication between hospital allied health and primary care practitioners: a systematic narrative review. J. Multidiscip. Healthc. **14**, 493–511 (2021). https://doi.org/10.2147/JMDH.S295549

From Strategy to Action: A Methodological Framework for Digital Transformation in Manufacturing SMEs

Marco Spaltini(✉), Federica Acerbi, and Marco Taisch

Department of Management, Economics and Industrial Engineering, Politecnico di Milano, via Lambruschini 4/b, 20156 Milan, Italy
{marco.spaltini,federica.acerbi,marco.taisch}@polimi.it

Abstract. The manufacturing industry is characterized by challenges including global competitiveness, unpredictable market evolutions, lack of skills, and supply chain disruptions. To gain competitiveness, manufacturing Small-Medium Enterprises (SMEs) shall pursue Digital Transformation (DT) which enables significant benefits such as operational excellence and long-term business sustainability. However, SMEs usually struggle to successfully implement DT due to several impediments like lack of consistency between business strategy and digital plans, internal cultural resistance, and limited resources. Thus, the present research aims to develop a roadmapping methodology that supports manufacturing SMEs to define and implement a structured DT plan. The authors first investigate the SMEs-tailored existing tools, their role, and their limits through a systematic literature review: detailing existing Maturity Models (MMs) and digital roadmapping approaches and evaluating how they intertwined. Today, digital roadmapping methodologies neglect some SME's barriers, appearing as too generic and more adequate for large companies. They do not provide pragmatic support to ensure the strategic alignment between digital initiatives and corporate goals, as well as they do not offer practical guidelines to select and prioritize actions. To fill this gap, the authors conjectured a roadmapping methodology tailored on SMEs' challenges. Three maps are elaborated to determine market, business, and strategic targets, and to assess their consistency with the identified digital initiatives. Furthermore, a methodological approach is provided to support SMEs during the evaluation of effort and benefits. The methodology was validated through application cases involving Italian manufacturing SMEs.

Keywords: Digital Transformation · Maturity Model · Roadmapping · Plan · SME · Strategy

1 Introduction

Manufacturing companies operate in a very dynamic environment, facing multiple challenges including economic uncertainty, global competitiveness, supply chain disruptions, and competences shortages [1]. Furthermore, evolutions of the market are difficult to predict and new trends such as product customization, servitization, and ESG sustainability

Published by Springer Nature Switzerland AG 2025
P. Sureephong et al. (Eds.): PLM 2024, IFIP AICT 741, pp. 265–279, 2025.
https://doi.org/10.1007/978-3-031-93323-3_20

affect the way in which organizations do business and capture value [1]. To successfully compete in this field, companies cannot neglect the importance of digital transformation, also called Industry 4.0 (I4.0). It is mainly described as the integration of technologies into production and operations, enabling visibility and communication among products, machines, and partners along the whole value chain [2]. DT generates great benefits for manufacturing enterprises, enhancing competitiveness as well as sustainability. It guarantees operational excellence, by shortening lead time, reducing complexity, minimizing costs, and increasing flexibility, quality, and effectiveness on the market. Nevertheless, I4.0 maturity and implementation by companies are still quite low, especially for SMEs [3]. In fact, SMEs usually lack consistency between business strategy and digital plan, resulting in unsuccessful digital transformation roadmap. Furthermore, SMEs reportedly have limited resources, both in terms of digital skills as well as money and time available for innovation [3]. Moreover, SMEs find more difficulties to evaluate benefits, risks, and effort's implications of I4.0 actions [3]. Recent concern for sustainability also made explicit how companies struggle to exploit I4.0 to foster environmental sustainability goals [4]. All these barriers obstacle them to successfully embracing DT and clarify the need for a SMEs-tailored and supporting tool. Thus, the author investigates the existing digital approaches with the goal of acquiring deep knowledge about the topic and elaborating a concrete methodology that allows SMEs to successfully overcome the unfaced challenges.

2 Theoretical Background

Literature describes many benefits and opportunities for companies and industries which undertake a digital transformation path. The appropriate use of I4.0 technologies enables huge innovation and competitive growth, enhancing at the same time the enterprise's sustainability [5]. Great potential of I4.0 emerges in manufacturing as well as in economics and society [6].

According to literature, DT enhances operational improvements and excellence [7]. Companies' operativity improves also through the continuous monitoring of production processes, which is supported by I4.0. This allows firms to reduce downtime and set up time, thanks to the anticipation of production issues, such as machines failures [7]. Moreover, [8] highlights the fact that the I4.0 technologies push industrial developments and allow companies to reduce complexity of the operations, simultaneously improving efficiency and effectiveness with a long-term cost reduction target. Also, [9] mention that digital technologies support decision making process through gathering and analysis of relevant information, enhancing operational efficiency, lowering costs, and raising profitability.

In this context, IoT and cloud computing enable full sharing, free circulation, on-demand use, and optimal configuration of information needed to the company, while big data analytics extract valuable insights from large amounts of data. In addition, DT is associated with highly flexible mass production, creation of new products, services and business models and optimization of value chains [10]. A more detailed list of opportunities enabled by DTs is provided below (Table 1).

Table 1. Overview of the main benefits of digital transformation

Domain	Opportunity	Reference
Business sustainability	• Long-term business sustainability and competitive advantage	[5, 9]
Innovation	• Better capabilities to innovate	[8, 10, 11]
Efficiency	• Costs reduction • Shortening of time • Decrease of complexity	[7, 12]
Effectiveness	• Improvement of quality conformance • Higher customization of the offer	[13, 14]
Flexibility	• Greater responsiveness and agility	[10, 13]
Integration	• Higher integration, transparency and collaboration	[14, 15]
Supply Chain	• Greater integration and visibility	[11, 14, 15], [16, 17]
	• Higher responsiveness and flexibility	
	• Collaboration and co-innovation	
	• Shortening time	
Environment	• Improve on environmental impact and green performance • Green sustainability	[18, 19]
Social	• Greater human safety and better work environment	[8, 11, 13]

3 Research Methodology

The present study's research objective is to "*develop a methodology to support manufacturing SMEs to define a feasible and valuable digital transformation path by concretely overcoming their I4.0 challenges and obstacles*". This objective is divided into two sub-objectives and is tackled by answering two key research questions.

- RQ1: *Which are the main tools that support Industry 4.0 transformation of manufacturing SMEs and how they intertwined?*
- RQ2: *Which methodology can be developed to concretely support SMEs in defining a feasible and valuable digital transformation path by overcoming their existing challenges and barriers?*

To answer these questions, a systematic literature review was adopted, led by the following query: ("Tool*" OR "Framework*" OR "Guideline*" OR "Methodology*" OR "roadmap*" OR "plan*") AND ("Industry 4.0" OR "Digital Transformation") AND ("SME*" OR "Small and Medium Enterprises").

Concerning the database selection, the authors relied on Scopus database as it offers a wider range of results than other databases yet offer peer-reviewed contributions. Some eligibility criteria have been posed. Only English conference and contributions since 2011 have been considered.

The authors have performed two screening activities. At first, the abstract and the conclusion were studied for each paper, while the complete reading characterised the second screening. Finally, papers coming from queries have been complemented with additional ones, identified through a "snowballing approach" thus enlarging the number of relevant publications included. Overall, 76 papers have been selected although a smaller, yet representative, sample has been cited for the sake of length limitations (see Table 2).

Table 2. Number of papers analysed through the systematic review implementation

Keywords	Initial	Filtered	First Screening	Second Screening	Snowball	Total
("Tool*" OR "Framework*" OR "Guideline*" OR "Methodolog*" OR "roadmap*" OR "plan*") AND ("Industry 4.0" OR "Digital Transformation") AND ("SME*" OR "Small and Medium Enterprises")	909	849	378	58	18	76

4 Literature Review Results

Organizations, especially SMEs, face numerous challenges to implement digital transformation; failing to deal properly with these challenges can lead to an Industry 4.0 path that slows down and takes longer than expected, reducing benefits, and eventually bringing economic losses. Strategic alignment between digitalization initiatives and business goals is key [20] and successful DT can be achieved when companies make long term plans, starting from their current situation [21]. To support manufacturing companies, a proliferation of dedicated tools and methodologies appears. [3] highlighted the presence of maturity and readiness models, framework and guidelines, and roadmapping methodologies. Unlike their structural differences, large companies as well as SMEs usually need to refer to similar tools [22]. From the analysis of literature emerged that 2 main research streams are well reported and address the topic (supporting SMEs in achieving Digital Transformation (DT). On the one hand, literature proposed multiple maturity Models (MMs) which aim at depicting the status quo of a firm and comparing it with the state of the art. On the other hand, literature has also investigated roadmapping methodologies which extend the boundaries of MMs to provide a structured plan to achieve

DT. Therefore, the remainder of this chapter is structured to present methodologies and frameworks belonging to these 2 macro-domains.

4.1 Maturity Models

The measurement of **digital maturity** is a necessary activity for companies which want to innovate and digitalize, bringing value to process design and target definition. I4.0 maturity assessment and comprehension represent the first step in planning and developing a successful digitalization strategy [23]. Digital maturity consists of two main elements: i) investments in technology to improve how the enterprise operates, i.e. digital intensity, ii) and development of capabilities needed to drive the DT [24]. To quantitatively measure it, digital Maturity Models (MMs) have been developed over the years [24].

MMs are well known in academia and present some key common characteristics. They are defined by a sequence of Maturity Levels, they delineate a development path, and they are characterized by **multiple stages**, ordered sequentially [25, 26]. Most of MMs in literature are characterized by 5 stages and different **dimensions of analysis**, which represent the areas assessed [27]. As DT doesn't just encompass technical factors, MMs shall enable a holistic view to consider also organizational and cultural changes [28]. In this direction, [27] underlines the following dimensions: (i) Strategy and Management, (ii) Corporate Culture, (iii) Organization and Processes, (iv) Products and Services, (v) Employees, (vi) Technology, (vii) Data, (viii) Customer, and (ix) Corporate Environment. Similarly, [29] highlighted 9 dimensions: (1) Strategy, (2) Leadership, (3) Customers, (4) Products, (5) Operations, (6) Culture, (7) People, (8) Governance, and (9) Technology. [24] included also data security; while [30] investigated also, risk management.

Despite the dimensions analysed, MM also varies in terms of processes assessed. In this direction [31] mapped the main MMs addressing manufacturing environment from a DT standpoint. The results of this mapping are shown in Table 3.

4.2 Digital and Twin Transformation Roadmapping

Starting from the assessment of the current digital maturity degree through a **maturity model**, several measures aiming to raise this level can be identified and a roadmap can be developed [32]. A roadmap is the outcome of a structured and broader approach called **roadmapping**, which includes the use of a maturity model and helps technological investment decisions through the identification of critical solutions and gaps [33]. Narrowing the scope, DT Roadmapping consists of steps for the elaboration of a DT roadmap, and it can be defined as a method finalized to identify, delineate and map technological strategies and innovation-based actions [2]. Its role is increasingly valuable, and it has been employed with higher frequency to support **strategic planning of digital innovation** by involving also managerial and organizational aspects of technology [21]. A roadmap appears as a set of consecutive steps to perform to reach a new digitalization degree [31] for which understanding needs and AS-IS situation is essential. With time, roadmapping methodologies evolved intending to support companies towards a holistic realization strategy which includes timelines and cost-benefit analysis [34]. As a matter of fact, from a graphical perspective, roadmaps are usually timebased represented,

Table 3. Most documented MM focused on manufacturing

Model	Processes and dimensions of Analysis	
DREAMY 4.0 Suite	§	**8 processes:** design & engineering, production, maintenance,
		quality, logistics, supply chain, data, people
	§	**4 dimensions:** Execution, Monitoring & Control (M&C), Tech-
		nology, Organization
Test I4.0	§	**8 processes + products + strategy**;
	§	**4 dimensions**: Plan, M&C, Technology, Organization
I4.0 readiness check	§	**6 dimensions:** Strategy and Organization, Smart factory, Smart operations, Smart products, Datadriven services, Employees
I4.0 maturity in-	§	**4 structural areas**: Resources, IT, Culture, Organizaion;
dex	§	**5 Functional areas**: development, Production, Logistics, Services, Marketing & Sales
I4.0 Readiness self-assessment	§	**5 dimensions:** Technology, Manufacturing, Economy & Market, Organization, Society & Culture
BEinCPPS I4.0 Maturity Method	§	**6 dimensions**: General, Strategy, People, Processes, Technologies, Integration
FAR-EDGE MM	§	**3 dimensions:** Technological, Operational, Human
6Ps MM	§	**6 dimensions:** Product, Process, Platform, People, Partnership, Performance

involving a series of layers that typically include business and technology perspectives [2].

As a set of sequential activities, DT roadmapping is characterized also by a number of mandatory macro-phases that must be followed consecutively. In this direction, regardless the many proposals available in literature, 5 main phases can be identified [23]:

1. *Goals and scope definition*: defining goals and vision to accomplish through DT. Strategy and environment are considered with a cross-functional mindset.
2. *As-Is digital maturity assessment*: conducting a digital MM aligned with the scope
3. *Target digital maturity assessment*: drawing the future state by combining business goals, market trends, industry benchmarks, and digital potentials.
4. *Initiatives identification, selection, and prioritization*: analysing the gap between As-Is and Target scenarios and identifying the initiatives to fill this gap. It determines

the actions to perform and prioritises them according to parameters including cost, benefits, and interdependency.
5. *Roadmap generation, implementation, and iteration of the framework*: placing the initiatives along a timeline on a graphical roadmap. The roadmap starts being implemented and periodical monitoring allows the company to ensure it is on track.

5 Literature Gap

Literature emphasizes that several topics are still not deeply explored in the analyzed field, especially DT roadmapping for manufacturing SMEs. This gap contributes to their low level of I4.0 implementation. The authors distinguish three main limits [6, 8]:

- DT roadmapping methodologies are too generic and not fully applicable by SMEs. They are more adequate for large companies and rarely empirically developed and validated with SMEs.
- DT roadmapping methodologies do not suggest practical steps to ensure the strategic consistency of the digital initiatives; this results in roadmaps misaligned with SMEs' objectives, and difficulties to accomplish the strategic vision.
- DT roadmapping methodologies rarely help SMEs to define *what* and *how* to do during the selection and prioritization phases; the importance of evaluating costs, risks, and benefits is highlighted, but a methodological support is not provided.

The main issue related to these three limits is the conflict with the peculiarities of the SMEs. SMEs (i) require tangible outcomes, even over perfection, (ii) lack suitable digital objectives aligned with business vision and I4.0 initiatives, and (iii) present difficulties in the evaluation of the parameters typical of the prioritization stage [6, 8]. Therefore, on one hand, existing tools are often unsuccessful in delivering actionable insights to SMEs, on the other hand, they are hardly able to face these phases and overcome the related obstacles autonomously. Thus, the authors developed a DT roadmapping methodology tailored to SMEs' peculiarities and unmet challenges.

6 Development of the Model

The roots of the conjectured digital roadmapping methodology consists of the five steps. However, innovative contributions are introduced and described step by step in the next sections.

6.1 Goals and Scope Definition

SMEs involve cross-functional people to define direction, targets, and scope of the DT path and, as reportedly documented, even a Sustainability Transition enabled by I4.0, namely Twin Transition [33]. To support this strategic vision, often inconsistent and sometimes even neglected, the authors elaborated three practical and methodological maps: (1) Market Perspective map (2) Business Target map (3) Digital Vision map.

The Market Perspective map helps SMEs in determining the market priorities, discussing also how they link with the operational levers of cost, time, and quality within the organization. The market preferences are translated into market drivers, which are classified in five macro-categories: (1) bread of the offer, in terms of variety and uniqueness provided to the customer, (2) value, considered as the relation between the perception about the specifications and the price of selling, (3) availability, representing the need of customers to have the solution available in the right moment and with the right time, (4) after sales service, which addresses requirements of post-sales, including re-calls and issues management, and (5) Sustainability as it has become a qualifier in most of industries and an increasingly source of competitive advantage (Fig. 1).

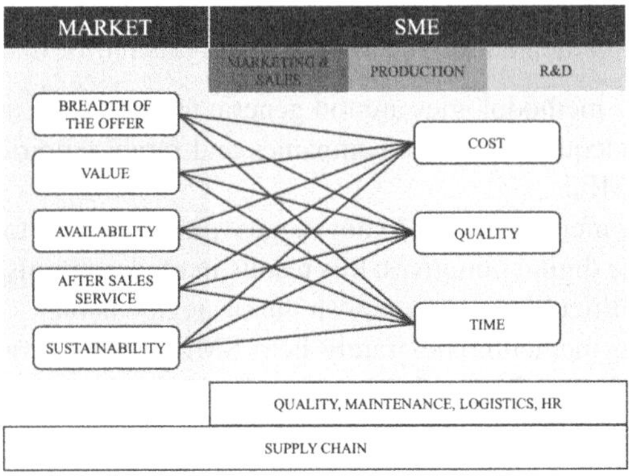

Fig. 1. Market Perspective Map

At this stage, the main aim of this map is to point out the key market priorities to guide SMEs in the next phases and avoid neglecting them during the identification of the initiatives to implement.

Business Target map supports SMEs in considering the corporate goals along the whole digital roadmapping process. At this step, SMEs need to define their business target, which is broken down into four main categories to practically simplify the discussion within the companies: (1) revenue growth, (2) costs decrease, (3) asset and structure management, and (4) long-term business sustainability (Fig. 2).

At this stage, the goal of this view is to make clear where business strategies focus and where they will focus in the future, grasping the strategic nature of digital transformation.

Digital Vision map helps SMEs to clarify the digital direction they want to pursue. This map consists of three main levels the organization must analyze: aim, type (i.e. incremental or radical), and focus (i.e. product, process, approach to market) of the digital transition (Fig. 3).

The framework moves from SME's market to its business strategy and digital vision, aiming to create a linkage along the whole company and stimulate a holistic vision, crucial to delineating a successful digital transformation roadmap.

6.2 As-Is Digital Maturity Assessment

The assessment of the As-Is digital maturity requires two main activities: (i) selection of the digital maturity model, and (ii) compilation of the chosen tool. Both these activities

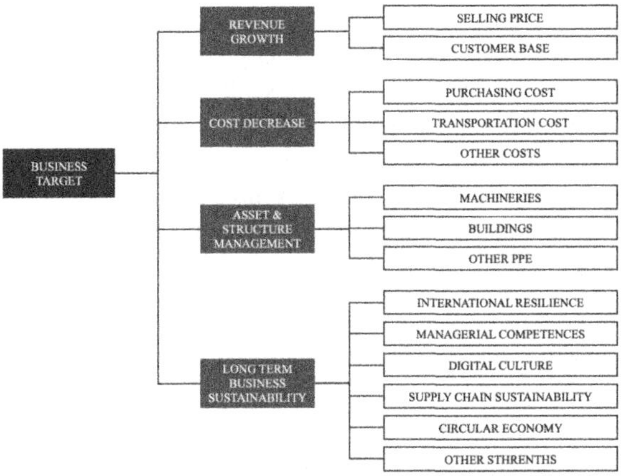

Fig. 2. Business Target Map

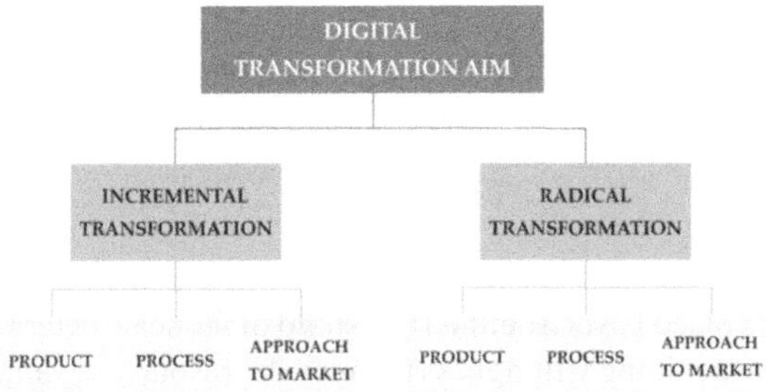

Fig. 3. Digital Vision map

can benefit from innovation introduced by the author in the stage before, since the previously defined goals and scope help the SME to select a digital MM consistent with its purposes. Then, this chosen MM is used to analyze the digital current status.

6.3 Target Digital Maturity Assessment

Once evaluated the As-Is digital readiness, the SME must assess the digital target, delineating the desired To-Be picture. To do that, the previously selected MM is employed. In this case the Market Perspective, the Business Target, and the Digital Vision maps introduced in the first stage can support the company to delineate a feasible and consistent Target picture.

6.4 Initiatives Identification, Selection, and Prioritization

To fill the gap between current and desired digital levels, several initiatives are identified. Thus, assessments' differences generate a list of actions. The SMEs must evaluate the alignment of these digital actions with market requirements, business goals, and digital vision, with the aim to select and prioritize the most consistent ones. The usage of the Market Perspective, Business Target, and Digital Vision maps concretely supports SMEs to perform this activity. A five-levels scale is used to quantify the alignment degree of each identified digital initiative with SMEs' market drivers, business targets, and digital vision (calculated as the average of alignments with digital aim, transition type, and transition focus). To guide SMEs in this phase, several supporting questions are provided (e.g. How much consistency does the digital initiative have with the key market drivers?) (Table 4).

Table 4. Alignment between the identified digital initiatives and the strategy of the SME

List of identified initiatives	Alignment with market priorities	Alignment with business target	Alignment with digital aim	Alignment with SME's strategic direction
In.1	5,00	4,00	4,00	4,33
In.2	3,00	2,00	4,33	3,11
In.3	4,00	5,00	2,33	3,78
In.4	1,00	1,00	1,67	1,23
In.5	5,00	4,00	4,00	4,33

Then, initiatives placed over a defined threshold of the scale determined by each SME through parameters including willingness to innovate, resources availability, and number of actions are selected. Therefore, once reached this step, the initial list of identified initiatives has been shortened. Now, SMEs need to prioritize the selected actions with the purpose of distributing them within the desired time horizon. To perform this activity, the author suggests considering benefits and broadening the common concept of cost to the broader concept of effort. Both efforts and benefits are faced at a high level in the model, since the focus is providing a structured and methodological approach rather than a punctual estimation of the economic impacts.

Regarding the efforts, four main parameters are presented:

1. Complexity: it exploits the Acatech I4.0 path development to calculate the number of leaps between the actual digital readiness of a process and the degree necessary to implement a single action over that process [52]. Then, the number of leaps is translated into a five-levels scale. Moreover, the interdependency between distinct initiatives is emphasized. Complexity analysis guarantees logical prioritization.
2. Costs: multiple typologies of expenses, including investments, scouting, integration, implementation, and updating costs, should be considered to evaluate the economic effort.
3. Duration: temporal constraints should be analyzed in terms of planning, development, integration, re-organization, etc. with the goal of identifying the most time-consuming actions.

4. Internal resources: it relates to the availability in terms of number and competences of the people needed to implement a digital initiative. The more the resources required and the more advanced their skills, the higher the effort.

The four values representatives of each group are averaged to estimate a high-level effort for each initiative.

Regarding the benefits, the model distinguishes six main categories, which are: (1) flexibility, (2) effectiveness, (3) process efficiency, (4) integration, (5) digital culture, and (6) business sustainability. To concretely support and guide SMEs during the assessment and the monitoring of the advantages brought by each distinct initiative, multiple KPIs for each class of benefits are suggested (e.g. time efficiency, OEE, etc.). The benefits of each initiative are calculated by combining the results of the six categories in a five-levels scale.

To conclude, digital initiatives are depicted within a graph that relates efforts and benefits (Fig. 4).

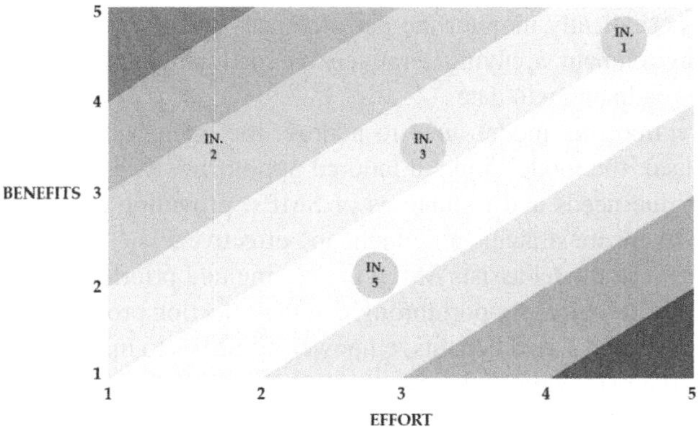

Fig. 4. Benefits-Effort graphical view

The prioritization of the initiatives happens in three further steps:

- Moving from top left to bottom right, select the actions belonging to the same bar.
- Prioritize the selected actions belonging to the same bar according to the highest level of alignment.
- Check the presence of interdependency relationships and, eventually, adjust priorities.

In this way, a right balance between low and high effort and benefits activities is guaranteed.

6.5 Roadmap Generation, Implementation, and Iteration of the Framework

At this stage, the SME graphically depicts the prioritized digital initiatives in a **roadmap**, composed by a **timeline of 4/5 years** and by the **main functions involved**. Actions for improvement with the highest priority are placed in the first phases, while other ones are scheduled in subsequent stages (Fig. 5).

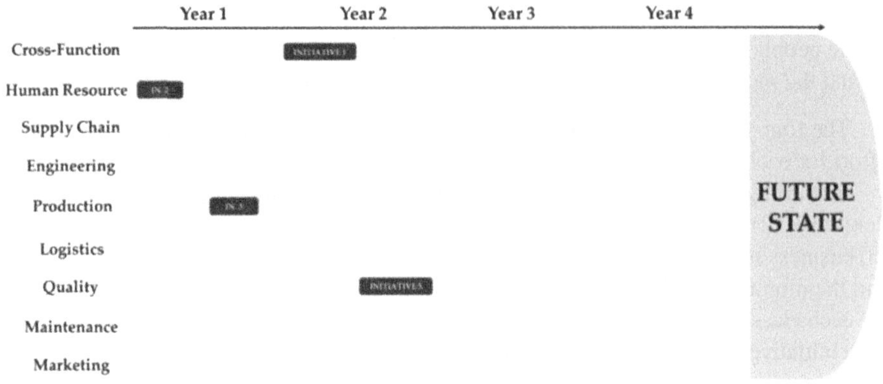

Fig. 5. Illustrative outcome of stage 5

Thus, the SME must start implementing the actions positioned in the first year. Then, it must periodically monitor the performances and the evolution of the internal and external environment, applying iteratively the roadmapping procedure with the goal of keeping the roadmap up to date.

In this direction, the model aims to address the main barriers to exploiting DT roadmaps. Indeed, the model offers a tailored approach which is designed to accommodate the unique needs and limitations of SMEs, providing concrete guidance and practical steps to ensure strategic alignment and effectiveness.

Furthermore, the model assists SMEs in defining and prioritizing digital initiatives by offering methodological support throughout the selection process. It emphasizes the evaluation of costs, risks, and benefits, empowering SMEs to make informed decisions that align with their strategic goals and resource constraints.

7 Conclusions

Manufacturing SMEs are increasingly pushed to carry out a profound DT of product and processes in order to keep up with business objectives, market requirements, regulations, and sustainability issues. Nowadays, the technological offer to achieve such goal is wide yet known. Nevertheless, such a variety tend to freeze manufacturers which struggle to set the right direction to navigate the DT journey. In this regard, DT roadmapping are well reported and recognized methods to support firms in overcoming those barriers. However, these methods have found limited applicability in reality as 1) they usually are too generic or designed for large companies, 2) they do not suggest practical steps to ensure the strategic consistency of the digital initiatives, and 3) they rarely help SMEs to define *what* and *how* to do during the selection and prioritization phases.

The developed methodology actively contributes fixing such literature gaps by focusing on SMEs. It provides concrete managerial support to ensure consistency between digital, market, business, and strategy. The verified alignment between corporate goals and digital actions can positively affect the I4.0 involvement of entrepreneurs and top management. In turn, this can be translated to bottom levels, establishing a widespread and holistic digital culture within the SME. Moreover, the framework suggests a methodological approach to perform initiatives' prioritization based on effort and benefits as well as to generate a feasible and structured digital transformation roadmap. To conclude, the developed model suggests parameters and indicators to evaluate the relevance and the complexity of each initiative, which could guide managers to maximize both shor tand long-term results. The main limits of the model deal with the high-level estimation of effort and benefits, which are not faced in real economics terms, and with the need to further validate the iterative nature of the model after the implementation of the first digital initiatives. As a consequence, further studies could integrate the model with detailed economic analysis of efforts and benefits as well as verify the iterative nature of the methodology by re-applying the procedure to the same companies.

Acknowledgment. This study has been partially funded by the Horizon Europe project FLASH-COMP – GA number: 101058458.

Co-funded by
the European Union

References

1. Paritala, P.K., Manchikatla, S., Yarlagadda, P.K.D.V.: Digital manufacturing applications past, current, and future trends. Procedia Eng. **174**, 982–991 (2017). https://doi.org/10.1016/j.proeng.2017.01.250
2. El-Warrak, L., et al.: Analyzing Industry 4.0 trends through the technology roadmapping method. Procedia Comput. Sci. **201**, 511–518 (2022). https://doi.org/10.1016/j.procs.2022.03.066
3. Schmitt, P., Schmitt, J., Engelmann, B.: Evaluation of proceedings for SMEs to conduct I4.0 projects. Procedia CIRP **86**, 257–263 (2019). https://doi.org/10.1016/j.procir.2020.01.007
4. Despeisse, M., Acerbi, F.: Toward eco-efficient and circular industrial systems: ten years of advances in production management systems and a thematic framework. Prod. Manuf. Res. **10**(1), 354–382 (2022). https://doi.org/10.1080/21693277.2022.2088634
5. Bai, C., Dallasega, P., Orzes, G., Sarkis, J.: Industry 4.0 technologies assessment: a sustainability perspective. Int. J. Prod. Econ. **229**, 107776 (2020). https://doi.org/10.1016/j.ijpe.2020.107776
6. Glova, J., Sabol, T., Vajda, V.: Business models for the Internet of Things environment. Procedia Econ. Finan. **15**, 1122–1129 (2014). https://doi.org/10.1016/S2212-5671(14)00566-8
7. Galati, F., Bigliardi, B.: Industry 4.0: emerging themes and future research avenues using a text mining approach. Comput. Ind. **109**, 100–113 (2019). https://doi.org/10.1016/j.compind.2019.04.018
8. Santos, C., Mehrsai, A., Barros, A.C., Araújo, M., Ares, E.: Towards Industry 4.0: an overview of European strategic roadmaps. Procedia Manuf. **13**, 972–979 (2017). https://doi.org/10.1016/j.promfg.2017.09.093
9. Khan, I.S., Ahmad, M.O., Majava, J.: Industry 4.0 and sustainable development: a systematic mapping of triple bottom line, circular economy and sustainable business models perspectives. J. Clean. Prod. **297**, 126655 (2021). https://doi.org/10.1016/j.jclepro.2021.126655

10. Hofmann, E., Rüsch, M.: Industry 4.0 and the current status as well as future prospects on logistics. Comput. Ind. **89**, 23–34 (2017). https://doi.org/10.1016/j.compind.2017.04.002
11. Reddy, K.R.K., Gunasekaran, A., Kalpana, P., Sreedharan, V.R., Kumar, S.A.: Developing a blockchain framework for the automotive supply chain: a systematic review. Comput. Ind. Eng. **157**, 107334 (2021). https://doi.org/10.1016/j.cie.2021.107334
12. Müller, J.M., Buliga, O., Voigt, K.-I.: Fortune favors the prepared: how SMEs approach business model innovations in Industry 4.0. Technol. Forecast. Soc. Chang. **132**, 2–17 (2018). https://doi.org/10.1016/j.techfore.2017.12.019
13. Albukhitan, S.: Developing digital transformation strategy for manufacturing. Procedia Comput. Sci. **170**, 664–671 (2020). https://doi.org/10.1016/j.procs.2020.03.173
14. Nosalska, K., Piątek, Z.M., Mazurek, G., Rządca, R.: Industry 4.0: coherent definition framework with technological and organizational interdependencies. JMTM **31**(5), 837–862 (2019). https://doi.org/10.1108/JMTM-08-2018-0238
15. Pereira, A.C., Romero, F.: A review of the meanings and the implications of the Industry 4.0 concept. Procedia Manuf. **13**, 1206–1214 (2017). https://doi.org/10.1016/j.promfg.2017.09.032
16. Ng, T.C., Lau, S.Y., Ghobakhloo, M., Fathi, M., Liang, M.S.: The application of Industry 4.0 technological constituents for sustainable manufacturing: a content-centric review. Sustainability **14**(7), 4327 (2022). https://doi.org/10.3390/su14074327
17. Ortt, R., Stolwijk, C., Punter, M.: Implementing Industry 4.0: assessing the current state. JMTM **31**(5), 825–836 (2020). https://doi.org/10.1108/JMTM-07-2020-0284
18. Beltrami, M., Orzes, G., Sarkis, J., Sartor, M.: Industry 4.0 and sustainability: towards conceptualization and theory. J. Clean. Prod. **312**, 127733 (2021). https://doi.org/10.1016/j.jclepro.2021.127733
19. Chauhan, A., Jakhar, S.K., Chauhan, C.: The interplay of circular economy with industry 4.0 enabled smart city drivers of healthcare waste disposal. J. Clean. Prod. **279**, 123854 (2021). https://doi.org/10.1016/j.jclepro.2020.123854
20. Grooss, O.F., Presser, M., Tambo, T.: Surround yourself with your betters: recommendations for adopting Industry 4.0 technologies in SMEs. Dig. Bus. **2**(2), 100046 (2022). https://doi.org/10.1016/j.digbus.2022.100046
21. Phaal, R., Farrukh, C.J.P., Probert, D.R.: Technology roadmapping—a planning framework for evolution and revolution. Technol. Forecast. Soc. Chang. **71**(1–2), 5–26 (2004). https://doi.org/10.1016/S0040-1625(03)00072-
22. Mittal, S., Khan, M.A., Romero, D., Wuest, T.: A critical review of smart manufacturing & Industry 4.0 maturity models: implications for small and mediumsized enterprises (SMEs). J. Manuf. Syst. **49**, 194–214 (2018). https://doi.org/10.1016/j.jmsy.2018.10.005
23. Spaltini, M., Acerbi, F., De Carolis, A., Terzi, S., Taisch, M.: Toward a technology roadmapping methodology to enhance sustainable and digital transition in manufacturing. Prod. Manuf. Res. **12**(1), 2298572 (2024). https://doi.org/10.1080/21693277.2023.2298572
24. Brozzi, R., Riedl, M., Matta, D.: Key readiness indicators to assess the digital level of manufacturing SMEs. Procedia CIRP **96**, 201–206 (2021). https://doi.org/10.1016/j.procir.2021.01.075
25. Schumacher, A., Erol, S., Sihn, W.: A maturity model for assessing Industry 4.0 readiness and maturity of manufacturing enterprises. Procedia CIRP **52**, 161–166 (2016). https://doi.org/10.1016/j.procir.2016.07.040
26. Hein-Pensel, F., et al.: Maturity assessment for Industry 5.0: a review of existing maturity models. J. Manuf. Syst. **66**, 200–210 (2023). https://doi.org/10.1016/j.jmsy.2022.12.009
27. Amaral, A., Jorge, D., Peças, P.: Small medium enterprises and Industry 4.0: current models'ineptitude and the proposal of a methodology to successfully implement Industry 4.0 in small medium enterprises. Procedia Manuf. **41**, 1103–1110 (2019). https://doi.org/10.1016/j.promfg.2019.10.039

28. Ribeiro Da Silva, E.H.D., Angelis, J., De Lima, E.P.: In pursuit of Digital Manufacturing. Procedia Manuf. **28**, 63–69 (2019). https://doi.org/10.1016/j.promfg.2018.12.011
29. Angreani, L.S., Vijaya, A., Wicaksono, H.: Systematic literature review of Industry 4.0 maturity model for manufacturing and logistics sectors. Procedia Manuf. **52**, 337–343 (2020). https://doi.org/10.1016/j.promfg.2020.11.056
30. Schroderus, J., Lasrado, L.A., Menon, K., Kärkkäinen, H.: Towards a PayPer-X maturity model for equipment manufacturing companies. Procedia Comput. Sci. **196**, 226–234 (2022). https://doi.org/10.1016/j.procs.2021.12.009
31. Spaltini, M., Acerbi, F., Pinzone, M., Gusmeroli, S., Taisch, M.: Defining the roadmap towards Industry 4.0: the 6Ps maturity model for manufacturing SMEs. Procedia CIRP **105**, 631–636 (2022). https://doi.org/10.1016/j.procir.2022.02.105
32. Colli, M., Madsen, O., Berger, U., Møller, C., Wæhrens, B.V., Bockholt, M.: Contextualizing the outcome of a maturity assessment for Industry 4.0. IFACPapersOnLine **51**(11), 1347–1352 (2018). https://doi.org/10.1016/j.ifacol.2018.08.343
33. Spaltini, M., Terzi, S., Taisch, M.: Development and implementation of a roadmapping methodology to foster twin transition at manufacturing plant level. Comput. Ind. **154**, 104025 (2024). https://doi.org/10.1016/j.compind.2023.104025
34. Schumacher, A., Nemeth, T., Sihn, W.: Roadmapping towards industrial digitalization based on an Industry 4.0 maturity model for manufacturing enterprises. Procedia CIRP **79**, 409–414 (2019). https://doi.org/10.1016/j.procir.2019.02.110

PLM Traceability Approaches Along the V-Model – A Systematic Literature Review

Detlef Gerhard(✉) and Jannick Bayar

Institute for Product and Service Engineering, Digital Engineering Chair, Ruhr-University Bochum, Universitaetsstr. 150, 44801 Bochum, Germany
{detlef.gerhard,jannick.bayar}@rub.de

Abstract. Given the complexity of products and multidisciplinary product creation processes, the ability to systematically capture and track logically linked information artefacts in both the forward and backward directions of MBSE processes along the V-model is crucial. Traceability is essentially about establishing a clear and transparent context of relevant information for decisions in the product development process. This extends from recorded requirements, through specification and development, to the final release and the ongoing refinements and iterations in each of these phases. Traceability solutions are usually implemented in PLM systems but there is a wide range of solution approaches and still challenges to overcome. The paper provides a categorization and systematization of approaches for traceability based on a systematic literature review, addresses challenges and how to deal with them.

Keywords: Traceability · MBSE · V-Model · PLM

1 Introduction

Within the Product Lifecycle Management (PLM) domain and Model-Based Systems Engineering (MBSE), various concepts go hand in hand to enable seamless communication and collaboration between different stakeholders in various engineering domains involved in the product creation process and even beyond, i.e. production and operations. The term "digital thread" refers to a concept for providing a seamless flow of data and information that connects all relevant artefacts to control the digital engineering processes in product design and development as well as in manufacturing and operations. The main goal of a digital thread is to enable consistency, traceability and transparency of information throughout the entire product life cycle and thus facilitate well-informed decision-making processes.

MBSE is an approach that employs models to define, design, analyze, and document complex engineered systems. MBSE and the digital thread are closely linked, as both underline the importance of digital models and simulations for understanding, optimizing, and managing product lifecycles. By establishing MBSE processes, organizations

Published by Springer Nature Switzerland AG 2025
P. Sureephong et al. (Eds.): PLM 2024, IFIP AICT 741, pp. 280–294, 2025.
https://doi.org/10.1007/978-3-031-93323-3_21

can create accurate and comprehensive models that feed into the digital thread, ensuring consistency and accuracy across various lifecycle stages.

In contrast to the digital thread, which focuses on the information flow, the "digital twin" focuses on a holistic model respectively a virtual replica of a physical product or system, that captures its characteristics, behavior, and performance. The digital twin can be understood as a backbone for the digital thread, providing a real-time representation of the physical product and enabling data exchange and analysis throughout its lifecycle. Therefore, the two concepts digital thread and digital twin are strongly interconnected, particularly in order to enable traceability, which is one of the foremost issues in MBSE. Nonetheless, implementing the digital thread and digital twin as part of a PLM solution requires to overcome certain challenges, such as integrating disparate systems, ensuring data security and privacy, managing complex data flows, and fostering a culture of digital transformation within organizations.

Despite the widespread acknowledgment of traceability as a critical research area and its pivotal role in PLM and MBSE, there is a notable lack of consensus regarding the precise definition of traceability. Multiple and sometimes even contradictory definitions of traceability exist, ranging across international standards, EU regulations, scientific articles and dictionary entries. Traceability needs and obligations heavily depend on the risk level of a product. For example, in the context of the aerospace industry, since even minor flaws can have catastrophic consequences, traceability has a particular focus on ensuring the complete and verifiable history of a component, material, or product. This goes beyond simply knowing the current location or function but encompasses the entire lifecycle, from the origin of raw materials to the manufacturing process, all the way to the final assembly, operation, and even potential future maintenance or disposal. Information about a component's history must be documented, accurate, and accessible. This allows for verification at any point in the supply chain. Furthermore, there are different views on traceability in different sectors [1]. With focus on engineering, traceability can be defined as an "association among logical entities such as requirements, system elements, verifications, or tasks." [2].

This diversity in definitions underscores the complexity of establishing a universally accepted understanding of traceability. It reflects the multifaceted nature of traceability as a concept that spans various industries, regulatory environments, and stages of the supply chain, each with its unique requirements and challenges. Therefore, this paper narrows the scope and puts the focus on traceability approaches along the V-model, which is a fundamental procedural guideline of the MBSE concept in product development and comprises engineering tasks and processes from requirements and early concept phase up to verification and validation. Production and operation phase with predominantly material and order-related processes are intentionally not taken into consideration apart from processes and information flows directly linked to engineering.

The paper provides a categorization and systematization of approaches for traceability based on a systematic literature review (SLR) in order to identifying gaps for future research. The SLR was conducted using a thorough four-step methodology. The initial phase of the involved identifying pertinent literature by applying a predefined set of filter criteria on leading scientific databases, namely Scopus, Google Scholar and Web of Science. These platforms were chosen for their comprehensive coverage across various disciplines. The criteria for selection included the time frame of publication, set from 2016 to 2024, to ensure the inclusion of the most recent studies reflecting current trends and technological innovations. Additionally, the search was refined using specific keywords within the article title, abstract, or keywords section, focusing on terms such as "traceability," "PLM" (Product Lifecycle Management), "v-model," and combinations thereof, emphasizing the context of product information and lifecycle. This search yielded a total of over 200 publications after eliminating duplicates. The third step "abstract review" was performed to match the publications content to the object of investigation. It was analyzed weather the publication tackles the topic of traceability focused on MBSE or if later stages of the product lifecycle in areas such as manufacturing or operation were in focus. The abstract review resulted in 45 papers as a base of the full paper review [3–47]. [14–47] are not further referenced in this paper.

2 Terminology Clarification

With the focus stated above, traceability refers to the ability to track and document relationships between various artefacts in engineering models, and thus the ability to track the origin, rationale and impact of decisions throughout the lifecycle of a system. It involves linking requirements to design decisions, simulations and the final system behavior. Traceability helps to recognize and rectify errors at an early stage in order to prevent problems from propagating across different development stages. This is crucial for ensuring consistency, integrity, and completeness of the model and digital engineering processes. Understanding the impact of changes on various aspects of systems through traceability enables engineers to make informed decisions during the product development process and manage changes effectively.

In terms of clarifying the terminology, this paper adopts the perspective that the two different but synergetic functionalities of tracking and tracing are crucial for achieving the above-mentioned goal. While distinct in their focus, both tracking and tracing involve data collection and analysis, serving as fundamental tools for gaining valuable insights. However, a key differentiator lies in their temporal focus, as shown in Table 1:

Table 1. Tracking and Tracing in the context of MBSE

Term	Temporal Focus
1. Tracking	opens a window into the model's present state that provides real-time information on the current state of a product
2. Tracing	offers a retrospective understanding by examining past information and unveiling the model's origins and relationships

Tracking accentuates the current state of the MBSE model and its constituent elements to provide real-time awareness. Version control systems offer a granular view of modifications made to models, meticulously recording who implemented the changes and the timestamps associated with them. Tracking empowers the identification of the current status of design elements, such as the latest revision of a crucial engine component within a complex aircraft model. The ability to track the progress of simulations or analyses being conducted on the model grants valuable insights into the ongoing engineering processes. Collaborative platforms facilitate real-time monitoring of model elements, enabling engineers working concurrently to maintain shared awareness of updates. By providing a shared understanding of model updates, tracking fosters seamless collaboration among engineers working on various aspects of the system.

Tracing as a retrospective look delves into the origins and intricate relationships between various elements encompassed within MBSE models. It centers around establishing explicit linkages and enabling access to past data, e.g., linkages between requirements, the corresponding design choices, and simulations that effectively fulfill them. Tracing enables the thorough tracking of how design decisions made in one part of the model propagate and influence downstream analyses, ensuring consistency throughout the development process. It empowers the identification of the origin of specific model elements for downstream analysis impact analyses, along with the rationale behind their creation. In MBSE, this retrospective view empowers investigations into root cause analysis, i.e. when faced with an issue (e.g., a malfunctioning component), tracing facilitates pinpointing the origin of the problem by examining past actions and decisions. Tracing allows for reconstructing the development trajectory of a product or process, highlighting the changes and influences that shaped its current state and, by examining historical records, tracing can reveal the maintenance history of a product, providing valuable context for its current condition.

The complementary nature of tracking and tracing particularly leads to enhanced transparency, reduced errors, and streamlined change management. A clear picture can be established regarding how different elements within the model are interconnected, fostering better understanding of the system's overall structure. The ability to comprehend the impact of changes throughout the model allows for the proactive identification and rectification of inconsistencies that might arise due to unreflected modifications. Engineers gain a comprehensive understanding of how proposed modifications impact various aspects of the system, enabling informed decision-making during the development process.

3 Trace Artifacts and Trace Links

Effective requirement traceability within MBSE hinges on efficiently managing trace artifacts and trace links [3], together with their associated functionalities. Trace artifacts are pivotal elements that form the backbone of effective traceability mechanisms. These artifacts that constitute the model representation are inherently diverse in nature, encompass a wide array of tangible or intangible elements such as requirements, design elements, verification and validation outputs, and other critical documentation integral to the engineering process. They serve as the foundational entities which, when interlinked, create a network of traceability that provides comprehensive visibility across the entire system development lifecycle. Thus, trace artefacts can be understood as the building blocks for understanding the intricate relationships within a system. The quintessential role of trace artifacts lies in their ability to encapsulate and convey essential information about the characteristics and evolution of the system under development. This includes the articulation of requirements, the embodiment of design decisions, the encapsulation of test cases and outcomes, as well as various other forms of documentation and data that substantiate the development process. Trace artefacts are not static, they evolve and change over time as system development progresses. This dynamic nature requires robust management strategies to ensure their relevance and accuracy. It is essential that these artefacts are subject to rigorous version control, change management and validation processes to maintain their integrity and applicability.

Trace links establish explicit directional or bidirectional relationships between two or more trace artifacts to capture their interactions and mutual influences. They are the backbone of a robust traceability framework, facilitating an in-depth understanding of the system's components and their interplay. Functioning as the relational threads within the MBSE development framework, trace links connect disparate elements, such as linking specific requirements to their respective design elements or associating test cases with their validating requirements. This connectivity enables engineers to navigate the complexities of system development, tracing a coherent path from abstract requirements to tangible implementation and subsequent verification and validation. Trace links also have a dynamic character, which is crucial for adapting to changes in system development. They accommodate evolving requirements, modified designs, and emerging test results, ensuring that the traceability framework remains current and reflective of the system's state (Fig. 1).

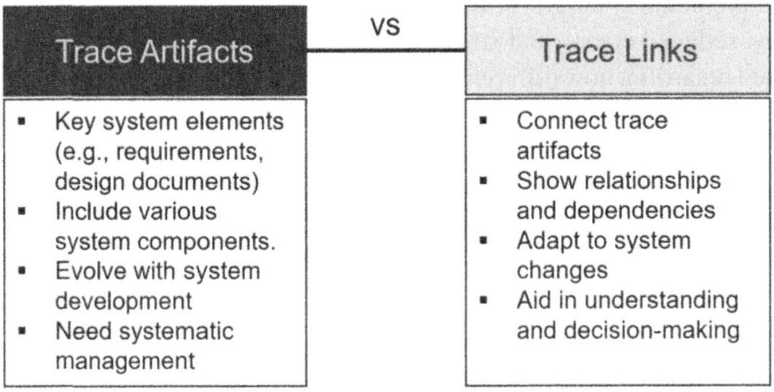

Fig. 1. Difference between trace artefacts and trace links

By thoroughly linking trace artefacts, a multidimensional and coherent traceability matrix is created that enables engineers to navigate through the complexity of the system and gain a comprehensive understanding of how e.g. requirements are translated into design decisions, respectively that the chosen design elements actually fulfil the specified requirements, how changes to the design are reflected in the simulations performed to validate system behavior, and, in turn, how the results of simulations and analyses affect subsequent design decisions. In such a traceability matrix, the rows and columns represent different types of trace artifacts. For instance, rows could represent requirements, while columns could represent design elements. Each cell in the matrix represents the relationship, or trace link, between the artifact in the row and the artifact in the column. A cell in the matrix is marked or filled if there is a direct relationship or link between the corresponding row and column artifacts. For example, if a particular requirement (row) is addressed by a specific design element (column), the intersecting cell indicates this relationship. The markings can vary depending on the complexity of the information being represented. Cells often contain additional information such as the nature of the relationship, its status, or links to more detailed documentation. While a basic traceability matrix is two-dimensional, multidimensionality is achieved through multiple matrices or layered matrices, each focusing on different aspects of the system (e.g., one matrix for requirement to design traceability, another for requirement to test cases traceability, etc.). In some advanced tool implementations, the matrix is interactive, allowing users to drill down into specific cells for more detailed information, effectively navigating through additional dimensions of data.

The underlying storage and management of trace links and system artifacts often rely on relational database technology to store traceability relevant data and support complex queries, enabling the retrieval and analysis of trace relationships. Trace links are also represented and managed using graph-based data structures, where artifacts are nodes, and links are edges. This approach is particularly effective for visualizing and analyzing the relationships and dependencies between numerous interconnected elements.

4 Traceability Types in MBSE

In the context of MBSE and the V-model, seven different types of traceability can be identified from the SLR and have to be distinguished with respect to challenges and particular support in PLM-systems. By exploring these traceability types, this paper aims to provide a more granular and nuanced understanding of traceability, highlighting its multifaceted role in modern MBSE and PLM environments (Table 2).

1. **Requirement Traceability**

   Challenges: Maintaining an up-to-date and accurate linkage transparent and verifiable connection between requirements and system elements across all phases of development can be complex, especially in large-scale projects. Changes in requirements can be difficult to propagate throughout the system particularly as requirements evolve and the model undergoes modifications.

   PLM Support: PLM systems can automate the tracking of requirements and their relationships with design elements. Standards like ISO/IEC/IEEE 29148 can provide guidance on requirements engineering. Methodical stakeholder analysis techniques, such as Quality Function Deployment (QFD), can be employed to ensure that stakeholder requirements are systematically captured and addressed. Another methodical approach is the systematic change management processes to update the trace links whenever requirements change.

   - Requirement management modules: Facilitate the definition, storage, and version control of requirements, enabling traceability to downstream model elements.
   - Impact analysis tools: Assess the influence of requirement changes on associated design elements and simulations within the MBSE model.

Table 2. T types of traceability

Type	Description
1. Requirement Traceability	Establishes a clear linkage between high-level system requirements and the corresponding design choices, simulations, and analyses that address them
2. Design Traceability	Tracks the relationships between various design elements within the MBSE model. This encompasses the connections between components, subsystems, and the overall system architecture
3. Verification and Validation Traceability	Links test cases, simulations, and analyses conducted during the V&V process to the specific requirements they are intended to verify or validate
4. Domain Spanning Traceability	Establishes linkages between information and concepts across various engineering disciplines involved in the MBSE process to bridge the disciplinary divide. This encompasses disciplines like mechanical, electrical, software, and control systems
5. Type-Instance Traceability	Provides linkages between general product type information (e.g., specifications, design rules) and the specific information pertaining to individual product instances (e.g., manufacturing lots).
6. Aggregation Level Traceability	Builds clear linkages between information and functionalities at various levels of the system hierarchy, encompassing the entire system, its constituent subsystems, and individual components
7. Safety and Compliance Traceability	Establishes linkages between safety requirements, regulations, and the corresponding design choices, analyses, and verification activities within the MBSE model to ensure adherence to regulations as well as compliant products and processes

2. **Design Traceability:**

   Challenges: Ensuring a consistent and transparent link between design decisions, requirements and implementation can be difficult, especially when there are multiple design alternatives. Managing the intricate web of interconnected elements, especially in complex systems with multiple parts and functions as well as documenting and understanding the impact of changes across the system, maintaining an audit trail for decision-making can also be difficult.

PLM Support: Utilizing PLM tools that support design traceability by linking design artifacts to requirements and downstream elements. Adopting design standards like ISO 10303 (STEP) can facilitate design data exchange and traceability. Methodologically, employing design rationale documentation can support traceability and decision-making. Change management methodologies, aligned with standards like ISO 9001, can provide structured processes for handling changes.

- Digital Mockup (DMU) functionalities: Allow for the creation and visualization of 3D assemblies, fostering a clear understanding of how design components interact and relate to each other.
- Bill of Materials (BOM) management: Provides a structured record of components and their hierarchical relationships within the assembly structure.

3. **Verification and Validation Traceability:**

   Challenges: The complexity of tracing each verification and validation activity to the corresponding system requirements and ensuring comprehensive coverage is a major challenge, particularly ensuring that tests effectively address the intended requirements and that the results are demonstrably linked back to the model elements they validate. Furthermore, linking performance metrics back to design decisions and requirements, ensuring that the system meets performance criteria is difficult. *PLM Support:* Integrating V&V activities into the PLM environment to ensure direct linkage with system requirements. Using standards such as ISO/IEC 25040 on software quality evaluation helps in defining V&V processes. Methodologically, adopting systematic V&V planning and execution frameworks can enhance traceability. Performance modeling and simulation techniques can be used to predict and analyze system performance early in the design process.

   - Test management tools: Facilitate the creation, execution, and recording of test results, enabling traceability to the corresponding requirements.
   - Simulation data management: Provides a central repository for storing and managing simulation data, ensuring clear links between simulations and the requirements they validate.

4. **Domain Spanning Traceability:**

   Challenges: Tracing elements across different domains (e.g., software, hardware) can be hampered by domain-specific languages and tools, leading to integration issues. Data exchange and communication barriers often exist between different engineering domains, hindering a holistic understanding of how decisions in one discipline impact others. Ensuring coherence and compatibility between multiple system interfaces, particularly in multi-vendor or multi-team environments.

   PLM Support: Implementing PLM systems with multi-domain integration capabilities. Methodically, cross-domain coordination and integration frameworks are essential.

   - Interoperable data models: Usage of standards like ISO 10303 (STEP), particularly AP 243 (MOSSEC), or SysML that provide a common modeling language for different domains to facilitate the exchange of information between different engineering tools used by various disciplines.

- Model-based collaboration tools: These tools enable engineers from different disciplines to work on a shared MBSE model, fostering real-time visibility into the impact of changes across domains.
- Interface management: Adopting interface management standards like ISO/IEC/IEEE 24765 can provide a common framework for defining and managing interfaces.

5. **Type-Instance Traceability:**

   Challenges: Linking generic models (types) to their specific instances in a way that facilitates understanding and managing variability poses significant challenges.

   Ensuring that specific product instances are manufactured in accordance with the broader product type definitions and managing various system configurations, handling version control, and maintaining consistency across multiple iterations is also difficult.

   *PLM Support:*Utilizing PLM tools that support type-instance relationships and variant management. Methodological approaches might include variant management techniques and systematic instantiation processes.

   - Product variant management tools: Management of a product type, variants, and end item configurations ensuring that each instance inherits the relevant specifications and design rules.
   - Manufacturing data management: Provision of a central repository for storing and managing manufacturing instructions and associated documentation specific to each product instance.

6. **Aggregation Level Traceability:**

   Challenges: Keeping track of how high-level system architectures are realized in lower-level components and vice versa can be complex, especially when dealing with hierarchical system structures. Maintaining a comprehensive understanding of how decisions and modifications at one level (e.g., component design) impact the overall system behavior is crucial.

   PLM Support: Adopting hierarchical modeling and decomposition standards like IEEE 1220 can be beneficial. Methodologically, employing top-down and bottomup approaches in system design can ensure coherence across different levels of aggregation.

   - Hierarchical MBSE models: Models that explicitly represent the system breakdown structure, clearly depicting the relationships between systems, subsystems, and components.
   - Change impact analysis tools: PLM modules that can assess the ripple effects of modifications made at a lower level (e.g., component) on the performance and functionality of higher levels (e.g., subsystem or entire system).

7. **Safety and Compliance Traceability:**

   Challenges: Guaranteeing that the developed system adheres to all relevant safety standards and regulatory requirements throughout the entire development lifecycle.
   PLM Support: PLM systems can maintain safety requirements and compliance documentation, ensuring traceability throughout the system lifecycle. Adherence to safety

and compliance standards, e.g. ISO 26262 for automotive safety, is critical in this regard.

- Regulatory compliance management tools: PLM systems can integrate with regulatory databases, enabling engineers to identify applicable standards and ensure their incorporation into the MBSE model.
- Risk assessment and mitigation tools: These functionalities within PLM systems facilitate the identification of potential safety hazards and trace their origin back to specific design elements or decisions within the model.

5 Resulting MBSE Traceability Challenges

Several studies (e.g. [4–8]) report that the heterogeneity of artefacts and tools in the development lifecycle pose the major challenge to be overcome when creating traceability links. The studies indicate that one of the solutions to this challenge is tool integration, where all existing tools are interconnected so that it becomes possible to exchange information about the heterogeneous artefacts and establish traceability links between them. Exchanging and merging trace information between different systems or organizations is often impeded by data incompatibility. This challenge can be approached through adopting interoperability standards such as OSLC (Open Services for Lifecycle Collaboration) that enable different systems, potentially from different vendors, to exchange and merge data effectively. To bridge gaps between different systems, especially when dealing with legacy systems, data transformation and mapping tools can be instrumental in converting trace data from one format to another [9]. Effective version control and change management processes are also crucial for maintaining the integrity and currency of trace information when it is reused or merged, particularly important for tracking the evolution of trace information over time [10].

ISO 10303-243 (MoSSEC) plays a significant role in ensuring traceability within the context of MBSE. MoSSEC establishes a common format for exchanging data between various modeling and simulation tools. This ensures consistency and reduces the risk of misinterpretations during data transfer, which can lead to traceability issues. The standard facilitates capturing information about the origin and history of data used in models and simulations. This allows engineers to track how requirements translate into design decisions, simulation inputs, and ultimately, the system's functionality. MoSSEC also enables the definition of relationships between different models used in the engineering process. This allows engineers to understand how changes in one model impact others, ensuring that the overall system design remains consistent with requirements. General challenges across all types of traceability include manual work in creating and updating traceability links, preparing for traceability perceived as overhead work, and lack of visualization tools.

The manual creation and maintenance of trace links, especially in complex systems involving numerous heterogeneous artifacts, are inherently labor-intensive and prone to human error [11]. This inefficiency arises from the difficulty in consistently identifying and maintaining relevant relationships across a wide array of artifacts. Automated approaches are suggested because they can significantly reduce the time and effort required for these tasks while enhancing accuracy. Automated techniques like event-based and model-driven approaches can systematically capture changes and updates in artifacts, thereby dynamically updating the trace links [12].

In many development environments, traceability is often viewed as a secondary task compared to the primary objective of developing functional solutions. This perception stems from the disconnect between those who create trace links (often for compliance or documentation purposes) and those who primarily benefit from them (such as quality assurance teams or project managers). Automating the generation of trace links is proposed to minimize this perceived overhead. By integrating traceability tasks into the normal workflow of development activities, the burden on developers is reduced.

Automated traceability ensures that developers are not sidetracked from their core tasks while still contributing to the creation of valuable trace links. This approach can bridge the gap between creators and users of trace links, making traceability a more integral part of the development process.

Effectively reusing trace artefacts and trace links across different projects can be difficult due to varying project contexts, specific functionalities, and unique requirements. Trace links and artefacts established in a previous project might not directly map onto the specific needs of a new project. Nonetheless, standardization is key to enabling reusability. Formats such as ReqIF (Requirements Interchange Format) provide a universal standard for representing and exchanging requirements, facilitating their reuse across diverse projects [9]. Establishing common ontological frameworks within an organization or across the industry can promote a shared understanding of traceability elements, enhancing reusability even amidst diverse project requirements [13]. Standardized traceability meta-models can capture the generic structure of traceability relationships. This allows for easier adaptation to different project context. Certain industries have their own standards that include traceability meta-model aspects. For instance, in the automotive industry, standards like Automotive SPICE and ISO 26262 incorporate traceability requirements specific to automotive systems development. Many organizations develop custom traceability meta-models tailored to their specific project or organizational needs. These are especially common in industries with unique requirements, such as aerospace, defense, or automotive. These custom meta-models are designed to align with the organization's specific processes, tools, and data formats, offering a more bespoke solution to traceability.

For overcoming reusability and interoperability challenges, research efforts are particularly ongoing to develop standardized metamodels for traceability information. These metamodels aim to capture the generic structure of traceability relationships while allowing for adaptation to specific project contexts. New frameworks are being developed that consider the specific context of each project. These frameworks can reason about the applicability of existing traces from past projects and suggest adaptations for reuse in new contexts [15]. With respect to enhancing interoperability and data exchange, research and industry is exploring the adoption of standardized data formats like Requirements Interchange Format (ReqIF). This format with a focus on requirements also facilitates the exchange of traceability information in a neutral and interoperable manner, enabling seamless data sharing between different MBSE tools and PLM systems.

Model transformation techniques are subject to investigations as well. These techniques can convert trace information from one system's format to another, allowing for effective merging of traceability data from different sources. In some implementations, formal grammars, ontologies and rule-based systems are utilized to define the rules and constraints for establishing trace links. Formal grammars are explored to define the rules and syntax governing the creation of valid trace links. Semantic representations using ontologies allow capturing the meaning and context associated with trace links, enabling richer understanding and reasoning capabilities. These approaches enhance consistency and adherence to predefined traceability criteria. Research in semantic representation of traceability information using ontologies holds promise for enhancing the understandability and interoperability. By capturing the meaning and context associated with traces, these approaches can facilitate better reasoning and analysis of traceability relationships.

6 Conclusion and Outlook

The concept of traceability has been explored within the selected literature with varying emphases on the nature of information that requires tracking and tracing. In one strand of literature, traceability is connected to the synergy between MBSE and PLM, particularly along the V-model. Addressing the multifaceted challenges associated with traceability in MBSE requires a combination of established technologies and ongoing research efforts. PLM systems and MBSE tools provide a foundation for managing trace artifacts, links, and their associated functionalities.

Standardized formats, semantic frameworks, and advanced change management processes are crucial for ensuring traceability across different contexts and maintaining consistency during changes. Major research directions include establishing a more universally adopted standardized metamodel for traceability to enhance interoperability and streamline data exchange across different tools and projects, developing frameworks that can adapt to the specific needs of diverse projects and facilitate the effective reuse of existing traceability information, and enriching traceability data with meaning and context using ontologies to enable deeper understanding and reasoning about the relationships within the system. Furthermore, research is exploring the potential of utilizing machine learning for tasks such as trace link recommendation, automated impact analysis, and anomaly detection within traceability data.

By addressing these research areas, traceability in MBSE developed in a direction where traceability information can be seamlessly exchanged and effectively reused across diverse projects and organizational boundaries. Advancements in standardization, interoperability, and innovative techniques will empower engineers to establish more robust, reliable, and adaptable traceability frameworks. This will ultimately lead to a more efficient and effective MBSE development process for complex engineering projects.

Disclosure of Interests. The authors have no competing interests to declare that are relevant to the content of this article.

References

1. Stark, J.: Product Lifecycle Management (Volume 1): 21st Century Paradigm for Product Realisation, 4th edn. Springer (2020)
2. Singh, V., Willcox, K.E.: Decision-making under uncertainty for a digital thread-enabled design process. J. Mech. Design (2021)
3. Rasiman, R., Dalpiaz, F., España, S.: How effective is automated trace link recovery in model-driven development? In: Gervasi, V., Vogelsang, A. (eds.) International Working Conference on Requirements Engineering: Foundation for Software Quality, 1st edn., p. 35. Springer (2022)
4. Maro, S., Steghöfer, J.-P., Staron, M.: Software traceability in the automotive domain: challenges and solutions. J. Syst. Softw. **141**, 85 (2018)
5. Parizi, R.M., Lee, S.P., Dabbagh, M.: Achievements and challenges in state-of-the-art software traceability between test and code artifacts. IEEE Trans. Rel. **63**, 913 (2014)
6. Regan, G., McCaffery, F., McDaid, K., Flood, D.: The barriers to traceability and their potential solutions: towards a reference framework. In: 38th Euromicro Conference (2012)
7. Kannenberg, A., Saiedian, H.: Why software requirements traceability remains a challenge. J. Defense Softw. Eng. **22**, 14 (2009)
8. Winkler, S., von Pilgrim, J.: A survey of traceability in requirements engineering and model-driven development. Softw. Syst. Model. **9**, 529 (2010)
9. Rose, L.M., García-Domínguez, A., Williams, J.R., Kolovos, D.S., Paige, R.F., Polack, F.A.: Saying hello world with epsilon – a solution to the 2011 instructive case (2011)
10. Conradi, R., Westfechtel, B.: Version models for software configuration management. ACM Comput. Surv. **30**, 232 (1998)
11. Ehring, D., Pluhňau, R., Nagarajah, A.: Concept development of a consistently traceable process and system solution for ensuring the requirements of engineering and functional safety. In: Exchange, DSMIE-2020 (2020). https://doi.org/10.1007/978-3-030-50794-7_23
12. Galvao, I., Goknil, A.: Survey of traceability approaches in model-driven engineering. In: 11th IEEE International Enterprise Distributed Object Computing Conference (EDOC 2007), p. 313 (2007)
13. Jackson, K., Dick, J.: Requirements Engineering. Springer, London (2011)
14. Allen, J.L.: An overview of model-based development verification/validation processes and technologies in the aerospace industry. In: 2016 AIAA Modeling and Simulation Technologies Conference (2016)
15. Bajzek, M., Fritz, J., Hick, H.: Systems engineering processes. In: Hick, H., Küpper, K., Sorger, H. (eds.) Systems Engineering for Automotive Powertrain Development. Powertrain. Springer (2020)
16. Bajzek, M., Fritz, J., Hick, H., Maletz, M., Faustmann, C., Stieglbauer, G.: Model based systems engineering concepts. In: Hick, H., Küpper, K., Sorger, H. (eds.) Systems Engineering for Automotive Powertrain Development. Powertrain. Springer (2020)
17. Bitzer, M., Eigner, M., Faißt, K.G., Muggeo, C.: Framework of the evolution in virtual product modelling and model management towards digitized engineering. DS 87-6 Proceedings of the 21st International Conference on Engineering Design (ICED 17) Vol 6: Design Information and Knowledge, p. 345 (2017)
18. Bougain, S., Gerhard, D.: Integrating environmental impacts with SysML in MBSE methods. Procedia CIRP (2017)
19. Bougain, S., Gerhard D.: Supporting product design decision with a SysML design history Assistant. Procedia CIRP (2020)
20. Dickopf, T., Eigner, M., Apostolov, H.: The KSCM as part of a holistic methodology for the development of cybertronic systems in the context of engineering 4.0. In: DS 92: Proceedings of the 15th International Design Conference (2018)

21. Ebert, C., Ray, R.: Toward a formal traceability model for efficient security validation. Computer (2021)
22. Eigner, M., Dickopf, T., Apostolov, H.: The evolution of the V-model: From VDI 2206 to a system engineering based approach for developing cybertronic systems. In: Product Lifecycle Management and the Industry of the Future, p. 382 (2017)
23. Eigner, M., Gilz, T., Denger, A., Fritz, J.: Applicability of model-based system lifecycle management for cyber-physical systems. In: Proceedings of the 14th Mechatronics Forum International Conference (2014)
24. Forte S, Weber S, Göbel JC. Strategic Planning in the Digital Engineering of Smart Product-Based System of Systems Reconfiguration. Proceedings of the 19th International Conference on Product Lifecycle Management, Grenoble 2022
25. Gerhard, D., Cordero, S.S., Vingerhoeds, R.: MBSE-PLM integration: initiatives and future outlook. In: Proceedings of the 19th International Conference on Product Lifecycle Management, Grenoble (2022)
26. Gräßler, I., Wiechel, D., Roesmann, D., Thiele, H.: V-model based development of cyber-physical systems and cyber-physical production systems. Procedia CIRP (2021)
27. Heber, D.T.: A framework to foster traceability of E/E artifacts during automotive development in consideration of model-based systems engineering within distributed engineering collaboration by means of the Blockchain. University of Twente, Enschede (2022)
28. Holt, J., Perry, S., Payne, R., Bryans, J., Hallerstede, S., Hansen, F.O.: A model-based approach for requirements engineering for systems of systems. IEEE Syst. J. **9**, 252 (2015)
29. Karban, R., Dekens, F.G., Herzig, S., Elaasar, M., Jankevičius, N.: Creating system engineering products with executable models in a model-based engineering environment. In: Proceedings of the SPIE 9911, Modeling, Systems Engineering, and Project Management for Astronomy VII, 99110B (2016)
30. Kim, E.H., An, K.I., Song, D.Y.: A study on the traceability analysis between nonstandardized documents. Korean J. Comput. Design Eng. **20**(4), 328 (2015)
31. Maro, S., Steghöfer, J.P., Staron, M.: Software traceability in the automotive domain: challenges and solutions. J. Syst. Softw. (2018)
32. Maro, S., Steghofer, J.-P.: Capra: a configurable and extendable traceability management tool. In: 2016 IEEE 24th International Requirements Engineering Conference, p. 407 (2016)
33. Mengist, A., Buffoni, L., Pop, A.: An integrated framework for traceability and impact analysis in requirements verification of cyber-physical systems. Electronics **10**, 983 (2021)
34. Nigischer, C., Gerhard, D.: Lightweight visualization of SysML models in PDM systems. In: DS 87-3 Proceedings of the 21st International Conference on Engineering Design (ICED 17) Vol 3: Product, Services and Systems Design, p. 61 (2017)
35. Rahimi, M., Cleland-Huang, J.: Evolving software trace links between requirements and source code. Empir. Softw. Eng. **23**, 2198 (2018)
36. Rasiman, R., Dalpiaz, F., España, S.: How effective is automated trace link recovery in model-driven development? In: Gervasi, V., Vogelsang, A. (eds.) International Working Conference on Requirements Engineering, p. 35. Springer (2022)
37. Rieger, E., Oleszek, S.:. The digital thread concept for integrating the development disciplines for mechatronic products. In: IFIP International Conference on Advances in Production Management Systems, p. 261 (2023)
38. Ríos, J., Bernard, A., Bouras, A., Foufou, S.: Product lifecycle management and the industry of the future. In: 14th IFIP WG 5.1 International Conference, PLM 2017, Seville, Spain, 10–12 July 2017, Revised (2017)
39. Stark, R., Figge, A.: Eco tracing-a systems engineering method for efficient tracelink modelling. In: DS 68-4: Proceedings of the 18th Conference on Engineering Design (ICED !!) Vol. 4: Product and Systems Design, p. 21 (2011)

40. Stolt, R., André, S.: Requirements handling in multidisciplinary product development–a company study. Transdiscip. Future Eng. **115** (2022)
41. Tsui, R., Davis, D., Sahlin, J.: Digital engineering models of complex systems using model-based systems engineering (MBSE) from enterprise architecture (EA) to systems of systems (SoS). INCOSE International Symposium (2018)
42. Viapiana, D., Riggio, G., Barbieri, L., Bruno, F.: An integrated approach to ensure requirements traceability during the product development process. In: International Conference on Design Tools and Methods in Industrial Engineering (2021)
43. Vosgien, T.: Model-based system engineering enabling design-analysis data integration in digital design environments: application to collaborative aeronautics simulation-based process and turbojet integration studies. Ecole Centrale Paris (2015)
44. Yaman, O.: Model-based approach for product requirement representation and generation in product lifecycle management (2020). search.proquest.com
45. Zamazal, K., Denger, A.: Product lifecycle management in automotive industry. In: Hick, H., Küpper, K., Sorger, H. (eds.) Systems Engineering for Automotive Powertrain Development. Powertrain. Springer (2020)
46. Zheng, C., Bricogne, M., Le Duigou, J., Eynard, B.: Survey on mechatronic engineering: a focus on design methods and product models. Adv. Eng. Inform. **8**(3), 241 (2014)
47. Zheng, C., Hehenberger, P., Le Duigou, J., Bricogne, M., Eynard, B.: Multidisciplinary design methodology for mechatronic systems based on interface model. Res. Eng. Design **28**, 333 (2017)

Data-Driven Digital Twin Model Construction and Stress State Monitoring of Grasping Robot

Wei Wei[1,2](✉) and Wenjie Gong[1]

[1] School of Mechanical Engineering and Automation, Beihang University, Beijing 100191, China
wei-wei@buaa.edu.cn
[2] Jiangxi Research Institute of Beihang University, Nanchang 330096, China

Abstract. As more and more robots are used in industrial production, real-time state monitoring during robot running process is attracting widespread attention. Aiming at the problem that conventional Finite Element Analysis costs long calculating time, making it difficult to meet the demand of real-time monitoring, the grasping robot is taken as the research object in this paper to construct its digital twin model for simulating the real motion state of the robot itself, and to realize a real-time monitoring in stress state of mechanical grasper based on Deep Learning method. The 3D model of the grasping robot is firstly constructed and then configured in Unity 3D environment. The running data of the robot is collected and transmitted to drive the digital twin model moving as same as the robot itself. In order to realize the real-time monitoring of mechanical grasper's stress state, a quick generating method for images based on Deep Learning is studied in this research. Finally, high-speed loading of image data and animation generation of mechanical grasper's stress state are realized in Unity 3D.

Keywords: Grasping robot · Digital twin · Deep learning · Stress · State monitoring

1 Introduction

With the rapid development of new technologies, people's life is undergoing great changes all over the world. As the cutting-edge technologies are combining closer and closer to conventional manufacturing technologies in manufacturing industry, it has become a chasing goal for each step in production that making digitalization, automation, informatization and intellectualization come true, which drives the enterprises strive to optimize and upgrade the production and management model [1]. During the transforming process from manufacturing towards intelligent manufacturing, Cyber Physical System is becoming a hot research direction [2].

Cyber Physical System is informational virtual space where digital objects share similar features with real ones, data interacts efficiently with one another and running process mirrors accurately from the realistic one by means of organic integration of signal

P. Sureephong et al. (Eds.): PLM 2024, IFIP AICT 741, pp. 295–306, 2025.
https://doi.org/10.1007/978-3-031-93323-3_22

acquisition, high-speed computing, data interaction and automatic control technologies [3]. Within the framework of Cyber Physical System, Digital Twin(DT) technology is developing fast these years and attracting widespread attention for its capacity of constructing virtual model of real object from many dimensions, based on which the real entity or process will be analyzed and optimized through virtual calculation and simulation. Besides, with the help of the high-speed and powerful computing performance, some problems that cannot be handled in real scenarios would be solved [4, 5].

The concept of digital twin was first mentioned in an article published by Dr. Michael Grieves, University of Michigan, USA. He advocates constructing virtual maps in the virtual space to represent the physical devices during the whole phases of the product, where the connection is bidirectional and dynamic [6]. In 2010, NASA applied digital twin technology in the research of spacecraft, which defined the digital twin paradigm of future aircraft. In 2011, a speech at the U.S. Air Force Laboratory also mentioned digital twins, solving the problem of fighter airframe maintenance through digital twin technology [7]. GE has developed corresponding digital twin models for the condition monitoring and maintenance guarantee of its equipment operation process, and has carried out practical application in the field of industrial production operation and maintenance [8]. Eric J. Tuegel et al. applied digital twin technology to aircraft health monitoring and management, and constructed a digital twin model of the aircraft entity for structural life prediction. Base on the multi-dimensional model of a certain type of aircraft, the structural deformation and temperature field are simulated and analyzed, and then the predictions of structural damage and material performance changes are realized [9]. Scholars from Vanderbilt University applied digital twin technology to aircraft wing health management, and achieved probability prediction of wing cracks through mathematical analysis [10]. Glaessgen E et al. applied the digital twin technology to the aircraft to establish a multi-dimensional model, and realized the health management combining the collected data in the working process [11].

With the development and iteration of technology, industrial robots are developing towards the direction of realizing more complex control, trickier angles, and faster response. For the purpose of real-time monitoring of industrial robots to ensure the smooth and safe production process, the multi-dimensional data of the operation process is collected and integrated to provide state monitoring service, avoiding possible manufacturing risks, and summarizing historical faults. At present, monitoring technology has been widely used in many links and scenarios of industrial manufacturing, and has become an important technology for production process management. ABB has built a remote state monitoring platform for equipment, which sends and receives real-time operating data through a processing box with communication functions to achieve data acquisition, data analysis, data transmission and other functions, thereby realizing remote operating state monitoring. With this platform, the cost for equipment operation and maintenance can be reduced [12]. By collecting torque signals and temperature signals, Lee et al. realized remote supervision of the operation status of industrial robots, so that they could timely warn and deal with possible faults during the operation process [13].

When it comes to stress calculating, FEA (Finite Element Analysis) would first appear in our mind. To accelerate the result-generating speed, much effort has been devoted in this field. Raza N et al. aimed at the simulation situation of complex-shaped objects with additional mass property, and studied the use of unstable models and variable mesh methods for finite element transient simulation of complex-shaped objects. The purpose of calculating the additional mass of complex-shaped objects is achieved by solving the difference of the solution results under the condition of variable speeds in the equilibrium state [14]. Brandvik T et al. invested in GPU-accelerated computing. Since GPU is specially developed for image processing, its capability in floating-point operations far exceeds that of CPU. Whether it is 2D or 3D solving, the performance of GPU is several times better than that of CPU [15]. W.D. Smith proposed the use of FPGA to realize distributed computing in the field of computational fluid dynamics, and realized the coordination between multiple computing platforms and the same computing subject [16]. Lifu Wang [17] et al. proposed a novel Pyramidal Efficient U-Net (PyEf-U-Net) network to predict the stress field distribution of the umbilical. The experimental research demonstrates that the proposed PyEf-U-Net can accurately predict the stress field of the umbilical in real time with a prediction accuracy of 94.2%.S. Gupta [18] et al. conducted a machine learning model that it takes the microstructural image as input with a different range of Young's modulus of carbon fibers and neat epoxy, and obtains output as visualization of the stress component S11 (principal stress in the x-direction). Anindya Bhaduri [19] et al. predicted the local stress distribution in fiber reinforced matrix composite materials under mechanical loading using deep learning models. They also predicted the stress field for systems with larger number of fibers using CNNs that are pertrained on data from relatively cheaper systems with smaller

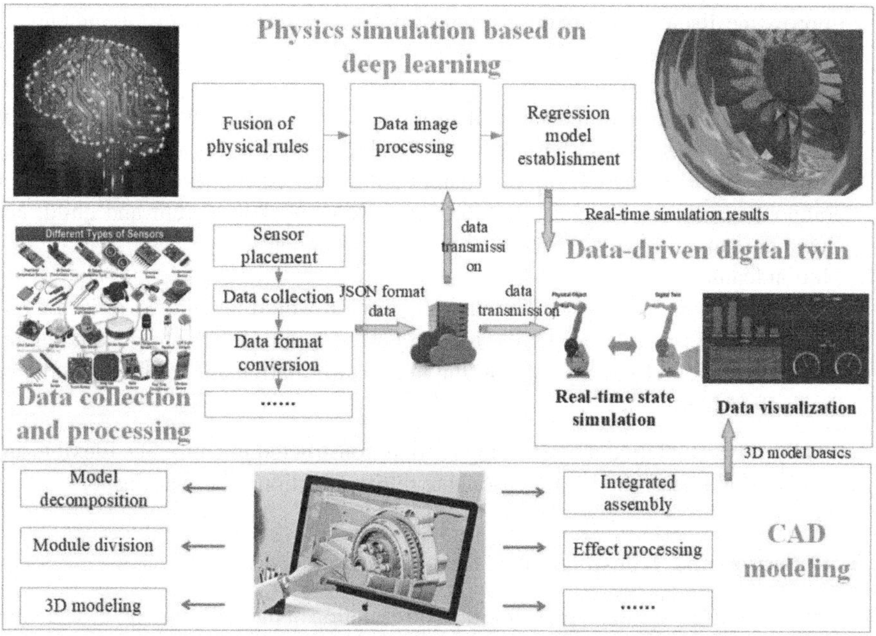

Fig. 1. Technological process

fiber number. Although these methods have improved the calculation efficiency to some degree, however, it is still difficult to realize the instantaneous calculation and output of finite element analysis results under the requirements of real-time state monitoring of grasping robots.

In this study, we build the digital twin model of a certain grasping robot and realize real-time simulation based on rapidly collected data. To generate the accurate stress figure at the same time of running, deep learning technology is used, instead of FEA, achieving quick calculating and relatively accurate results (Fig. 1).

2 Digital Twin Model Construction

2.1 The Preparation of 3D Model

According to the concept of five-dimensional framework in digital twin technology, the model plays the role of engine [20]. The first problem to be solved in the research on the state monitoring of the grasping robot based on digital twin is the model construction. The following requirements should be met:

(1) The models are simplified.

The software used in this study to realize digital twin model construction and datadriven model motion is Unity 3D. When building 3D models, it is necessary to consider the computer resource occupancy and performance allocation problem of subsequent import into the Unity 3D environment. Therefore, selecting standard parts, reducing unnecessary model details, optimizing rendering memory, etc., can improve the display effect of the grasping robot model, so as to ensure the authenticity of the model and reduce the occupation of computer resources.

(2) The structure is hierarchical.

In order to facilitate the subsequent motion of the digital twin model based on realtime data, it is necessary to set hierarchical relationships for all the components, including parallel relationship and parent-child relationship, based on which one component can automatically link to the movement of its parent and child, and shield the movement influence on its parallel components.

(3) Motion features are considered.

After obtaining a static 3D model, the motion would be applied to it in Unity 3D. It is necessary to consider the motion law of the grasping robot, including the motion direction of each axis joint, the maximum rotation angle, the limit motion range of each arm, etc.

Finally, the virtual model of grasping robot is constructed in Unity 3D. Flexible and efficient control is feasible due to reasonable setting mentioned above (Fig. 2).

2.2 Collection and Transmission of Real-Time Data

Arduino Uno is used to collect real-time data information of the grasping robot. The robot has 6 degrees of motion freedom and is driven by 6 servos that work independently. Each servo has three leads for power supply, PWM signal transmission and grounding. Under

Fig. 2. Grasping robot model in Unity 3D

the condition of stable power supply, the PWM signal line is connected to the programmable Arduino Uno board, through which the servo's running attitude information is read.

Then ESP8266, a WiFi transmission MCU, is used to transmit the data to Unity 3D for further motion application. The ESP8266 is set to STA mode to connect to the local area network to realize wireless cross-domain data transmission.

A new thread is used in Unity 3D to perform data reception. Since the servo attitude information collected from the Uno is relatively simple, for the convenience of reading, a script code is written in Unity 3D to convert the received data information into a string form for unification, and then disassemble it into ready-to-use data type for model motion.

2.3 The Motion of the Digital Twin Model

The Transform component is composed of three basic properties: position, rotation, and scale, with which basic movement, rotation and scaling of the virtual model can be performed. After the running information is transmitted from the ESP8266 to the Unity 3D environment, it has become a data type that can be directly used after processing. If substitute these real-time updated data of the pose information of the grasping robot into the parameters of the Transform component, the virtual model can be controlled to perform pose transformation. To meet the requirements of real-time update, the Transform component is placed in the loop to timely control the model's attitude, achieving the visual effect of virtual motion simulation.

3 Stress State Monitoring Based on Deep Learning

In this chapter, the stress state monitoring of the grasping robot would be studied. To simplify the calculation process and focus on the research method itself, the mechanical grasper of the robot is selected as research object, which directly contacts with outer entity, and has the most complex load distribution during the process of movement when grasping things.

3.1 Finite Element Analysis of the Grasper

By constructing a three-dimensional model of a grasping robot in CAD software and assembling it, the mechanical grasper can be imported into ABAQUS for finite element analysis. Since the bottom of the grasper is connected to the robot body, allowing for transformations and rotations, it can be considered as fixed at the base. After gripping an object, the grasper attains a state of equilibrium. In the static simulation, the opening and closing angles of the grasper can be fixed, considering only the load impact on the grasper model due to the clamped heavy object during motion. The load is applied as a surface load on both ends of the grasper. By altering the load conditions at the ends of the grasper, the stress states of the gripper under different object-gripping scenarios can be simulated.

The stress figure generated by finite element analysis under a certain working instant is shown in Fig. 3. It can be seen that the most concentrated part of the stress is the connection at the bottom of the mechanical grasper. Due to the small size and light working conditions of the robot, the material limit is rarely reached when grasping objects, but the influence on the tangential direction of the robot arm cannot be ignored, which may cause the fracture of the bottom parts and cause the overall failure.

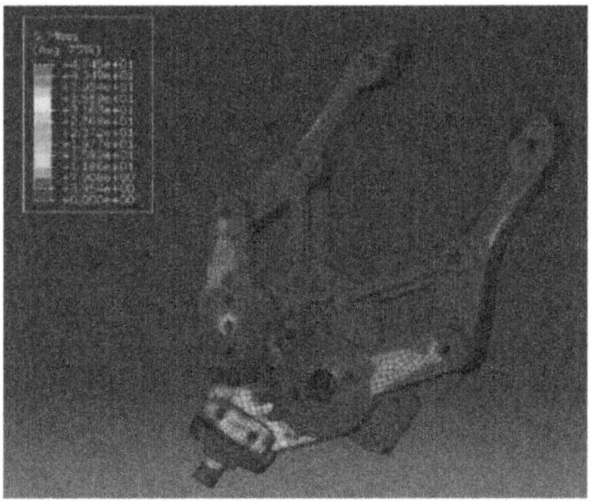

Fig. 3. The stress figure of the grasper under a certain working instant

3.2 Training Samples Collection Tool Development

The training samples of the deep learning model come from the stress figures generated in ABAQUS under different working situations. Since deep learning model training requires many samples to ensure the precision of the model, we prepared 3000 samples for training, which means we ran 3000 times of finite element simulation in ABAQUS. To improve efficiency, a sample collection tool is developed based on ABAQUS in Python language. The function of such tool is to automatically adjust the loads and submit the job for calculation. After the FEA result is generated, the tool will make sure the stress figure shows in the same view as the formerly collected figures and save it, as well as the value of all the parameters. In this way, the mechanical grasper stress figures with the size of 512 × 512 pixels are finally obtained, and the value of parameters under all the corresponding working situations is saved in the format of txt file. Figure 4 shows 20 figure samples collected by the tool.

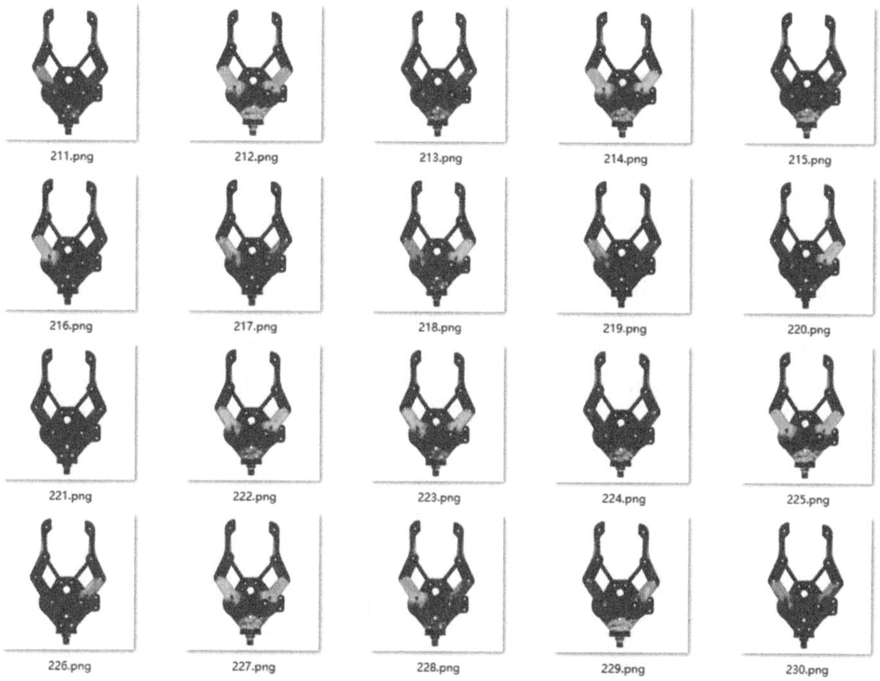

Fig. 4. Some figure samples collected by the tool

3.3 Stress State Monitoring Deep Learning Model Building

This study constructs a deep learning model based on Conditional Generative Adversarial Nets (CGAN), where the input is a matrix of mechanical parameters, and the output is stress result images. The CGAN model comprises two main components: a generator built with a fully connected neural network and a discriminator constructed using a convolutional neural network.

The generator network includes an input layer, three hidden layers, and an output layer. For specific robotic gripper loads and constraints, the stress prediction output is deterministically defined, eliminating the need for random load inputs at the output layer; instead, dimensional expansion of load parameters suffices. Within the hidden layers, data undergoes normalization and is processed through the LeakyReLU activation function, then mapped between (−1, 1) via the tanh nonlinear activation function. After reshaping, this process generates the stress result output.

The discriminator receives two three-channel images: one stress image from the generator and another from finite element analysis results. These images are merged into a six-channel image through a connector, then processed through convolution, pooling, and LeakyReLU activation, with a final mapping to a two-dimensional array in the (0, 1) range via the sigmoid function. Each element in this array represents the confidence level of pixel area information.

3.4 The Comparison of Stress Figures Generated by Deep Learning Model and ABAQUS Simulation

In order to verify the accuracy of the prediction results generated by the deep learning model, the deep learning method and the finite element method are compared. Select a random state during the robot running process, determine the load application form on the grasper model, and perform finite element analysis in ABAQUS to obtain the stress distribution figure. Meanwhile, the same load condition is input into the trained deep learning model, and a stress figure of the mechanical grasper is generated as well.

The comparison is shown in Fig. 5. It can be seen that the stress prediction image of the mechanical grasper generated by the deep learning model is highly consistent with the stress result calculated by the finite element method, and the stress concentration area can be predicted extremely well, meaning high accuracy of deep learning method.

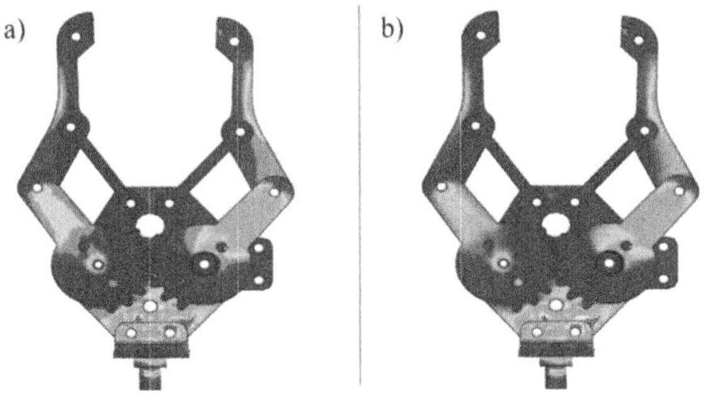

a) Finite element analysis results b) Deep learning output results

Fig. 5. The comparison of stress figures generated by these two methods

While ensuring the accuracy, the deep learning method is faster than the conventional FEA method. The finite element software ABAQUS and the deep learning model are put on the same machine to obtain the stress distribution figure of the mechanical grasper under the same load condition. By actual measurement, it takes about 20 sonds for ABAQUS to obtain the static stress figure of the mechanical grasper, while under the same load and the same operating state, the deep learning method only requires 0.003 s to generate a figure. The time cost comparison for five testing conditions are shown in Table 1.

It should be pointed out that the time consumed by the FEA method to generate the stress figure includes more computing contents such as grid node calculation, image rendering, etc., which are not computed by deep learning model. However, it is out of

Table 1. Time comparison of five testing conditions

Serial number	ABAQUS stress image generation time (s)	Deep learning prediction image generation time (s)
1	21	0.003
2	21	0.003
3	20	0.003
4	20	0.003
5	20	0.003

question that FEA method takes more time. For scenarios such as digital twin model mapping reality which have extremely high real-time requirements, deep learning methods have great research and application value.

3.5 Conversion Between Pose Parameters and Load Parameters

During the gripping process, the spatial position and orientation of the grasper continuously change, resulting in variations in the load applied by the gripped object on both ends of the grasper. Therefore, it is necessary to convert the servo motor posture information into load parameters, which will then serve as input parameters for the deep learning model.

As shown in Fig. 6. When the manipulator holds the object, the force analysis of the object is:

$$F_f = Gsin\alpha = Gsin(\theta 1 + \theta 2 - \theta 3) \tag{1}$$

$$F_{ver} = Gcos\alpha = Gcos(\theta 1 + \theta 2 - \theta 3) \tag{2}$$

In this context, G represents the weight of the heavy object, and F_f is the frictional force provided by the robotic grasper, preventing axial sliding of the object. F_{ver} denotes the force exerted by the robotic grasper perpendicular to the grasper plane. The angles $\theta_1, \theta_2, \theta_3$ represent the pose angles of the robotic hand. At this moment, the robotic arm experiences an equal but opposite force F'_f, with the load parameter in the x-direction denoted as F_x.

$$F'_f = -Gsin\alpha = Gsin(\theta_1 + \theta_2 - \theta_3) = F_{1x} + F_{2x} \tag{3}$$

$$F_{1x} = F_{2x} = -\frac{1}{2}Gsin(\theta_1 + \theta_2 - \theta_3) \tag{4}$$

After excluding the role of F_f in axial slip, the mechanical gripper is observed along. the mechanical gripper plane, as shown in Fig. 7.

$$F_{1y} = F_{2y} = \frac{1}{2}F_{vera} = \frac{1}{2}Gcos(\theta_1 + \theta_2 - \theta_3)cos\theta_4 \tag{5}$$

$$\begin{cases} F_{1z} = -Gcos(\theta_1 + \theta_2 - \theta_3)sin\theta_4 \\ F_{2z} = 0 \end{cases} or \begin{cases} F_{1z} = 0 \\ F_{2z} = -Gcos(\theta_1 + \theta_2 - \theta_3)sin\theta_4 \end{cases} \quad (6)$$

The resultant force of F_{vera} and F_{verb} is F_{ver}, F_{vera} is composed of F_{1y} and F_{2y} on the two forearms, F_{verb} is provided by the lower arm F_y, and F_z is the load parameter in the y and z directions.

θ_1, θ_2, θ_3, θ_4 are directly related to the attitude Angle of the steering gear. Through the above conversion, the mapping from the real-time attitude data of the steering gear to the real-time load parameters is realized.

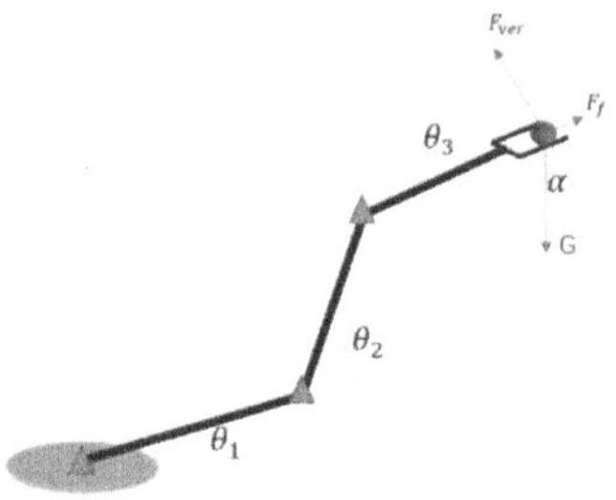

Fig. 6. Attitude and force analysis of grasping robot

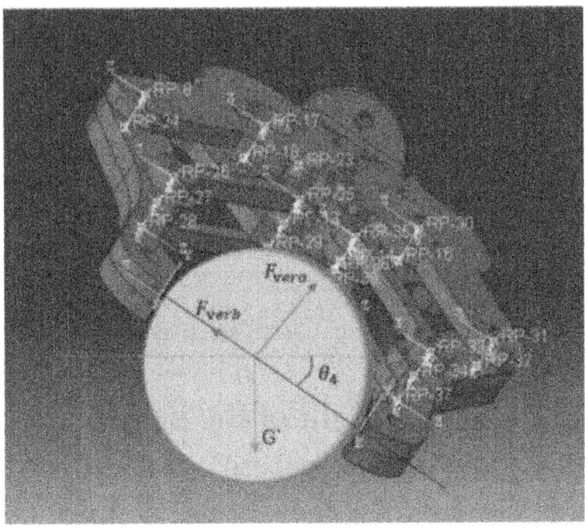

Fig. 7. Grasper force analysis diagram in local coordinate system

3.6 Stress State Animation Generation

To enhance real-time stress monitoring in robotic grippers, static stress images are converted into video animations within the Unity 3D environment. In the process of grabbing objects, the pose parameters are collected and converted into the load parameters on the grasper, which are used as the input of the deep learning model.Here, deep learning model outputs are serialized into byte arrays and stored in memory streams, enabling efficient data transfer and integration across different software environments. This approach, leveraging byte arrays, is space-efficient and fast-loading, meeting digital twin real-time requirements. Despite each animation frame being generated from unique load parameters, the continuous nature of robotic operation ensures smooth transitions between frames, avoiding abrupt visual changes. This results in a coherent animation that effectively visualizes stress changes in robotic grippers, facilitating realtime monitoring and analysis.

4 Conclusions

In this research, the digital twin model of the grasping robot is constructed and the stress state monitoring of the mechanical grasper component based on deep learning is studied. In view of the lack of cross-regional, cross-equipment, high-fidelity and vivid monitoring of the running state of the grasping robot in the conventional method, the digital twin method is used by building virtual model in Unity 3D and driving the model moving with the real-time data. Aiming at the problem that the finite element method has high hardware requirements and long calculation time for computing equipment, and cannot meet the real-time output of stress state during operation, the deep learning method is used to realize the real-time response of stress figure output during the grasper operating through pre-training in advance and calling when needed. The stress monitoring animation is generated based on Unity 3D.

5 Discussion

In this paper, the construction of the digital twin model of the grasping robot and the monitoring of the stress state based on deep learning method are carried out. Due to the limited time and energy, the research of this paper can still be further expanded.

Firstly, the research object of this paper is a relatively simple physical entity. Even so, in order to build its virtual model, more than 100 3D models were built and assembled in the early stage. It is foreseeable that modeling physical entities in the research and development of digital twins is a time-consuming and labor-intensive process. It is a valuable development direction in the field of digital twins that how to achieve rapid modeling of equipment and scenes. Some existing technologies such as 3D scanner rapid modeling are good methods to be combined with digital twins.

Secondly, the real-time stress state prediction of the grasper of the grasping robot is based on the deep learning method, and the training samples come from the finite element simulation. In this research we only focus on the stress figures of the mechanical grasper, and use the figures as the training input of the deep learning model. In the follow-up research, the node data information of the model can be devoted more effort into, which includes multi-dimensional data such as stress and strain of the model, with which deep learning model can give more accurate prediction results, and even give deformation prediction, further improving the monitoring effect. In addition, combing node data with 3D model makes dynamic display possible, which is also a promising research direction.

References

1. Leitao, P., Kamouskos, S., Ribeirio, L.: Smart agents in industrial cyber-physical systems. Proc. IEEEoc. IEEE **5**(104), 1086–1101 (2016)
2. Pirvu, B., Zamfirescu, C., Gorecky, D.: Engineering insights from an anthropocentric cyber-physical system: a case study for an assembly station. Mechatronics **34**(2), 147–159 (2016)
3. Kuo, S., Chou, Y., Chen, C.: Quantum-inspired algorithm for cyber- physical visual surveillance deployment systems. Comput. Netw. **117**(10), 5–18 (2017)
4. You, M., Liu, Q., Sun, H.: New communication strategy for spectrum shar- ing enabled smart grid cyber-physical system. IET Cyber-Phys. Syst.: Theory Appl. **2**(3), 136–142 (2017)
5. Vachálek, J., Bartalský, L., Rovný, O., et al.: The digital twin of an industrial production line within the industry 4.0 concept. In: 2017 21st International Conference on Process Control, pp. 258–262 (2017)
6. GRIEVES M. Digital twin: manufacturing excellence through virtual factory replication [EB], pp. 12–20 (2016)
7. Tuegel, E.: The airframe digital twin: some challenges to realization. In: 53rd AIAA/ASME/ASCE/AHS/ASC Structures, Structural Dynamics and Materials Conference 20th AIAA/ASME/AHS Adaptive Structures Conference 14th AIAA (2012)
8. Warwick. G.: GE advances analytical maintenance with digital twins. Aviation Week and Space Technology, 10–19 (2015)
9. Tuegel, E.J., Ingraffea, A.R., Thomas, G., et al.: Eason reengineering aircraft structural life prediction using a digital twin. Int. J. Aerosp. Eng. **8**(2), 1–14 (2011)
10. Li, C., Mahadevan, S., Ling, Y., et al.: Dynamic Bayesian network for aircraft wing health monitoring digital twin. AIAA J. **55**(3), 930–941 (2017)
11. Glaessgen, E., Stargel, D.: The digital twin paradigm for future NASA and US Air Force vehicles. In: 53rd Structures, Structural Dynamics and Materials Conference (2012)
12. Bjerkeng, M., Falco, P., Natale, C., et al.: Stability analysis of a hierarchical architecture for discrete-time sensor based control of robotic systems. IEEE Trans. Rob. **30**(3), 745–753 (2017)
13. Alejandro, E., Jesus, R., Yosimar, J.: Vibration suppression for improving the estimation of kinematic parameters on industrial robots. Shock. Vib. **15**(23), 178–182 (2016)
14. Raza, N., Mehmood, I., Rafiuddin, H., et al.: Numerical simulation of added mass determination of standard ellipsoids. In: Proceedings of 2012 9th International Bhurban Conference on Applied Sciences & Technology (IBCAST), pp. 270–273 (2012)
15. Brandvik, T., Pullan, G.: Acceleration of a 3D Euler solver using commodity graphics hardware. In: 46th AIAA Aerospace Sciences Meeting and Exhibit **22**(7), 443–456 (2008)
16. Smith, W.D., Schnore, A.R.: Towards an RCC-based accelerator for computational fluid dynamics applications. J. Supercomput. **30**(3), 239–261 (2004)
17. Wang, L., et al.: Real-time stress field prediction of umbilical based on PyEf-U-net convolutional neural network (2023)
18. Gupta, S., Mukhopadhyay, T., Kushvaha, V.: Microstructural image based convolutional neural networks for efficient prediction of full-field stress maps in short fiber polymer composites. Defence Technology **24**, 58–82 (2023)
19. Bhaduri, A., Gupta, A., Graham-Brady, L.: Stress field prediction in fiber- reinforced composite materials using a deep learning approach. Compos. Part B: Eng. **238** (2022)
20. Tao, F., Xiao, B., Qi, Q., Cheng, J., Ji, P.: Digital twin modeling. J. Manuf. Syst. **64**, 372–389 (2022)

MESLedger: Investigating Manufacturing Execution Improvements Through the Integration of AI and Blockchain Technologies

Abdelhak Belhi[1](✉), Khaled Benfriha[2], and Abdelaziz Bouras[3]

[1] Joaan Bin Jassim Academy for Defence Studies, Al Khor, Qatar
abdelhak.belhi@jbj.edu.qa
[2] Arts et Métiers Institute of Technology, Paris, France
khaled.benfriha@ensam.eu
[3] CSE, College of Engineering, Qatar University, Doha, Qatar
abdelaziz.bouras@qu.edu.qa

Abstract. Digitalized manufacturing processes necessitate a shift from traditional production control systems to more intelligent frameworks. Existing systems reliant on Programmable Logical Controllers (PLCs) fall short in handling the influx of data generated by the Internet of Things (IoT) layer and the array of IT systems integral to modern operations. This paper proposes the development of a novel manufacturing execution environment capable of dynamically analyzing data, orchestrating operations, and making informed decisions, especially in response to malfunctions. Furthermore, such environment aims to optimize production activities based on diverse priorities such as cost, energy efficiency, and production time. This necessitates a reevaluation of manufacturing operations, incorporating both standard and parametric design elements. The success of such environment is crucial for industries motivated to achieve optimal performance through intelligent data utilization. From a scientific perspective, the challenge lies in devising scalable algorithms capable of autonomously driving digital production systems while maintaining a high level of adaptability and efficiency. This paper presents MESLedger, a collaborative project aiming at improving manufacturing execution through the integration of blockchain technology to secure communication in collaborative environments and artificial intelligence to optimize major aspects of production. It also emphasizes on the significance of this effort in meeting industrial demands and outline the scientific advancements required to realize the project's objective.

Keywords: Manufacturing Execution System · Blockchain · Artificial Intelligence · Smart Manufacturing · Industry 4.0

P. Sureephong et al. (Eds.): PLM 2024, IFIP AICT 741, pp. 307–316, 2025.
https://doi.org/10.1007/978-3-031-93323-3_23

1 Introduction

The aim of an intelligent system, also known as "Smart Manufacturing System", is to enhance productivity and resource efficiency in real-time production environments. Achieving this objective necessitates flexible responses to changes in the manufacturing landscape, as well as real-time control, scheduling, and optimization capabilities. The integration of Internet of Things (IoT) and Cyber-Physical Systems (CPS) mark the transition to the fourth industrial revolution, making factories more inclusive and intelligent [1]. Cyber-physical systems combine computational decision-making and communication capabilities with integrated physical activities, such as coordinating, monitoring, and controlling physical operations and engineering systems. This integration opens up new avenues for addressing industrial challenges. Furthermore, the concept of "Industry 4.0" establishes a strong connection between the real and virtual worlds, resulting in increased flexibility, cost savings, and enhanced customization. Jackson et al. distinguish between dynamic and static scheduling, where static scheduling assumes all processing jobs are ready and does not require adjustments once established, while dynamic scheduling continuously adapts to changing circumstances [2]. Dynamic events, categorized into four classes, present challenges such as machine damage, operational delays, changes in delivery dates, and production instability. While dynamic scheduling is complex, advancements in real-time data accessibility have increased its relevance. Various scheduling approaches exist, from entirely reactive to predictive-reactive, with the latter modifying schedules in response to current events. Incorporating real-time information into decision models across manufacturing processes enables prompt responses to disturbances without disrupting operations. Real-time connectivity, monitoring, and control enhance production systems, with scheduling and planning playing crucial roles. Recent research explores the impact of this manufacturing paradigm shift on production planning and control, highlighting the potential contributions of scheduling research to intelligent manufacturing. The term "real-time" in production decisions signifies immediate responses to disturbances and unexpected events, ensuring uninterrupted operations [3].

Security and reliability of information communication are also important aspects to consider especially with the context of Industry 4.0. Manufacturing Execution Systems (MES) which encounters numerous security challenges necessitating attention to safeguard manufacturing operations' integrity and confidentiality. These challenges include maintaining data integrity to thwart unauthorized modifications or tampering, implementing stringent access controls to limit system and data access to authorized personnel, mitigating cyber threats like malware and ransomware through robust cybersecurity measures such as network segmentation and intrusion detection, addressing supply chain risks by conducting thorough security assessments of third-party systems and implementing supply chain security measures, and ensuring physical security measures such as access control systems and surveillance to prevent unauthorized access or tampering with MES systems and infrastructure. Addressing these challenges comprehensively through technical controls, cybersecurity best practices, and robust risk management strategies is vital for securing MES systems and ensuring uninterrupted manufacturing operations [4].

The integration of blockchain with a Manufacturing Execution System (MES) offers manufacturers a multitude of benefits, including enhanced transparency, traceability, security, and efficiency in production operations. By leveraging blockchain technology, manufacturers can establish an immutable and transparent ledger to track every step of the manufacturing process, from raw material sourcing to final product delivery, ensuring greater visibility and accountability. Smart contracts enable automation of various tasks, such as inventory management and supplier payments, streamlining processes and reducing administrative overhead. Additionally, blockchain facilitates decentralized manufacturing networks, enabling direct interaction between multiple stakeholders and fostering collaboration and interoperability across the manufacturing ecosystem. Overall, blockchain integration with MES holds promise for driving innovation and competitiveness in the manufacturing industry [4].

The remainder of this paper is organized as follows. In section two, we present works related to AI and blockchain integration in the context of manufacturing execution. In section three, we present MESLedger, a research initiative aiming at investigating the benefits of AI and optimization in addition to the blockchain technology with MES systems. In Sect. 4, we draw our final remarks and present future research challenges.

2 Related Work

Recent years have witnessed significant advancements in technology, profoundly impacting industrial performance. One such innovation is the workshop production system, strategically designed to facilitate the simultaneous production of diverse products within manufacturing plants. Its primary objectives include minimizing cycle times, ensuring high schedule reliability, optimizing capacity utilization, and maintaining low inventory levels. Central to this system is production scheduling, a pivotal process that involves planning, resource allocation, and sequencing operations to achieve various objectives such as maximizing efficiency and minimizing costs and production cycle times [5].

However, the complexity of real-world Job Shop Scheduling Problems (JSSP) often poses significant challenges. These problems, typically characterized as non-deterministic (NP) hard combinational optimization, require careful consideration. Traditional optimization algorithms, primarily designed for centralized scheduling and smaller-scale problems, struggle to ensure computational stability regarding cost and time optimization in JSSP. Thus, a more comprehensive analysis of the problem is imperative [6, 7].

Recognizing the need for enhanced productivity in the era of Industry 4.0, industries have increasingly focused on smart distributed scheduling (SDS) methods for modeling and optimization. Industry 4.0, also known as Smart Manufacturing, emphasizes advanced manufacturing techniques aimed at creating adaptable systems capable of automatically adjusting production processes to varying product requirements and conditions. This approach not only enhances quality, efficiency, and flexibility but also facilitates the sustainable production of customized items on a large scale with minimal resource consumption [8].

Despite the successes achieved through digitalization and AI in Industry 4.0, the rapid evolution of technology underscores the importance of continued innovation. AI applications, with their potential to contribute significantly to the global economy, are increasingly being adopted across various industrial sectors to drive competitiveness, innovation, cost reduction, and performance improvement.

However, the widespread adoption of Industry 4.0 principles faces challenges, primarily due to the time-consuming nature of the transition process. Nevertheless, machine learning techniques, such as reinforcement learning, are already demonstrating their utility in optimizing manufacturing processes and detecting anomalies, contributing to the industry's advancement [2].

Furthermore, security remains a pressing concern in the production industry. Blockchain technology, with its decentralized and tamper-proof nature, offers a promising solution to address information security challenges. Initially developed for maintaining public ledgers in monetary transactions, blockchain finds applications across diverse domains, including healthcare, finance, and supply chains. Its adoption in Qatar's economic sector, particularly within supply chains, holds significant potential for enhancing transparency, traceability, and efficiency while reducing costs and time [9, 10].

The emergence of Blockchain 2.0 introduces programmable infrastructure capable of executing self-triggered contracts known as Smart Contracts. These contracts, residing within the blockchain, offer a versatile means of automating agreements based on predefined conditions, further streamlining operations across various sectors.

3 MESLedger Approach

Current methodologies are only partially conventional, lacking real-time data consideration and robust decision-making capabilities. MESLedger project's approach aims to equip the control system with blockchain and artificial intelligence, enabling continuous learning and introducing new security paradigms. Through MESLedger, our aim is to investigate the integration of AI and blockchain for addressing scheduling, malfunctions, and security challenges in modern manufacturing execution scenarios. Our first major task consists of studying the integration of AI in the context of smart manufacturing aiming at optimizing the manufacturing execution process. To address issues related to malfunctions and scheduling, we propose employing AI techniques for anomaly detection and optimization, aiming to minimize process anomalies and defects through various monitoring approaches. Reinforcement learning, a subset of machine learning, offers promise in enhancing efficiency, reducing costs, and optimizing production processes, particularly in tasks such as production scheduling and Automated Guided Vehicles (AGV) operations, which we plan to explore for our scenario [8, 11]. The second major task consists of using blockchain technology as a secure and reliable collaboration platform focusing on the aspects related to Industrial IoT (IIoT) and collaborative manufacturing where data integrity and smooth collaboration plays a great role. In the following, we present details about these two major tasks [12–14]. Blockchain technology holds immense potential to revolutionize manufacturing, offering benefits such as improved supply chain management, enhanced data security, transparency, and the facilitation of smart contracts and autonomous systems within Industry 4.0. In integrating blockchain with manufacturing execution, our objective is to explore its added value in decentralized manufacturing, particularly in enhancing security, transparency, traceability, and tamper-proofing. Our aim is to oversee critical manufacturing processes through blockchain smart contracts, transitioning from centralized to federated management of product manufacturing, particularly beneficial for large product lines.

Although the traditional MES will remain a component, the blockchain platform will take charge of critical processes, as depicted in Fig. 1. We have opted for Hyperledger Fabric as our blockchain development platform and a conventional MES for integration, aiming to delve deeper by exploring integration solutions and data exchange with SCADA/PLC platforms to leverage the full potential of decentralized architecture. This involves assessing the feasibility of deploying blockchain endpoints on edge devices, potentially requiring firmware adjustments. The overall research methodology is structured sequentially to meet project objectives, encompassing steps to transition existing production systems into intelligent and secure setups [15].

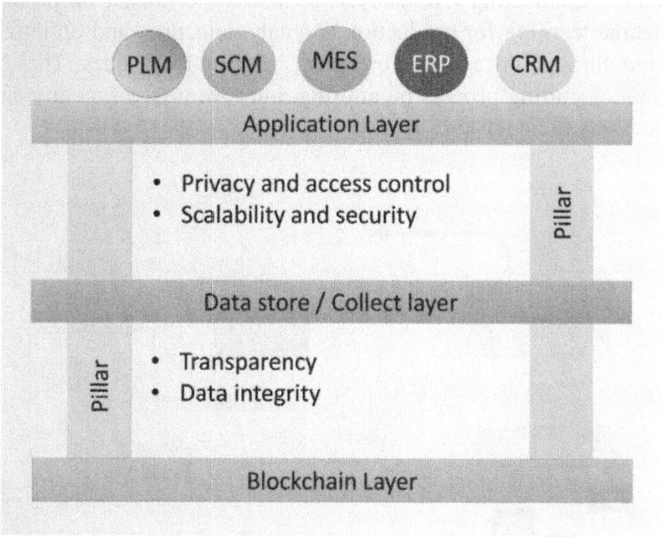

Fig. 1. The different integration layers of our scenario

3.1 AI Optimization of Manufacturing Execution

The primary aim of this task is to create an AI-driven solution for managing data orchestration and detecting product malfunctions within manufacturing processes, with a focus on monitoring both the execution processes and product quality. This entails employing manufacturing data flow analytics, where logs from Supervisory Control and Data Acquisition (SCADA)/PLC platforms are scrutinized by an AI-based event anomaly detection tool. AI offers many ways to enhance MES within the context of AGVs, PLCs, and SCADA systems [10, 16, 17]. Through predictive maintenance, AI can anticipate equipment failures by analyzing real-time data, thus allowing for proactive scheduling of maintenance activities to minimize downtime. Additionally, AI algorithms can optimize AGV routing and scheduling based on various factors such as production demands, traffic conditions, and equipment availability, resulting in improved efficiency and reduced waiting times. Additionally, Dynamic reconfiguration is another key area where AI can have a significant positive impact, as it enables production lines to adapt swiftly to changing demands, resource availability, and environmental conditions. This flexibility ensures agile and real-time responses to market fluctuations and unexpected disruptions. Moreover, AI-powered image recognition and machine learning algorithms enhance quality control by inspecting products in real-time to detect defects and deviations from quality standards, thereby reducing waste and rework (See Fig. 2 and Fig. 3).

In our context, triggers will be implemented to identify and understand the source and reasons behind any malfunctions. Additionally, to address product quality concerns, an AI visualization tool will be developed for inspecting product quality along manufacturing lines. The tasks within this package include identifying manufacturing anomalies using AI models, conducting risk analysis to assess the impact on production lines, utilizing machine learning for production anomaly detection, and optimizing production parameters through AI and reinforcement learning techniques. These efforts aim to enhance manufacturing process monitoring, improve product quality, and optimize overall production efficiency [8].

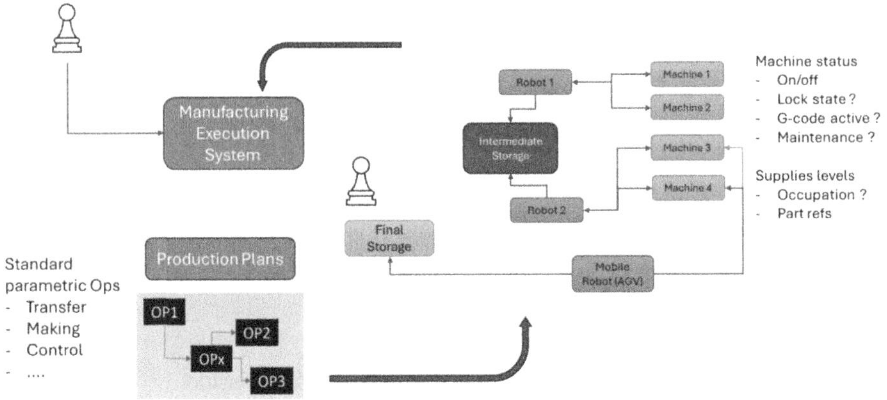

Fig. 2. Our manufacturing plant scenario

3.2 Blockchain in Manufacturing Execution

We have previously investigated the integration of blockchain with information systems within industrial settings, with a primary focus on a manufacturing supply chain use case. Rather than pursuing a conventional end-to-end integration at the transactional level between the supply chain and blockchain services, our innovative method centers on establishing a shared metamodel. This metamodel encapsulates data structures, business logic, transaction triggers, and more, serving as a blueprint for constructing and updating data models and business logic within both the information system and the blockchain platform. The core of our solution lies in the mechanisms required to develop and translate this metamodel seamlessly between the blockchain and information systems. In our previous work, we have explored the both synchronous and asynchronous data exchange between the two platforms [15, 18].

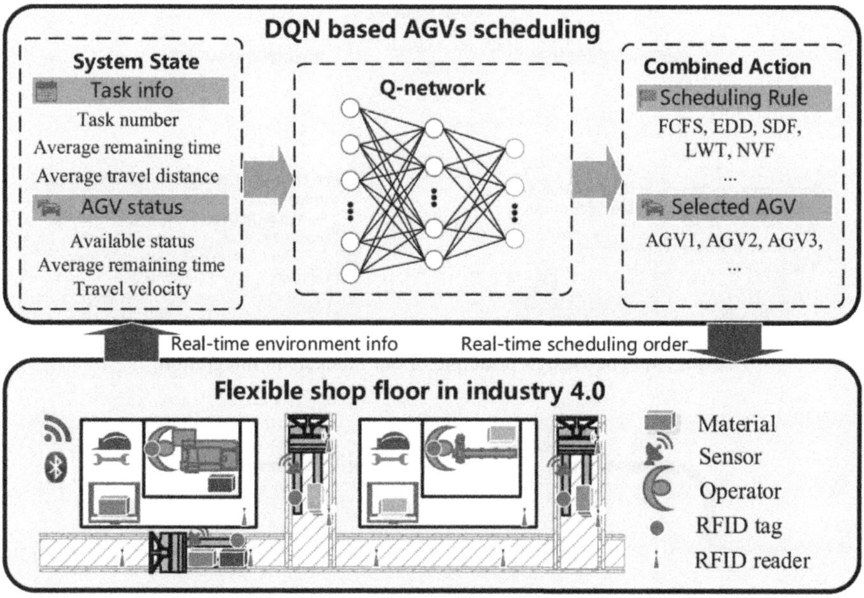

Fig. 3. AGV scheduling approach

The architecture of the proposed approach is streamlined, envisioning integration between two heterogeneous systems—the Odoo ERP Enterprise Resource Planning system and the Hyperledger Blockchain platform—at a higher architectural level. To facilitate communication between these systems, we advocate for two communication modes determined by the transaction criticality index. For less critical transactions not necessitating data locking, asynchronous mode is employed to exchange information between the ERP and the blockchain platform. Conversely, for more critical transactions mandating data locking, data is withheld from commitment until fully integrated into the blockchain platform, as this approach is deemed to better enforce business logic. At the heart of our strategy is the establishment of a shared metamodel, describing database table schemas, business rules ensuring data integrity, data triggers, and transactional details.

Building on this work, in MESLedger, our aim is to implement an integration between MES and blockchain platforms, leveraging the decentralized nature of blockchain for more effective management of manufacturing execution tasks. The focus will be on identifying the needs and possibilities of integrating MES with other platforms, particularly utilizing available APIs [15, 19, 20]. The objective is to harness the advantages offered by blockchain, such as enhanced security, tamper-proof features, and decentralization, within the context of manufacturing processes. This task involves identifying and characterizing potential constraints on production lines, emphasizing requirement analysis and identification for the integration of blockchain with MES systems, with a specific focus on SCADA/PLC devices. In addition, it proposes a decision decentralization plan based on blockchain to facilitate dialogue between components of Cyber Physical Systems (See Fig. 4 and Fig. 5).

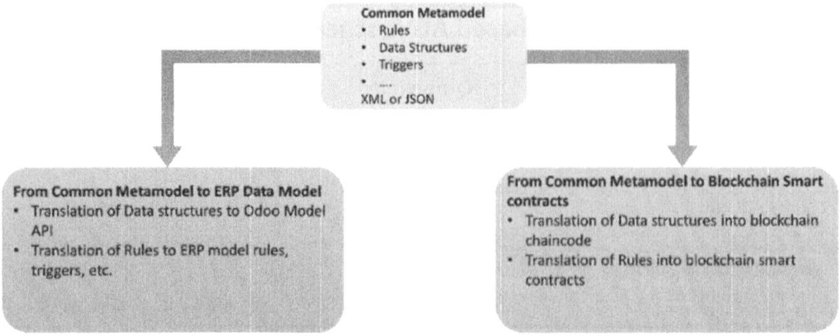

Fig. 4. The Design principle of our blockchain integration

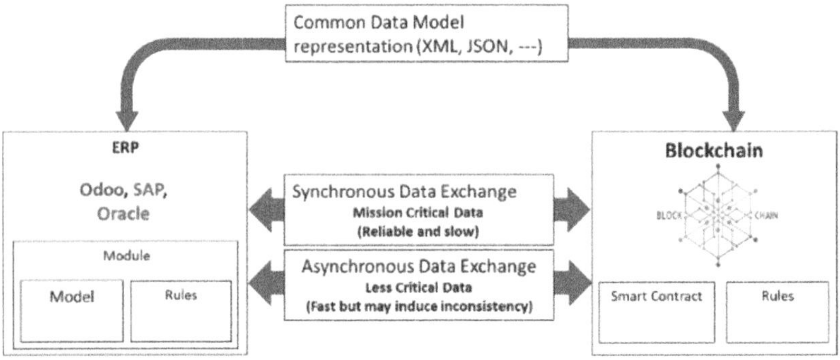

Fig. 5. The data exchange modes in pour blockchain integration

3.3 Anomaly Detection

During previous work, we investigated the detection and recovery from business process anomalies in the context of supply chain processes. Our approach consisted of modelling business processes and the establishment of SWRL rules to detect and classify aberrant processes. Illustrated in Fig. 6 is a depiction of a typical procurement workflow alongside various anomaly types. We conducted diverse anomaly simulations utilizing ProM 6.10, a renowned Process Mining tool offering extensive capabilities in simulating business processes and generating logs. Furthermore, we employed Deep Learning techniques like Long Short-Term Memory (LSTM) to forecast normal behavior, enabling comparison with real data to identify anomalies and issues within manufacturing data flows [21].

Our aim in MESLedger is to be able to detect anomalies at multiple stages of the manufacturing execution. The integration of our approach and the use of blockchain dataflow will ensure an accurate handling of manufacturing execution and a effective identification and correction of anomalies.

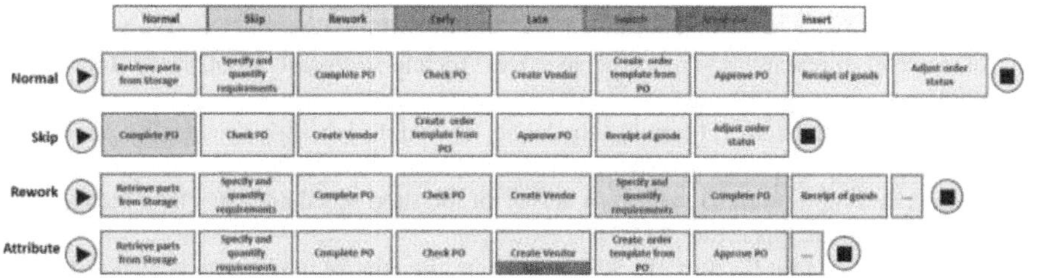

Fig. 6. Different business process anomalies

4 Conclusion

With Industry 4.0, there is a pressing need to transition from traditional production control systems to more intelligent frameworks. Traditional systems have a difficulty in working within highly digitized environments. Therefore, this paper presents a new initiative, called MESLedger, aiming for the development of a pioneering manufacturing execution system capable of dynamically analyzing data, orchestrating operations, and making informed decisions, especially in response to malfunctions. This system aims to optimize production activities considering diverse priorities like cost, energy efficiency, and production time. Achieving this requires a reevaluation of manufacturing operations, integrating both standard and parametric design elements. Its success is paramount for industries striving for peak performance through intelligent data processing. From a scientific standpoint, the challenge lies in making scalable algorithms capable of autonomously driving digital production systems while upholding adaptability and efficiency. Additionally, through the blockchain technology, we aim at securing the data exchange in IIoT environments to make smart manufacturing systems more secure and resilient. These innovations are aimed at addressing pressing challenges in the context of manufacturing execution and the expected results are set to make a new paradigm in smart manufacturing. These developments are made within the current QU-IRCC-CENG-24/25-517 MESLedger project, a joint applied research project between Qatar University and ParisTech Arts et Métiers university.

Acknowledgment. This publication was supported by Qatar University Internal Grant No. QU-IRCCCENG-24/25-517. The findings achieved herein are solely the responsibility of the authors.

References

1. Wang, B., Tao, F., Fang, X., Liu, C., Liu, Y., Freiheit, T.: Smart manufacturing and intelligent manufacturing: a comparative review. Engineering **7**, 738–757 (2021)
2. Magd, H., Jonathan, H., Khan, S.A., El Geddawy, M.: Artificial intelligence— the driving force of Industry 4.0. Roadmap Enabling Ind. 4.0 b Artif. Intell. 1–15 (2022)
3. Bousdekis, A., Lepenioti, K., Apostolou, D., Mentzas, G.: A review of data-driven decision-making methods for industry 4.0 maintenance applications. Electronics. 10, 828 (2021)
4. Shojaeinasab, A., et al.: Intelligent manufacturing execution systems: a systematic review. J. Manuf. Syst. **62**, 503–522 (2022)

5. Meindl, B., Ayala, N.F., Mendonça, J., Frank, A.G.: The four smarts of industry 4.0: evolution of ten years of research and future perspectives. Technol. Forecast. Soc. Change. **168**, 120784 (2021)
6. Fattahi, P., Fallahi, A.: Dynamic scheduling in flexible job shop systems by considering simultaneously efficiency and stability. CIRP J. Manuf. Sci. Technol. **2**, 114–123 (2010)
7. Xiong, H., Shi, S., Ren, D., Hu, J.: A survey of job shop scheduling problem: the types and models. Comput. Oper. Res. **142**, 105731 (2022)
8. Hu, H., Jia, X., Liu, K., Sun, B.: Self-adaptive traffic control model with behavior trees and reinforcement learning for AGV in industry 4.0. IEEE Trans. Ind. Informatics. **17**, 7968–7979 (2021)
9. Gupta, R., Tanwar, S., Kumar, N., Tyagi, S.: Blockchain-based security attack resilience schemes for autonomous vehicles in industry 4.0: a systematic review. Comput. Electr. Eng. **86**, 106717 (2020)
10. Mathew, D., Brintha, N.C., Jappes, J.T.W.: Artificial intelligence powered automation for industry 4.0. In: New Horizons for Industry 4.0 in Modern Business, pp. 1–28. Springer (2023)
11. Hu, H., Jia, X., He, Q., Fu, S., Liu, K.: Deep reinforcement learning based AGVs real-time scheduling with mixed rule for flexible shop floor in industry 4.0. Comput. Ind. Eng. **149**, 106749 (2020)
12. Tyagi, A.K., Dananjayan, S., Agarwal, D., Thariq Ahmed, H.F.: Blockchain— internet of things applications: opportunities and challenges for industry 4.0 and society 5.0. Sensors. **23**, 947 (2023)
13. Ekblaw, A., Azaria, A., Halamka, J.D., Lippman, A.: A case study for blockchain in healthcare: "MedRec" prototype for electronic health records and medical research data. In: Proceedings of IEEE Open & Big Data Conference, Vienna, Austria, p. 13. (2016)
14. Abeyratne, S.A., Monfared, R.P.: Blockchain ready manufacturing supply chain using distributed ledger. Int. J. Res. Eng. Technol. **5**, 1–10 (2016)
15. Belhi, A., Gasmi, H., Bouras, A., Aouni, B., Khalil, I.: Integration of business applications with the blockchain: odoo and hyperledger fabric open source proof of concept. IFAC-PapersOnLine. **54**, 817–824 (2021)
16. Kraus, N., Kraus, K., Shtepa, O., Hryhorkiv, M., Kuzmuk, I.: Artificial intelligence in established of industry 4.0. WSEAS Trans. Bus. Econ. 1884–1900 (2022)
17. del Real Torres, A., Andreiana, D.S., Ojeda Roldán, Á., Hernández Bustos, A., Acevedo Galicia, L.E.: A review of deep reinforcement learning approaches for smart manufacturing in industry 4.0 and 5.0 framework. Appl. Sci. **12**, 12377 (2022)
18. Belhi, A., Gasmi, H., Hammi, A., Bouras, A., Aouni, B., Khalil, I.: TMSLedger: a transactions management system through integrated odoo hyperledger smart contracts. In: Blockchain Driven Supply Chains and Enterprise Information Systems, pp. 201–220. Springer (2023)
19. Lee, J., Azamfar, M., Singh, J.: A blockchain enabled cyber-physical system architecture for industry 4.0 manufacturing systems. Manuf. Lett. **20**, 34–39 (2019). https://doi.org/10.1016/j.mfglet.2019.05.003
20. Lee, C.K.M., Huo, Y.Z., Zhang, S.Z., Ng, K.K.H.: Design of a smart manufacturing system with the application of multi-access edge computing and blockchain technology. IEEE Access **8**, 28659–28667 (2020). https://doi.org/10.1109/ACCESS.2020.2972284
21. Musa, T.A., Bouras, A.: Anomaly detection in blockchain-enabled supply chain: an ontological approach. In: IFIP International Conference on Product Lifecycle Management, pp. 253–266. Springer (2021)

Navigating the Future

Integrating Safety Passports for Enhanced Product Implementation on Machinery - Conceptualization and Use Case

Alexandra Saliger(✉) and Manfred Grafinger

Institute of Engineering Design and Product Development, TU Wien, Getreidemarkt 9, 1060 Vienna, Austria
alexandra.saliger@tuwien.ac.at

Abstract. As industry moves towards smarter and more connected processes, the need for efficient and secure implementation of safety information and product lifecycle information on manufacturing machines, industrial robots and in this representational use case a modular manipulator, becomes increasingly important. The article lights up the development of a safety passport tailored to manufacturing machines across their entire life cycle. Inspired by the digital product passports our conceptual framework explores the integration of such a safety passport on certified production machines throughout their lifetime.

The research emphasizes the importance of aligning the safety passport with existing development processes and presents a comprehensive approach to seamlessly integrate safety aspects from design implementation, production, usage, maintenance and dismantling. Case studies and stakeholder insights are considered and provide a real-world context, highlighting successful implementations and potential challenges. It concludes by recommending strategies for implementation beginning in the engineering department and outlining avenues for future research to continuously refine and optimize security passport systems.

Keywords: Safety · digital product passport · product life cycle

1 Introduction

In an era characterized by rapid technological advances and an evolving industrial landscape, the seamless integration of products into operational machinery is a critical determinant of overall efficiency and safety. Recognizing the need to accelerate and strengthen this integration process, this paper proposes the development of a specialized safety passport tailored to products originating from the engineering department. As the development department plays a pivotal role in shaping the products that drive industries forward, the implementation of an effective security passport system holds the promise of not only streamlining processes, but also strengthening security protocols. This paper undertakes a comprehensive exploration of the conceptual framework, requirements and integration strategies to contribute to the discourse on improving the efficiency and safety.

Published by Springer Nature Switzerland AG 2025
P. Sureephong et al. (Eds.): PLM 2024, IFIP AICT 741, pp. 317–328, 2025.
https://doi.org/10.1007/978-3-031-93323-3_24

In industry, increased safety standards are essential to ensure the physical integrity of employees. The use of machinery, heavy equipment and potentially hazardous materials poses a high risk of accidents and injuries. Safety measures also help to increase productivity by minimizing downtime due to accidents or injuries. A good safety record builds trust with stakeholders, customers and the community. In addition, high safety standards help companies meet global expectations, standards and facilitate international co-operation.

2 Conceptual Background

Crafting a security passport in accordance with established norms and standards necessitates a meticulous approach. Analogous to Digital Product Pass (DPP), a safety passport can be envisioned as a comprehensive entity encompassing a distinctive product identifier and extensive data gathered from diverse stakeholders in the value chain [1]. Adhering to recognized norms and standards, the safety passport should incorporate crucial product-related details, encompassing aspects such as product characteristics, disassembly instructions, and safety measures. Aligned with emerging policy directives, particularly those articulated by the European Commission, the passport should harmonize with static repositories of product-related data [2]. By DPP, for sectors like batteries, a predefined information set is integrated, with plans for expansion to encompass all product groups, excluding medical devices and food, in subsequent phases [3]. Given the potential for dynamic and cross-product data exchange throughout the lifecycle, the safety passport should be tailored to facilitate efficient and sustainable product management practices, aligning with the principles of the circular economy [4]. Ensuring alignment with the data needs and requirements of value chain participants is pivotal for the safety passport to function as a robust decision support tool in sustainable product management. Addressing concerns regarding the sensitivity of specific product lifecycle data is paramount, considering stakeholders' potential reluctance to share sustainability-related information. Furthermore, for specific products like electric vehicle batteries (EVB), careful consideration of potential drawbacks, including intellectual property infringement and competitive disadvantage, is essential to foster widespread acceptance and compliance with industry standards [5]. The development of a safety passport, grounded in these considerations, not only enhances product safety but also contributes to the overall sustainability and efficiency of product management practices.

The advent of safety product passports introduces a revolutionary approach to managing and documenting the entire life cycle of safety-critical products. These passports serve as comprehensive digital records, encapsulating crucial information from product development to disposal. However, a significant challenge arises concerning the handling of sensitive product life cycle data. As safety-related information becomes increasingly digitized and interconnected, ensuring the confidentiality and secure management of data, such as risk assessments, design specifications, and maintenance records, becomes paramount. Striking the right balance between transparency and data protection is essential. Robust cybersecurity measures, encrypted storage, and access controls are integral components to address these concerns. Organizations must navigate the delicate task of making pertinent safety information accessible for stakeholders while safeguarding against unauthorized access or data breaches. Achieving this delicate equilibrium is vital to fully realize the potential benefits of digital safety product passports in enhancing safety, traceability, and transparency across the product life cycle.

2.1 Underlying Challenges in the Practical Implementation of Safety Design Processes

In safety development, several challenges can hinder the effectiveness of security measures. One main issue is the insufficient incorporation of security considerations in the early stages of development, leading to sub-optimal protection mechanisms. Insufficient allocation of resources, such as time and budget, for security aspects can also result in vulnerabilities. Additionally, the lack of training for developers regarding current security practices and threats can lead to insecure implementations. Communication gaps between developers and security teams, as well as the absence of continuous monitoring and updating of security measures, are other factors that can contribute to security problems. Neglecting data protection requirements and failing to address evolving threats can also create security gaps. To address these challenges, effective security development requires early integration of security considerations, adequate resources, continuous training, transparent communication, and regular adaptation to current threats.

To overcome the challenges in safety development, a holistic approach is necessary, along with the integration of best practices. One key step is the early integration of safety. This means considering safety from the very beginning of the development process. Providing training to developers on security aspects can help create awareness of safety issues and promote integration from the outset. In addition to early integration, organizations should allocate adequate resources, both in terms of time and budget, to security development. This ensures that robust security measures can be implemented without compromise due to lack of resources.

Continuous monitoring is essential for safety development. Implementing monitoring tools and processes that regularly review security measures helps in the early detection of anomalies or threats. This enables a rapid response to potential security issues. Additionally, integrating data protection aspects from the outset and regularly reviewing data protection guidelines are crucial steps in maintaining security. These measures help avoid data breaches and maintain user trust.

Implementing these measures requires an organizational commitment to security and continuous improvement of security processes throughout the whole lifecycle. By following these steps, companies can enhance their security development practices and effectively address the challenges in this new field.

2.2 The Concept of Safety Passports and Their Relevance to Product Implementation

A safety passport for machinery is a digital customized document that contains essential technical data, operating features, and safety information. The passport must be included in the delivery documents and must be available throughout the entire life cycle of the machinery or equipment. The passport has to provide the main technical data and operating features of the machinery, including information on the manufacturer's warranty, technical parameters, product certification, and the disposal of the machine. It must also contain fields for any changes made to the machine that must be recorded in modernization, alterations, replaced parts, repair, and welding work. The passport must be prepared separately for each machine or device and serial number. As addition the passport must include information on the manufacturer's name and address, date of manufacture, factory number, estimated service life and provide background information on the need for enhanced safety standards in specific industrial settings.

In relation to machinery and industrial processes, safety passports should become important tools to minimize risks and ensure the well-being of workers and the machine itself. By integrating safety passports into product implementation, companies can systematically check whether machine operators and the products in question fulfil the required safety standards.

2.3 The Importance of Safety Passports

Safety passports go beyond the traditional approach to safety measures by incorporating a comprehensive documentation system. Safety passports are not limited to certifying the qualifications of individuals, but extend their scope to verifying the compliance of machines throughout their life cycle. At a time when technological progress is rapidly advancing and the complexity of machinery is increasing, safety passports could be a tangible solution to ensure that every aspect of the industrial process complies with the highest safety protocols (Fig. 1).

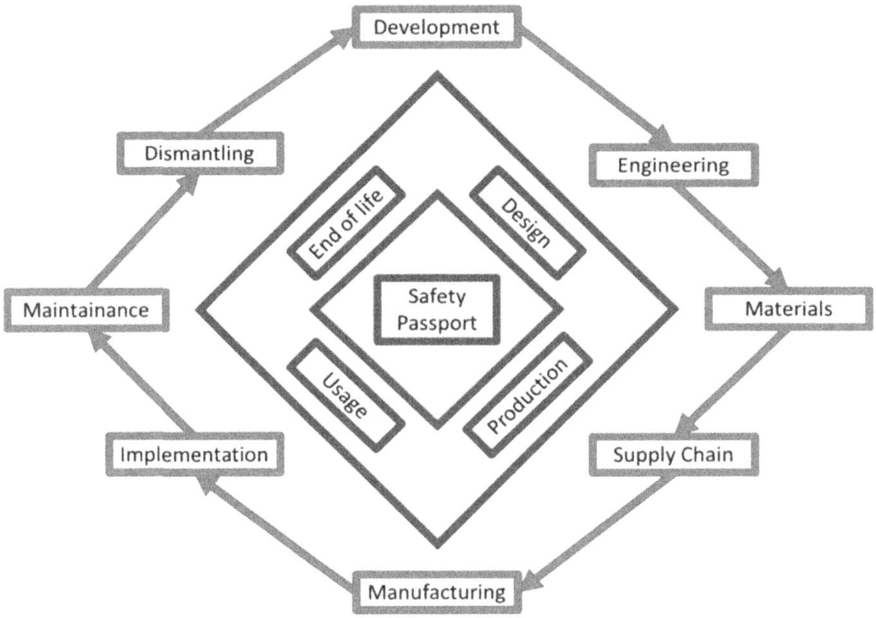

Fig. 1. Safety Pass and Product lifecycle Interaction

3 Conceptual Framework

The university pilot factories in Vienna, Graz, and Linz are aiming to develop a role model for networking by utilizing the tools of digitalization. We plan to use our own infrastructures, as well as those of other research centres, to enable better cooperation between companies. One key aspect of effective collaboration is data interoperability, which requires the use of cutting-edge information and communication technologies. This is particularly important for small and medium-sized enterprises, and the necessary data technology infrastructures should be standardized and specified for them.

3.1 Data Collection and Analysis

Collecting and analyzing data is crucial for ensuring safety throughout the life cycle of a product. Different types of data can be used to identify risks, enhance safety measures, and comply with safety standards. Key areas to focus on include the design and development phase, where failure analyses and FMEA (Failure Mode and Effects Analysis) can be used to identify potential design failures and safety gaps. In the production phase, quality control data and performance monitoring can help in monitoring safety performance and detecting deviations. During the utilization phase, data from accident reports, user feedback, and maintenance logs can provide insights into safety incidents, end-user experiences, and maintenance activities. Monitoring and surveillance can be achieved through real-time sensor and IoT data as well as log data for detailed analysis of security incidents. Compliance and certification can be ensured through audit logs and documentation tracking. Analysis of recall and repair data can help in identifying safety defects and improvements. Finally, data on recycling and disposal can ensure the safe dismantling and disposal of the product. The use of advanced analytics, machine learning, and statistical modelling can support the analysis of this data by identifying patterns, trends, and anomalies. The data collection and sources for our study encompass various types of data and their respective origins. Operational data, such as temperature, vibration, and speed, are collected from machinery, while environmental data, including ambient temperature and humidity, are monitored using IoT devices. Maintenance records detailing service dates and parts replaced are logged digitally, and incident reports, documenting safety breaches and near-misses, are entered through human-operated digital systems. Data analysis techniques employed include descriptive analytics, which summarize past safety incidents to identify common factors; for instance, it was found that 70% of incidents were due to improper machine handling over the past year. Predictive analytics utilize machine learning models to forecast potential safety issues based on historical data, leading to insights such as a 30% reduction in unexpected downtimes through predictive maintenance alerts. Prescriptive analytics recommend actions to mitigate risks and enhance safety, with suggestions like implementing specific safety protocols to potentially reduce incident rates by 25%. [11] The impact on safety improvements is validated statistically through correlation and regression analyses, identifying key predictors and quantifying the effects of safety measures. For example, a regression model indicated that a 10% increase in maintenance frequency correlates with a 15% decrease in safety incidents. Additionally, case studies from real-world implementations demonstrate the effectiveness of the safety passport, such as a pilot project where accident rates were reduced by 40% over six months [12].

4 Conceptualization

The first component of the concept is a holistic approach to safety. This involves integrating safety aspects into the design process early on, ensuring high-quality production, considering user experience during product use, monitoring safety parameters continuously, and ensuring safe disposal of the product. By incorporating safety into all stages of the life cycle, potential risks can be addressed and mitigated.

Lastly, the concept takes a broad view of security, encompassing not just physical safety but also data protection, compliance, and ethical considerations. Adopting a comprehensive approach to security helps create products that are not only safe but also reliable and trustworthy.

4.1 Use Case Analysis: Co-Design and Co-production

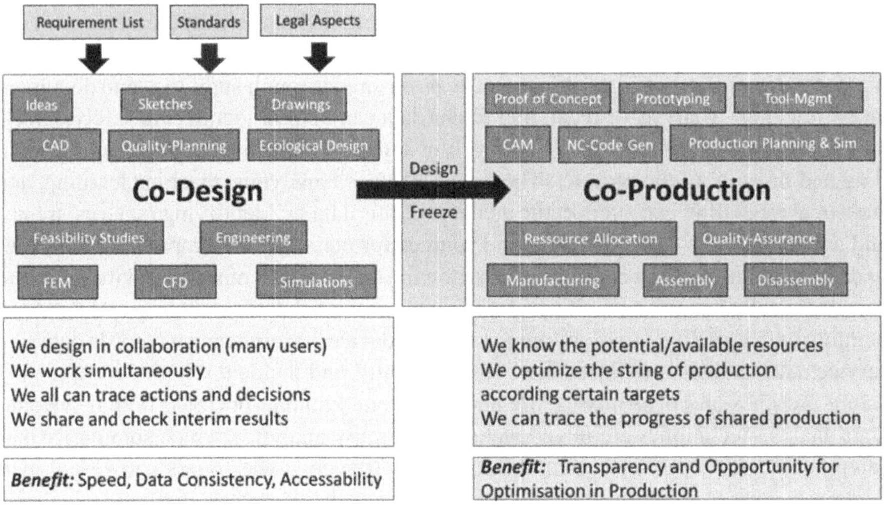

Fig. 2. Co-Design and Co-Production use case

Co-design is a design approach that embraces collective intelligence and diverse perspectives. It encourages active participation from users, allowing them to contribute their insights and experiences to the design process. The emphasis on traceability in codesign enables participants to track the progress of decisions and actions, fostering accountability and facilitating continuous improvement. Iterative refinement through the exchange and review of interim results ensures that the design aligns with user expectations and evolving requirements. This feedback loop also minimizes misunderstandings and leads to more effective and user-friendly designs. The end result of co-design is an innovative solution that reflect the diverse needs of users (Fig. 2).

Co-production is a collaborative approach to production that involves utilizing the resources available within our collective framework. Participants are actively involved in optimizing the entire production process towards the specific goals, defined for the use case. This emphasis on resource awareness allows for strategic allocation, maximizing productivity and minimizing waste. Traceability is a key component of co-production, ensuring transparency and accountability. It allows all parties involved to monitor development and results in real-time, promoting collective responsibility and continuous improvement. Co-production optimizes the production chain based on defined goals, such as efficiency gains, sustainability targets, or quality benchmarks. This targeted optimization ensures that joint efforts align with overarching goals and contribute to a more effective and targeted production process.

Both processes contribute to the creation of traceability records detailing the progress of development and production activities.

4.2 Use Case Selection

The chosen use case provides a representative illustration of the safety-related challenges in the specific application area. It highlights critical safety requirements in the product. The use case not only demonstrates relevant safety aspects but is also practical and realistic. By selecting this use case through evaluation, ensures an effective demonstration of safety principles and allows for the identification and validation of appropriate solutions and practices. The development in the field of industrial robotics is experiencing an innovative turn with the introduction of the modular industrial robot construction kit. A key aspect of this innovation lies in the use of a virtual project space and the product development process, supported by TU Graz in coordination with TU Wien (Fig. 3).

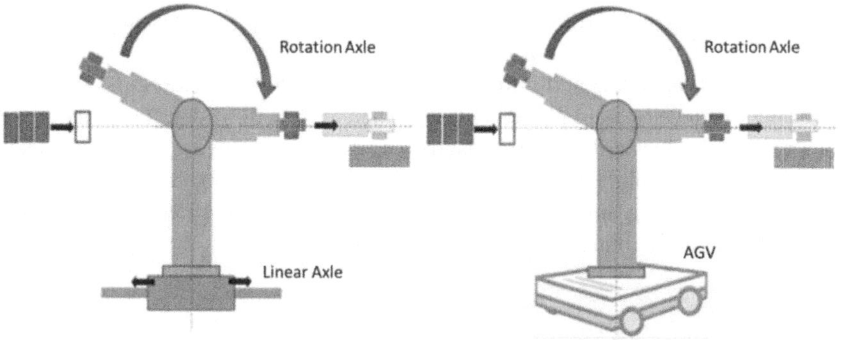

Fig. 3. Scheme of the modular manipulator

The key innovation is to detach from the conventional notion of a "mobile two-pronged manipulator with an end-effector." The Modular Manipulator (MM) offers the possibility to have lower requirements compared to conventional robots, such as a reduced number of axles. This results in a reduction in traditional complexity and associated costs, while creating space for innovations.

The application areas of the MM are diverse and particularly relevant at a high degree of specialization. This results in a catalogue of requirements that defines the functionalities of the MM. The mobility of the MM is considered in detail, from portable/portable to mobile. This includes aspects such as the type of control (joystick, AGV), the energy source, the control coupling of the chassis and manipulator axes.

A further key point is the modularity of the system, which enables the exchange of end-effectors and manipulator elements in order to handle different tasks. Safety is an integral part of the overall system. It is not only legally binding, but also essential for compliance with the new machine guidelines.

4.3 Relevant Data

To create a safety passport for our MM throughout its life cycle, an extensive amount of data and information is necessary. Firstly, detailed design specifications are required, outlining the product's design and any safety aspects that need to be considered. These specifications provide a clear understanding of how the product should be developed. Additionally, the results of a comprehensive risk assessment should be documented, highlighting potential hazards associated with the product. This allows for proactive measures to be taken to mitigate these risks.

Another key element is ensuring that the product complies with relevant standards and regulations. In our use case we use DIN ISO 12100 [6], DIN ISO/TR 14121-2 [5], ISO 10218-1 [7] and ISO 10218–2 [8]. It is essential to have a list of applicable standards and regulations that the product needs to meet to ensure its safety. Furthermore, detailed information on the materials and components used in the product is necessary, including their safety features and certifications. This information helps in assessing any potential risks associated with the usage of specific materials or components.

All the mentioned data and information form the basis for creating the safety passport during the development phase. This documentation will be continued and updated throughout the product's life cycle to ensure that all safety aspects are considered and that the product meets the highest safety standards. By maintaining a comprehensive and reliable safety passport, organizations can demonstrate their commitment to product safety and compliance throughout the entire life cycle.

4.4 Technical Specification and Operational Parameters

The technical specifications and operational parameters of safety passports are crucial for enhancing safety in production environments. The system employs cloud integration for scalable and secure data storage, complemented by edge computing to ensure real-time data processing and reduced latency. Data security is maintained through encryption both in transit and at rest, along with role-based access control to restrict data access to authorized personnel. Interoperability is achieved through APIs that enable seamless integration with existing ERP and MES systems, adhering to industry standards like ISO 13849 for machinery safety. The system's scalability is ensured by its modular design, allowing it to expand with the production line and handle data from up to 10,000 sensors and devices simultaneously. Adaptability is facilitated by customizable templates for various machinery and production processes, and machine learning models that continuously improve with new safety data. The implementation process is divided into three phases: initial assessment of current safety protocols and data infrastructure, integration with existing systems and data sources, and continuous monitoring and optimization of safety protocols.

Performance metrics highlight the system's efficiency, with real-time monitoring ensuring safety incidents are detected and addressed within milliseconds. Reliability is guaranteed with 99.99% uptime through redundant systems and failover mechanisms.

User feedback indicates positive reception from operators and maintenance personnel regarding ease of use and effectiveness.

Future enhancements include integrating advanced AI algorithms for predictive maintenance and anomaly detection, blockchain technology for tamper-proof logging of safety incidents and maintenance records, and augmented reality tools for on-site safety training and real-time maintenance guidance. This detailed analysis and technical specifications provide a comprehensive understanding of how the safety passport can improve safety outcomes in production environments, demonstrating its scalability, adaptability, and measurable impact.

4.5 Risk Assessment

Risk assessment is the method that we use for confirming safety in order to ensure the safety of workers and other individuals and to reduce to the absolute minimum the possibility of harm. Even if harm does not occur, there may exist potential for risk and harm. If these are not addressed, it can be said that the situation possesses a high possibility of industrial accidents occurring. To further reduce these accidents, it is necessary to implement safety and sanitation measures in a pre-emptive, not reactive, manner. The process for carrying out risk assessments is defined by the ISO basic safety standards. The following flowchart shows the risk assessment procedure established by ISO 12100 [6]. Generally speaking, safety is confirmed with the following procedure during a risk assessment and also the basis for the safety passport (Fig. 4).

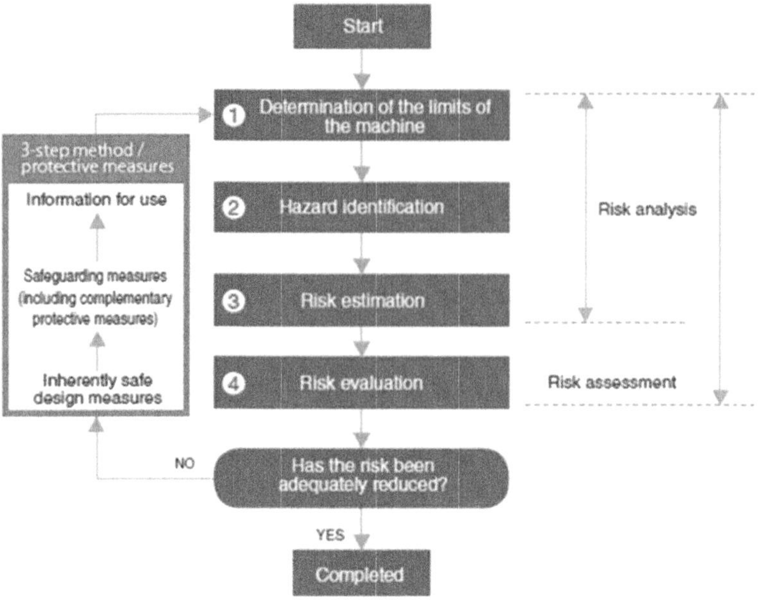

Fig. 4. Risk Assessment by ISO 12100 [6]

5 Overcoming Challenges and Limitations

The introduction of safety passports requires a strategic approach that harmonizes with existing workplace systems. Standardized protocols for documenting qualifications, ongoing training programs and collaboration with regulators are critical components. The integration of security passports in industrial processes is a promising development for enhancing security and efficiency. However, this integration is not without its challenges, which require thoughtful strategies for successful adoption.

One of the main challenges is resistance to change. Many employees and stakeholders are accustomed to traditional security protocols and may perceive the introduction of security passes as disruptive. To overcome this, proactive communication and comprehensive training programs are vital. Organizations need to clearly communicate the benefits of security passports, highlighting how they contribute to a safer working environment and more efficient processes.

Collaboration with regulators, industry experts, and other stakeholders is critical. Establishing an open dialogue with these bodies ensures alignment with broader industry standards and regulations. Collaborative efforts can lead to the development of standardized requirements for the security passport, enabling a smoother integration process. Regulatory support can significantly increase the adoption of security passports across the industry.

The successful integration of safety passports requires a multi-layered approach. By overcoming resistance, simplifying implementation, promoting education, collaborating with stakeholders, leveraging technology, and committing to continuous improvement, the industry can address and overcome the challenges associated with this transformative safety initiative.

6 Conclusion and Outlook

The integration of safety passports into the production, packaging and logistic industries heralds a new era of safety-conscious practices. It is a proactive step to not only fulfil legal requirements, but also to promote a workplace culture where safety is paramount. By adopting this new approach, the industry is embarking on a journey where the integration of safety passports is not just a compliance measure, but also its commitment to shaping a future where innovation and safety co-exist harmoniously.

Looking to the future, the successful implementation of this strategy has the potential to not only streamline processes but also improve the overall safety culture in industrial environments. The integration of safety passports is a point for improved product implementation that prioritizes safety in every operation.

This paper examines the conceptual framework of integrating safety passports as a strategic approach to accelerate and improve the introduction of products along the entire product lifecycle. By providing a standardized and verifiable proof of compliance, the integration of safety passports aims to streamline and accelerate the product implementation process, promoting not only a safer working environment, but also the handling of the products throughout their entire life cycle.

References

1. Lemos, P.: Towards a circular economy. Presentation on product passports. In: European Commission. Conference presentation Hamburg Sustainability Session, vol. 2 (2020)
2. Berger, K., et al.: Confidentiality-preserving data exchange to enable sustainable product management via digital product passports-a conceptualization. Procedia CIRP **116**, 354–359 (2023)
3. EU Commission. Proposal for a Regulation of the European Parliament and of the Council establishing a framework for setting ecodesign requirements for sustainable products and repealing Directive 2009/125/EC. COM (2022), 142 (2022)
4. Berger, K., Schöggl, J.P., Baumgartner, R.J.: Digital battery passports to enable circular and sustainable value chains: Conceptualization and use cases. J. Clean. Prod. **353**, 131492 (2022)
5. DIN ISO/TR 14121-2, DIN SPEC 33885. Safety of machinery—Risk assessment—Part 2: Practical guidance and examples of methods (ISO/TR 14121–2:2012). International Organization for Standardizationb (2012)
6. DIN EN ISO 12100. Safety of machinery—General principles for design—Risk assessment and risk reduction (ISO 12100:2010); German version EN ISO 12100:2010, Corrigendum to DIN EN ISO 12100:2011-03. International Organization for Standardization (2011)
7. International Organization for Standardization. ISO 10218-1:2011 Robots and robotic devices—Safety requirements for industrial robots—Part 1: Robots. International Organization for Standardization (2011)
8. International Organization for Standardization. ISO 10218-2:2011 Robots and robotic devices—Safety requirements for industrial robots—Part 2: Robot systems and integration. International Organization for Standardization (2011)

9. SICK AG. Leitfaden Sichere Maschinen (Publication No. 8008007) (n.d.). from https://www.sick.com. Accessed 25 Feb 2024
10. Deutsche Gesetzliche Unfallversicherung (DGUV). DGUV Information 209–074 (formerly BGI/GUV-I 5123): Industrieroboter (n.d.). https://medien.bgetem.de/medienportal/artikel/MjA5LTA3NA--
11. U.S. Bureau of Labor Statistics. Injuries, Illnesses, and Fatalities (n.d.). https://www.bls.gov/iif/. Accessed 28 May 2024
12. Health and Safety Executive. Work-related fatal injuries in Great Britain (n.d.). https://www.hse.gov.uk/statistics/fatals.htm?v=clpu&v=clpu. Accessed 28 May 2024

Towards a Multi-criteria Decision-Making Framework for R&D Project Prioritisation Driven by Innovation Incentives and Enterprise Strategies

Bruno Paludo, Anderson Luis Szejka(✉), Marcelo Rudek, and Osiris Canciglieri Junior

Industrial and Systems Engineering Graduate Program (PPGEPS), Pontifical Catholic University of Parana (PUCPR), Curitiba, Brazil
bruno.paludo@pucpr.edu.br, {anderson.szejka,marcelo.rudek, osiris.canciglieri}@pucpr.br

Abstract. Innovation and R&D projects usually hold a high risk of partial or total failure. Thus, considering that research and innovation are crucial to a country's economic development, governments worldwide tend to institute incentives and subsidy programs (i.e., financial aid and tax incentives) to facilitate these endeavours by enterprises. However, given the vast array of incentives available in different countries, the technical knowledge required to file for each specific subsidy is sparse. Whenever determining the priority of each project within an organisation's portfolio, along with this knowledge, managers must have a deep understanding of innovation and R&D projects and portfolio management, as well as the company's strategic guidelines. Hence, this paper focuses on the current issues and the state of the literature regarding methods for prioritisation of projects that encompass all three knowledge domains, considering the following research question: how can the prioritisation of innovation and R&D projects supported by tax and financial incentives be systematised considering companies' strategic organisational guidelines? This initial literature review focuses on the projects, incentives, and strategy domains, examining them by themselves and their integration. Finally, this research contributes to the research community with an overview of trends for incentives and subsidies, innovation and R&D projects and the company's strategy. Moreover, it contributes to the industry by proposing a framework to integrate all domains when prioritising projects, especially for those who adopt a product-service system business model.

Keywords: Innovation Projects Prioritisation · Multi-Criteria Decision-Making · Innovation Incentives · Enterprise Strategies

1 Introduction

Investments made in research and development (R&D) and innovation projects have been continuously increasing throughout the years; more developed countries, such as Israel, South Korea, China, and the United States of America (USA), tend to lead

Published by Springer Nature Switzerland AG 2025
P. Sureephong et al. (Eds.): PLM 2024, IFIP AICT 741, pp. 329–342, 2025.
https://doi.org/10.1007/978-3-031-93323-3_25

the charts when analysing gross domestic spendings on R&D [1]. However, R&D and innovation projects are reputed to be permeated by a high-risk and high-reward outcome system. According to [2], these projects may have up to a 90% total or partial failure rate, depending on their nature.

However, R&D and innovation projects are fundamental to the countries' economic and industrial development [3]. Different governments have been adopting the triple helix model to incite the integration of governmental policies with activities developed by academia and the industry [4]. The objective is to streamline project execution and mitigate associated risks in their development.

In Brazil, multiple laws (nation, state, and municipal-wide) establish fiscal and tax incentives for companies that invest in R&D and innovation projects. Also, there are other forms of fostering innovation, such as credit lines incentivised by the federal government to finance these projects and hiring researchers partially funded by governmental institutions [5, 6]. Although specific incentives are predominantly favoured by Brazilian enterprises [7], municipalities' authority to establish distinct programs results in various incentives for research and development (R&D) and innovation in Brazil. Under this scenario, companies and project managers leading these projects can become easily overwhelmed by all the different incentives and their bureaucracy. Hence, integrating each project's specific features with the best incentives available and considering the enterprise's strategic guidelines makes prioritising the execution of R&D and innovation projects arduous. Based on this context, this research presents a multi-criteria decision-making conceptual framework for a formal R&D and innovation project prioritisation according to (i) the project's features, (ii) the incentives available, and (iii) the enterprise's strategic plan and guidelines. The result will be a structured list of the order of execution of projects aligned to the budget, the company's guidelines, and the incentivised resources available.

The research herein aims to develop a conceptual framework to prioritise companies' R&D and innovation projects utilising a multi-criteria decision-making procedure. This decision involves gathering specific data on projects' features and conducting a cross-analysis with various parameters derived from available incentives, such as tax relief and incentivised credit lines, as well as the company's strategic plan and guidelines.

This paper is structured as follows: Sect. 2 delves into the problem statement when the project manager does not know all the available incentives and cannot integrate them into the R&D and innovation project's execution. Section 3 provides a literature review of the main issues involved in the range of incentives available and the integration with the enterprises' guidelines and the projects' features. Section 4 presents the proposed framework's key components and detailed expectations of results from its implementation. Section 5 addresses the practical significance of the present study for improved strategic decision-making, concludes this paper and presents perspectives for future studies.

2 Problem Statement

R&D and innovation projects may have both positive and negative ramifications: on the one hand, the outcome may be a revolutionary product or an incredibly more efficient process; on the other hand, it may result in a partial or complete failure with time and money invested in a project with little or no outcome [2, 8]. Although the risk may ward off investments, many factors can mitigate the risks inherent in these processes [25].

The decision-making process of prioritising these projects is very complex and multidisciplinary: not necessarily in this order, the portfolio manager and its team must (i) identify the project's features, such as scope, budget, staff, management methods (i.e. cascade, agile), urgency etc.; (ii) have profound knowledge on the regional and national incentives available for funding the project (partially or entirely); (iii) understand the company's internal strategic guidelines and consider its vision when establishing a list of priority for the portfolio execution.

Regarding topic (i), there are systems and frameworks already designed to manage innovation and open innovation projects for better efficiency and enhanced results, and the positive effects of good project management on the outcomes are nearly undisputed [9–11]. In [13], the authors have shown that, if applied correctly, the project management methods used at all stages of innovation projects increase the chances of success. However, project management is an extensive body of knowledge that the manager conducting the company's portfolio must deeply study and master.

When considering topic (ii), government-instituted tax reliefs and refunds are the primary sources of fostering innovation and R&D projects. As an example, it can be regarded as the Canadian Scientific Research and Experimental Development (SR&ED) [14], the USA Research and Experimentation Tax Credit [15], the Singaporean Productivity and Innovation Credit (PIC) Scheme [16], the French Research Tax Credit (CIR) [17], and so on. In Brazil, the program with the highest values invested and incentivised was instituted by Federal Law n. 11.196 of 2005 and is commonly known as "*Lei do Bem*" (or Law of Good, as per translation to English) [7, 18, 19]. Its main tax relief is a deduction from the basis of the calculation of corporate income tax, depending on the values invested in R&D and innovative projects by the applying company. However, due to the legal structure of how incentives are instituted, the Brazilian innovation ecosystem comprises many tax reliefs and incentivised grants and loans.

Additionally, there are other forms of fostering innovation, some conflicting with each other, compelling project managers to prioritise one over the other [5, 6]. Suppose the company seeks funding for R&D and innovation projects beyond its internal budget. In that case, the manager must comprehensively understand the multitude of incentives and subsidies available in Brazil for such endeavours.

It is also important to note how these incentives are operationalised. For example, after identifying projects susceptible to claiming tax relief, the company must inform all expenditures in R&D in its ancillary obligations, and those values will automatically be deducted from the basis of its corporate income tax. Although this accelerates the process, the company will be audited if the understanding of tax authorities differs from that of the project manager regarding which expenditures are subject to the incentive. Then, tax authorities may disallow, note, and charge the values with fines and interests. Thus, having to consider all these vast bodies of knowledge regarding (i) project management, (ii) incentives to R&D and innovation projects, and (iii) aligning these features with the company strategy to prioritise and compose the company's portfolio duly proves to be an arduous task for project managers.

This problem is illustrated in Fig. 1. The manager responsible for the project must delve into the different domains and utilise the knowledge in a non-structured prioritisation process. This method may result in other complications, as per the abovementioned tax audit, wrongful prioritisation of projects based on subjectivity, or the archiving of projects due to budget since the knowledge of all different incentives and subsidies is a challenge.

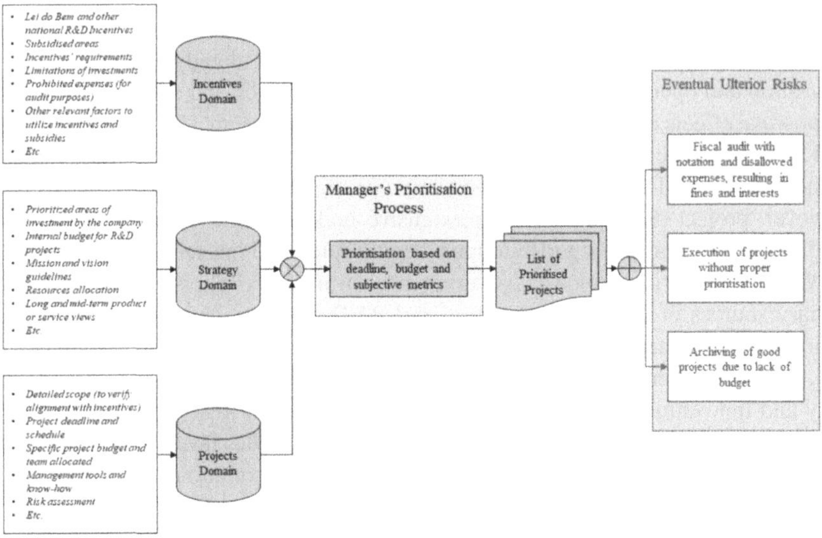

Fig. 1. Prioritisation process without formal and proper scientific method.

The illustration above offers a comprehensive view of the manager's pivotal role in prioritisation. However, it is imperative to delve deeper into the ramifications of this process on other stakeholders and their intricate interactions within the system. Adopting an enterprise-centric perspective broadens our scope to encompass internal and external stakeholders. Internally, this includes the manager and the project team, whose expertise and insights occupy a central role in informed decision-making. Furthermore, the company's strategic plan, formulated by shareholders or board members, exerts substantial influence over R&D endeavours, delineating preferred investment areas and allocating scarce resources.

Additionally, external stakeholders such as governmental and regulatory bodies wield significant influence over R&D initiatives: they are responsible for structuring, regulating, and inspecting available subsidies and their usage by the companies. Therefore, if an institution such as the Brazilian Ministry of Science, Technology, and Innovation (MCTI) changes its understanding regarding what is or is not an innovation project, the process above will be directly affected. Thus, a holistic framework that constantly considers and monitors the perspectives and needs of all stakeholders is imperative for effective R&D project prioritisation and successful innovation management and product or service development.

After an analysis of the literature, the authors identified two issues. The first issue is regarding the vast array of incentives currently available to foster innovation in Brazil – even though this is an issue of technical knowledge, it profoundly influences this study and the proposed framework. The main scientific issue identified is the literature gap and lack of established methods and barriers for integrating all three bodies of knowledge (i.e., project management, incentives, and company strategies) when prioritising and composing the company's R&D and innovation project portfolio. For example, if a feature of the project is poorly stated, or if a new incentive has been instituted and the project manager is unaware of its specific guidelines, the project portfolio may be poorly structured, with inefficient execution order and sub-optimal budget and resource composition. Considering this problem, the authors proposed the following research question: *how can the prioritisation of innovation and R&D projects supported by tax and financial incentives be systematised considering companies' strategic organisational guidelines?*

3 Related Works

To comprehensively examine the presence of a gap in the existing literature, the research for related works was structured considering the three bodies of knowledge abovementioned: (i) project management, (ii) incentive, and companies' (iii) strategic plan. When considering the research question, only the project management domain had frameworks and approaches proposed to create, monitor, and prioritise the projects. Other domains lacked the guidelines of a structured methodology to prioritise said projects. Moreover, although many scientific articles were published (especially regarding innovation and project management), the integration of the domains was poorly explored.

The research methodology follows four steps:

- **Step 1 – Definition of keywords:** the authors defined specific keywords for each domain. The selected keywords for the project management domain were innovation and R&D projects, multi-criteria prioritisation, project features, and their equivalent translated words into Portuguese. For incentives, the chosen words were innovation incentives, governmental subsidies, and their Portuguese counterparts, as well as specific Brazilian incentives, such as "*Lei do Bem*", "*Lei de Informática*", "*Lei das TICS*", and "*BNDES*". Lastly, the company strategy domain keywords were organisational strategy, strategic plan, and project portfolio strategy.
- **Step 2 – Selection of databases:** after defining the keywords, the researchers designated Google Scholar, Science Direct, and SciELO as the databases for the articles substantiating the literature review.
- **Step 3 – Exclusion of duplicates:** duplicate articles and papers were excluded with the initial research results gathered from the databases.[1] Also, articles eventually imported from the databases with empty titles or null and void content were excluded.

[1] The authors used the keywords in different search strings and downloaded the BibText files from the results. Then, these files were converted into.csv and unified in a single database for further analysis.

- **Step 4 – Title and abstract analysis:** An initial analysis excluded the articles with titles unrelated to the research's proposed problem. Likewise, the abstract analysis excluded articles that did not delve into the studied subjects.
- **Step 5 – Choosing the most pertinent and impactful articles:** the last step was to organise a list of the most relevant articles. This organisation was based on the publisher journal's impact factor, citation index, and the classification criteria below.

Initial searches conducted after the second step of the defined protocol resulted in a database summing up to 20.991 articles. The distribution of these articles among the studied domains is shown in Fig. 2.

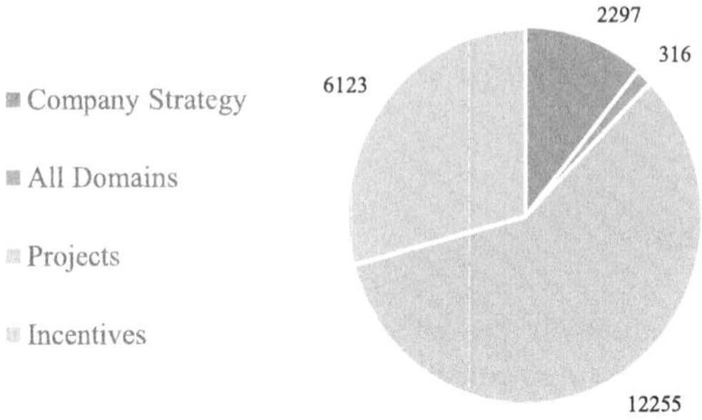

Fig. 2. Distribution of articles per domain.

After the initial composition, the research proceeded with the application of the exclusion criteria. First, exclude all duplicate and invalid articles. Then, articles were excluded based on title and abstract alignment. Finally, the remaining articles and papers underwent a journal impact analysis. This procedure is demonstrated in Fig. 3, which lists the remaining articles after each step.

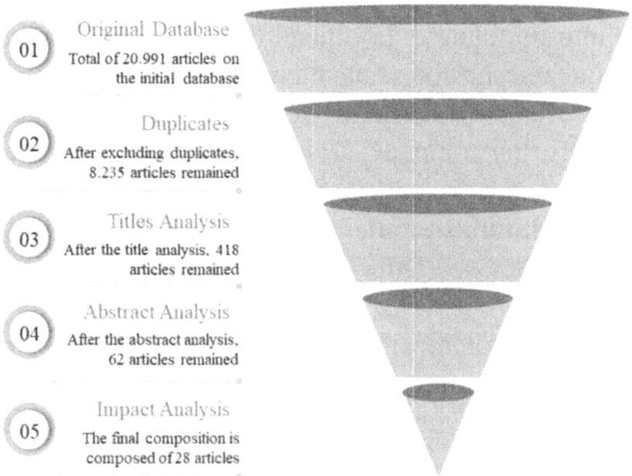

Fig. 3. Research articles funnel.

As done by [20], when analysing works related to requirements interoperability in complex systems engineering, the authors structured a literature review considering the three domains and bodies of knowledge studied by the related works and their integration or interrelation. In Table 1, it is proposed to summarise this analysis based on specific categories defined by the authors; the criteria utilised for this analysis were as follows:

- (P) *Project management*: papers/articles either proposed a framework or approach for innovation and R&D projects' management or studied their features.
- (S) *Strategic planning*: papers/articles that discussed the impacts of strategic planning on the company and its components.
- (I) *Incentives*: papers/articles that discussed fiscal and tax incentives or any other form of fostering innovation and R&D projects promoted in the analysed ecosystem.
- (PS) *Integration of project management and strategic planning*: papers/articles that studied these domains separately and analysed or discussed their impacts on one another.
- (PI) *Integration of project management and incentives*: papers/articles that analysed or discussed the integration of these bodies of knowledge.
- (SI) *Integration of strategic planning and incentives*: papers/articles that analysed or discussed the integration of these bodies of knowledge.
- (PSI) *Integration of project management, strategic planning, and incentives*: finally, papers/articles that discussed or analysed not only all domains separately but also the integration of all three domains.

Table 1. Summary of related works categorised by domains.

Authors and Publication Year	*Individual Domains*			*Integrations Amongst Domains*			
	P	*S*	*I*	*PS*	*PI*	*SI*	*PSI*
Rabechini Jr.; Maximiano; Martins, 2005	✓	✓		✓			
Dos Santos et al., 2023	✓	✓					
Bellucci; Pennacchio; Zazzaro, 2023			✓				
Kaveski; Lopes; Beuren, 2020			✓				
Shabir; Pazienza; de Lucia, 2023		✓	✓				
Tarapuez; Guzmán; Hernández,2016	✓	✓	✓	✓			
Jardim; Saes; Mesquita, 2013	✓	✓		✓			
Avagyan et al., 2022	✓	✓		✓			
Antolín-López et al., 2016		✓	✓			✓	
Guenduez et al., 2023	✓		✓		✓		
Sheyk; Azizi; Sobhiyah, 2013	✓	✓		✓			
Dos Santos; Rapini; Mendes, 2021		✓	✓				

(*continued*)

Table 1. (*continued*)

Authors and Publication Year	*Individual Domains*			*Integrations Amongst Domains*			
	P	*S*	*I*	*PS*	*PI*	*SI*	*PSI*
Porto; Memória, 2019		✓	✓				
Moctezuma; López; Mungaray, 2017			✓				
Zaman; Nawaz; Nadeem, 2020	✓	✓		✓			
Coccia, 2023	✓	✓		✓			
Gordon; Cassiolato, 2019			✓				
Duong et al., 2022	✓	✓					
Rudnik et al., 2021	✓						
Frezatti et al., 2010		✓					
Esfahani; Sobhiyah; Yousefi, 2016	✓						
Purnus; Bodea, 2014	✓	✓					
Sánchez et al., 2023	✓						
Kleine; Heite; Huber, 2022	✓		✓				
Guerrero; Urbano, 2017	✓	✓	✓		✓		
Cirani et al., 2016			✓				
Ommen et al., 2016	✓	✓					
Bressan, 2013		✓	✓			✓	

With the enumeration of the related works, it may be observed that the gaps above are coherent with the reality of academic literature: the bodies of knowledge, when individually analysed, have been extensively studied throughout the years. However, when it comes to the analysis of the integration of these bodies of knowledge (especially when involving *incentives*, such as PI, SI, and PSI), not much research has been published. Incentives (in Brazil and worldwide) are constantly changing, as are their rules for adherence to the different types of programs. In addition, companies' strategies will also vary, but according to market changes and constrained resources available for the projects' execution [21]; this shows the importance of the deep understanding and constant monitoring of the available incentives for the company's projects, since they may alleviate the latter problem regarding strategy changes.

Therefore, the authors deemed it necessary not only to explore and study the impacts of integrating the appointed bodies of knowledge but also to propose a framework to implement a scientific-based methodology to the decision-making process regarding the prioritisation of execution of innovation and R&D projects, considering the company's strategy and funds available. If implemented, the framework should help project managers and companies to integrate the knowledge across all studied domains.

4 Conceptual Framework for R&D Project Prioritisation Driven by Innovation Incentives and Enterprise Strategies

Considering the problem statement and literature gap, the authors hypothesised that it is possible, by establishing specific premises for an innovation or R&D project, to identify the most suitable set of subsidies and fiscal incentives available. Consequently, guided by the organisational guidelines of the company, it is feasible to determine, through a multi-criteria analysis, varying degrees of priority among the listed projects. Based on this context, Fig. 4 shows the core structure of the framework's functionalities and parameters.

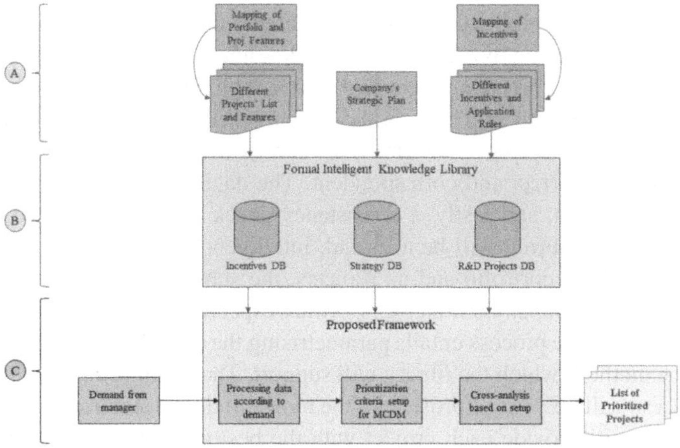

Fig. 4. Structure of the R&D project prioritisation driven by innovation incentives and enterprise strategies framework.

According to Fig. 4, the initial stage (A) will involve comprehensive data collection across all relevant domains under study, encompassing incentives, corporate strategies, and project attributes. Subsequently (B), the knowledge acquired will be inserted into a formal, intelligent knowledge library. Finally (C), whenever the manager demands the framework, it will process data according to the demand, resulting in a list of prioritised projects and indicating the best subsidies for each project.

A constant and comprehensive mapping of all available subsidies and incentives in the selected region will be conducted within the first phase. While the framework will be tailored explicitly for Brazil and its ecosystem, the data collection process and essential criteria for analysing incentives will be standardised. As a result, this phase will be designed so that researchers from other countries may readily adapt it to their local contexts based on the features they consider essential or of greater significance.

In parallel, analysing the company's strategies is crucial for the framework's functioning. Suppose the company's medium to long-term strategy involves investing in green energy under an environmental, social, and governance (ESG) guideline. In that case, projects aiming to develop new products and services or improve existing ones in this area should be prioritised. However, if the company is more aligned with providing top-quality healthcare services, reducing carbon dioxide emissions, or promoting social actions, these goals should be prioritised. Moreover, this analysis should also be extended to a financial assessment of the organisation to understand the scenario of resource allocation better and promote an MCDM prioritisation process with more accuracy [21]. Finally, within this first phase, the company's portfolio will be studied: all projects categorised as innovative or R&D projects within the company's scope will be mapped. The information gathered from each project will be used to fill out a standardised form, which will be constructed based on thorough research on project features, guided by scientific articles and the [23]. After that, the list of projects and their main features (i.e., but not limited to scope, deadline, budget, and personnel) will be utilised in the composition of the integrated database, considering data on incentives and strategies.

After the collection phase, the data will undergo a meticulous treatment procedure: initially, with data cleaning by removing duplicates, handling missing values with imputation or removal, and correcting identified errors; then, all data gathered will undergo a normalisation phase, and aggregation in different levels to be defined; finally, datasets will be merged and matched to ensure that records from other datasets that refer to the same entity are correct and correspondent. The database will then be analysed for any outliers' detection, and lastly, a consistency check will be performed. The result of this database structuring will be a formal, intelligent knowledge library comprised of all elements of the portfolio and its projects, the subsidies and incentives rules for application, and the company's strategic priorities, especially regarding investments. The subsequent stage of the process entails parametrising the criteria utilised for the MCDM prioritisation method, which the library will support. The result of thoroughly undergoing this process will be a list of projects in the most efficient execution order. Moreover, the framework will provide each project with the best possible mix of incentives and subsidies available, based on the mapping and the information input – the feasibility of incentives and subsidies will also be a determinant variable when considering the prioritisation system.

4.1 Expected Impacts on Sustainable Product and Services Development and Product-Services Systems

R&D projects and product and service development processes are intrinsically connected [26] and, when successful, usually result in new or vastly improved products or processes. Growth in R&D expenditure has been shown to positively influence more outstanding production and availability of renewable energy, which, in turn, is a crucial factor in reducing CO2 emissions [27]. In China, governmental subsidies have been vastly studied in terms of their positive impacts on corporate ESG and green innovation performance [28].

In this context, subsidies available within Brazil typically target strategic areas aligned with the principles of ESG, green production, and advancements in Industry

4.0 and 5.0. For instance, one prominent initiative under the Brazilian Federal Government's subsidy framework is the Funding Agency of Research and Projects (FINEP[2]). Presently, projects focusing on innovation in education, telecommunications, sustainability, Internet of Things (IoT), healthcare, green energy, or collaborations between companies and academia for their development are eligible for preferential rates and even non-repayable resources.

Thus, with the formalisation of incentives and subsidies knowledge, the proposed framework will enhance resource accessibility by connecting projects to the best and appropriate governmental subsidised resources – especially regarding projects with ESG, green production, and Industry 4.0 and 5.0. Moreover, companies located in developing countries are usually faced with more significant challenges regarding funding for R&D and innovation projects. However, these projects commonly result in new products and services, which must be developed from a customer and environmental perspective. It is also important to note that these outcomes may encourage companies to adopt a more sustainable product-service system (PSS) business model. In this perspective, this framework aims to bridge the gap between limited funding availability and the pressing need for innovation in developing countries. Addressing these challenges and facilitating access to resources have the potential to catalyse the development of new products and services, thus contributing to economic growth and societal advancement.

Finally, by implementing the proposed framework, companies should have more resilient and sustainable internal processes regarding R&D projects; this premise will be evaluated through a case study in a Brazilian industry to be conducted in 2025. After all knowledge required for the system is formalised, a company will utilise it to prioritise their R&D portfolio and identify the best subsidies for each project. Then, all results will be analysed, and the impacts of implementing the proposed framework will be published.

5 Discussion and Conclusion

This research has presented a framework for the multi-criteria prioritisation of R&D and innovation projects. This framework considers the projects' inherent characteristics and conducts cross-analysis with features from available incentives and subsidies, as well as the company's strategic plans and guidelines. Given the extensive array of incentives established within the Brazilian innovation ecosystem and the expansive bodies of knowledge within project management and strategic planning, it becomes essential to systematise this prioritisation process in alignment with established literature and scientific methodologies. This framework aims to mitigate biases and preempt the rejection of projects submitted for incentives.

Therefore, the proposition of this framework will bring significant theoretical and practical impacts for researchers and companies. The first impact lies in its attempt to enhance understanding of integrating the three studied domains. Given the scarcity of literature concerning the integration of innovation, strategy, and their collective interplay, we anticipate that this paper and the proposed framework will significantly contribute to a

[2] All information regarding the program's incentivized credit lines were extracted from the official governmental web page dedicated to this funding agency.

more profound comprehension of these integrations and their implications for both companies and R&D and innovation projects. Moreover, the framework will offer practical utility to companies, empowering them to apply it within their day-to-day projects and activities, enhancing their strategic decision-making capabilities. Our primary objective is to nurture the innovation ecosystem and streamline the execution of projects that might otherwise falter due to resource constraints or the absence of scientific methodologies for prioritisation, which can enhance the likelihood of project prioritisation occurring impartially, devoid of managerial biases.

Following [24], the adopted Design Science Research (DSR) methodology encompasses a pivotal concluding step: disseminating results from thoroughly demonstrating the designed solution within a practical context. Consequently, the authors expect this framework to be formulated, theoretically proposed, and subjected to empirical validation through a case study. The outcomes will be disseminated via scholarly and professional publications, enhancing the framework's significance for academic and industrial audiences. After delineating the problem and motivation for the research, the authors proceeded to conduct a literature review encompassing four principal areas: the integration between project domains, incentives domains, strategy domains, and the integration of all three domains collectively. The review revealed a notable gap in the literature concerning the interrelationships among these domains, particularly in integrating incentives with the other domains and analysing all three domains collectively.

Consequently, the presented research and proposed framework should be continued, beginning with identifying the most appropriate scientific methodologies to guide the design and implementation of the framework. Additionally, particular attention should be directed towards analysing the impacts of utilising incentives and subsidies on project prioritisation and execution and how the company's strategy influences this.

Acknowledgements. The authors thank the National Council for Scientific and Technological Development (CNPq), Coordination for the Improvement of Higher Education Personnel (CAPES) and Pontifical Catholic University of Parana for financially supporting this research.

References

1. OECD. Main Science and Technology Indicators (Edition 2022). In: OECD Science, Technology and R&D Statistics (database) (2023). https://doi.org/10.1787/2e2e1820-en. Accessed 4 Feb 2024
2. Rhaiem, K., Amara, N.: Learning from innovation failures: a systematic review of the literature and research agenda. Rev. Manag. Sci. **15**, 189–234 (2021). https://doi.org/10.1007/s11846-019-00339-2
3. Cainelli, G., Evangelista, R., Savona, M.: The impact of innovation on economic performance in services. Serv. Ind. J. **24**, 116–130 (2004). https://doi.org/10.1080/0264206041233130 1162
4. Ranga, M., Etzkowitz, H.: Triple helix systems: an analytical framework for innovation policy and practice in the knowledge society. Ind. High. Educ. **27**, 237–262 (2013). https://doi.org/10.5367/ihe.2013.0165
5. Martins, L.: Análise do Apoio Indireto à Inovação pelo BNDES. O Eco da Graduação **1**(2), 1–11 (2016)

6. Bueno, A., Torkomian, A.L.V.: Financiamentos à inovação tecnológica: reembolsáveis, não reembolsáveis e incentivos fiscais. Review of Administration and Innovation - RAI. **11**, 135 (2015). https://doi.org/10.11606/rai.v11i4.100276.
7. Zucoloto, G.F.: Lei do Bem: Impactos nas atividades de P&D no Brasil. http://www.ipea.gov.br. Accessed Feb 2010
8. Forsman, H.: Innovation failure in SMEs: a narrative approach to understand failed innovations and failed innovators. Int. J. Innov. Mgt. **25**, 2150104 (2021). https://doi.org/10.1142/S1363919621501048
9. Lappalainen, L., Aleem, M., Sandberg, B.: How to manage open innovation projects? An integrative framework. Project Leadersh. Soc. **4**, 100095 (2023). https://doi.org/10.1016/j.plas.2023.100095
10. Barbosa, A.P.F.P.L., Salerno, M.S., Nascimento, P.T.D.S., Albala, A., Maranzato, F.P., Tamoschus, D.: Configurations of project management practices to enhance the performance of open innovation R&D projects. Int. J. Project Manag. **39**, 128–138 (2021). https://doi.org/10.1016/j.ijproman.2020.06.005
11. Klaus-Rosińska, A., Pliński, W.: Management of R&D projects in the early phases of the project life cycle – empirical research. Procedia Comput. Sci. **219**, 1994–2002 (2023). https://doi.org/10.1016/j.procs.2023.01.500
12. Müller, R., Turner, R.: The influence of project managers on project success criteria and project success by type of project. Eur. Manag. J. **25**, 298–309 (2007). https://doi.org/10.1016/j.emj.2007.06.003
13. Vrchota, J., Řehoř, P.: Project management and innovation in the manufacturing industry in Czech Republic. Procedia Comput. Sci. **164**, 457–462 (2019). https://doi.org/10.1016/j.procs.2019.12.206
14. Government of Canada. Scientific Research and Experimental Development (SR&ED) Tax Incentives. https://www.canada.ca/en/revenue-agency/services/scientific-research-experimental-development-tax-incentive-program.html. Accessed 4 Feb 2024
15. Government of The United States of America. Internal Revenue Service - IRS. Research Credit. https://www.irs.gov/businesses/research-credit. Accessed 4 Feb 2024
16. Government of Singapore. Inland Revenue Authority of Singapore - IRAS. Productivity & Innovation Credit (PIC) Scheme. https://www.iras.gov.sg/taxes/corporate-income-tax/specific-topics/productivity-innovation-credit-(pic)-scheme. Accessed 4 Feb 2024
17. Government of France. République Française. Research Tax Credit (RIC). https://entreprendre.service-public.fr/vosdroits/F23533?lang=en. Accessed 4 Feb 2024
18. Santos, U.P.D., Rapini, M.S., Mendes, P.S.: Impactos dos incentivos fiscais na inovação de grandes empresas: uma avaliação a partir da pesquisa Sondagem de Inovação da ABDI. Nova econ. **30**, 803–832 (2020). https://doi.org/10.1590/0103-6351/5687
19. Government of Brazil. Law number 11.196, of November 21, 2005. Institutes the Special Tax Regime for the Export Platform of Information Technology Services - REPES, the Special Regime for the Acquisition of Capital Goods for Exporting Companies - RECAP, and the Digital Inclusion Program; provides fiscal incentives for technological innovation; amends various legal texts and provides other measures. Brasília, DF: Official Gazette of the Union (1997). Accessed 4 Feb 2024
20. Szejka, A.L., Aubry, A., Panetto, H., Júnior, O.C., Loures, E.R.: Towards a conceptual framework for requirements interoperability in complex systems engineering. In: Meersman, R., et al. (eds.) On the Move to Meaningful Internet Systems: OTM 2014 Workshops, pp. 229–240. Springer, Heidelberg (2014). https://doi.org/10.1007/978-3-662-45550-0_24
21. DuCoin, C., Kuo, P.C.: Creating an integrated strategic plan. Surgery **175**, 1240–1243 (2024). https://doi.org/10.1016/j.surg.2023.11.006.
22. PROJECT MANAGEMENT INSTITUTE. A Guide to the Project Management Body of Knowledge (PMBOK Guide) – Seventh Edition. Project Management Institute, 2021

23. Peffers, K., Tuunanen, T., Rothenberger, M.A., Chatterjee, S.: A design science research methodology for information systems research. J. Manag. Inf. Syst. **24**, 45–77 (2007). https://doi.org/10.2753/MIS0742-1222240302
24. Reisdorfer-Leite, B., Marcos De Oliveira, M., Rudek, M., Szejka, A.L., Canciglieri Junior, O.: Startup definition proposal using product lifecycle management. In: Nyffenegger, F., Ríos, J., Rivest, L., and Bouras, A. (eds.) Product Lifecycle Management Enabling Smart X, pp. 426–435. Springer, Cham (2020). https://doi.org/10.1007/978-3-030-62807-9_34
25. Otto, R.B., Szollosi, C.H., Terada, G.G., Hisamura, R.K., Zat, G., Pereira, J.A.: Methodology for product development in R&D companies. In: Hiekata, K., Moser, B.R., Inoue, M., Stjepandić, J., and Wognum, N. (eds.) Advances in Transdisciplinary Engineering. IOS Press (2019). https://doi.org/10.3233/ATDE190132
26. Paramati, S.R., Alam, M.S., Hammoudeh, S., Hafeez, K.: Long-run relationship between R&D investment and environmental sustainability: evidence from the European Union member countries. Int. J Fin Econ. **26**, 5775–5792 (2021). https://doi.org/10.1002/ijfe.2093
27. Xu, J., Liu, F., Shang, Y.: R&D investment, ESG performance and green innovation performance: evidence from China. K. **50**, 737–756 (2021). https://doi.org/10.1108/K-12-2019-0793
28. Na, C., Ni, Z., Shu, Q., Zhang, H.: Can Government Subsidies Improve Corporate ESG Performance? Evidence from Listed Enterprises in China (2024). https://doi.org/10.2139/ssrn.4699016. https://www.ssrn.com/ab-stract=4699016

Personalized Exercise Plans for the Elderly Using KNN Based on FITT Principle

Phachara Wuttisetpaiboon[1](✉), Suwit Wongsila[2], Pradorn Sureephong[2], Nauman Aslam[3], Shanfeng Hu[3], and Adnan Shahid[4]

[1] Graduated School of College of Arts, Media and Technology, Chiang Mai University, Chiang Mai, Thailand
phachara_w@cmu.ac.th

[2] College of Arts, Media and Technology, Chiang Mai University, Chiang Mai, Thailand
{suwit.w,pradorn.s}@cmu.ac.th

[3] Computer and Information Science, Northumbria University, New Castle Upon Tyne, UK
{nauman.aslam,shanfeng2.hu}@northumbria.ac.uk

[4] Engineering and Architecture, Ghent University, Flanders, Belgium
adnan.shahid@ugent.be

Abstract. Thailand is experiencing a significant demographic shift toward an aging population, leading to increased healthcare demands. This research aims to address personalized health management for the elderly by developing a reinforcement learning system based on the FITT (Frequency, Intensity, Time, and Type) standard. The system monitors and recommends personalized activities to improve health outcomes.

Data on elderly activities were collected using wearable devices and mobile apps, capturing the full FITT components. After preprocessing, the K-Nearest Neighbors (KNN) algorithm was used for classification, with hyperparameter tuning via Grid Search to optimize performance. Additional models, including Decision Trees, Random Forests, SVM, and Logistic Regression, were implemented and compared. Ensemble methods further enhanced predictive accuracy.

The KNN model demonstrated high classification accuracy, while the reinforcement learning-based recommendation engine effectively adapted to user-specific data, offering comprehensive and personalized activity plans. Challenges such as data variability and user compliance highlight the need for further innovation, including improved sensor technologies and strategies for engagement.

In conclusion, the integration of the FITT framework and advanced machine learning algorithms resulted in a robust system that improves health outcomes and user satisfaction. This technology-driven approach provides a promising tool for caregivers and healthcare providers, laying the foundation for enhanced elderly care through personalized exercise recommendations. Future work will focus on refining the system with real-world testing and enhancing user adherence.

Keywords: Machine Learning · FITT · Recommendation system · Elderly · Reinforcement Learning

P. Sureephong et al. (Eds.): PLM 2024, IFIP AICT 741, pp. 343–353, 2025.
https://doi.org/10.1007/978-3-031-93323-3_26

1 Introduction

Thailand, like many other countries, is experiencing a significant demographic shift towards an aging population. The proportion of elderly citizens is growing rapidly, leading to increased healthcare demands and challenges. Many elderly individuals face health issues such as chronic diseases, reduced mobility, and cognitive decline. Ensuring their well-being requires effective and personalized health management strategies. Traditional approaches to elderly care often lack the personalization needed to address the diverse health needs of older adults.

To address this, there is a need for innovative solutions that leverage technology to provide tailored health recommendations. One such approach is using the FITT (Frequency, Intensity, Time, and Type) standard to design and evaluate exercise programs specifically for the elderly. The FITT standard allows for structured and individualized exercise regimens that can cater to the varying capabilities and health conditions of older adults, thereby promoting their physical and mental well-being.

This research aims to develop a reinforcement learning system that utilizes the FITT standard to monitor and recommend personalized activities for the elderly. By focusing on Frequency and Time and incorporating Intensity and Type, the system aims to improve health outcomes through precise and adaptive recommendations. This approach promises to enhance the quality of life for elderly individuals in Thailand and beyond by providing personalized, data-driven activity recommendations, ultimately contributing to more effective and sustainable elderly care solutions.

2 Literature Review

Recommendation systems play an important role in today's digital world. They have found applications in various areas such as music platforms like Spotify and movie streaming services like Netflix. These systems use various algorithms to predict user preferences and provide personalized content, enhancing user experience and engagement. However, sedentary lifestyles have become the major driver of several diseases as well as healthcare costs, highlighting the need for recommendation systems in the health and wellness sector [1]. Exercise recommendation systems are significantly different from streaming recommendation systems. Unlike streaming services, where click feedback and user interaction data are readily available, exercise recommendation systems cannot rely on such immediate feedback. Instead, they must infer user engagement and effectiveness of recommendations through indirect measures such as activity logs, sensor data, and health outcomes. This makes the design and implementation of exercise recommendation systems more complex and challenging. The FITT standard provides a structured approach to exercise prescription, making it a suitable framework for developing personalized exercise recommendations. Incorporating the full FITT components (Frequency, Intensity, Time, and Type) ensures that recommendations are comprehensive and tailored to the individual's health needs and capabilities. Previous studies have shown the effectiveness of using machine learning and reinforcement learning in developing personalized health interventions, highlighting the potential for these technologies to improve elderly care. In conclusion, leveraging the FITT standard with advanced machine learning techniques presents a promising solution to the growing healthcare challenges faced by the elderly population in Thailand. By developing a system that monitors and recommends personalized activities, this research aims to enhance the health and well-being of elderly individuals through precise, data-driven interventions.

Introduction to Expert Systems in Health and Wellness

Expert systems have become increasingly significant in various domains, including health and wellness, to provide personalized recommendations and interventions. These systems utilize artificial intelligence (AI) and machine learning (ML) techniques to process vast amounts of data, infer patterns, and deliver tailored recommendations that can enhance user engagement and health outcomes. In elderly care, expert systems can address unique health challenges by providing personalized exercise recommendations, thereby improving overall well-being and quality of life [1].

Personalization in Elderly Care [2]

Personalized health interventions are crucial for the elderly due to their diverse and specific health needs. Traditional one-size-fits-all approaches often fall short in addressing individual differences in health status, physical capabilities, and preferences. Personalization in elderly care can be achieved through the integration of the FITT (Frequency, Intensity, Time, and Type) principles, which provide a structured framework for designing exercise programs. Studies have shown that personalized exercise recommendations can lead to significant improvements in physical and cognitive health outcomes for older adults.

Machine Learning and Reinforcement Learning in Health Interventions

Machine learning techniques, such as K-Nearest Neighbors (KNN), and reinforcement learning (RL) have been effectively utilized in developing personalized health interventions. KNN is a simple yet powerful classification algorithm that can categorize data points based on their similarity to other data points. This makes it suitable for classifying different types of physical activities based on frequency and duration. Reinforcement learning, on the other hand, is particularly useful for developing recommendation systems as it can learn and adapt based on user feedback and changing conditions. By defining a reward system that incentivizes desirable health outcomes, RL can provide personalized and adaptive exercise recommendations that optimize health benefits for the elderly.

The FITT Principle in Exercise Prescription

The FITT [3, 4] principle is a well-established framework in exercise science, encompassing four key components: Frequency (how often an activity is performed), Intensity (the effort level of the activity), Time (the duration of the activity), and Type (the kind of activity). Incorporating all four components ensures that exercise recommendations are comprehensive and tailored to the individual's needs. Research has demonstrated that applying the FITT principle can effectively enhance physical fitness, reduce the risk of chronic diseases, and improve overall health among the elderly.

Challenges and Considerations in Developing Exercise Recommendation Systems
Developing effective exercise recommendation systems for the elderly involves several challenges. These include handling data variability, ensuring user compliance, and accurately measuring the intensity and type of activities. Wearable devices and mobile apps can facilitate data collection by providing continuous and real-time monitoring of physical activities. However, ensuring the accuracy and consistency of the collected data remains a critical task. Additionally, motivating elderly users to adhere to recommended activities requires a user-friendly interface and consideration of individual preferences and limitations [5].

Case Studies and Applications
Several studies have explored the application of expert systems and ML techniques in elderly care. For instance, a study by Ordonez et al. (2013) utilized a decision tree algorithm to classify physical activities and provide personalized recommendations to older adults. Another study by Ristaniemi et al. (2018) employed reinforcement learning to develop an adaptive exercise recommendation system that improved user engagement and health outcomes. These case studies highlight the potential of technology-driven interventions in enhancing elderly care.

In summary, leveraging the FITT standard with advanced machine learning techniques presents a promising solution to the growing healthcare challenges faced by the elderly population. By developing a system that monitors and recommends personalized activities, this research aims to enhance the health and well-being of elderly individuals through precise, data-driven interventions. The pilot research outlined here sets the stage for future studies and practical applications, offering valuable insights into the integration of AI and ML in personalized elderly care.

K-Nearest Neighbors (KNN) in Exercise Recommendation
The K-Nearest Neighbors (KNN) algorithm is one of the simplest yet effective classification algorithms used in ML. KNN works by classifying a data point based on the majority class among its k-nearest neighbors in the feature space. This algorithm is particularly suitable for exercise recommendation systems for the elderly due to its simplicity and effectiveness in handling small to moderate-sized datasets.

KNN can classify exercise activities by analyzing features such as Frequency (number of times an activity is performed per week), Intensity (effort level of the activity), Time (duration of each activity session), and Type (category of activity). By finding the most similar historical data points, KNN can recommend exercises that are most likely to benefit the user, considering their current health status and activity preferences.

Reinforcement Learning (RL) for Adaptive Recommendations
Reinforcement learning (RL) is another powerful ML technique used in developing adaptive exercise recommendation systems. RL involves training an agent to make a sequence of decisions by rewarding desirable outcomes and penalizing undesirable ones. This approach is particularly useful in dynamic environments where user preferences and health conditions may change over time.

In the context of elderly exercise recommendations, an RL-based system can continuously learn and adapt its recommendations based on user feedback and health outcomes. For example, if a particular exercise regimen leads to improved health metrics, the system will reinforce this behavior by recommending similar activities in the future. Conversely, if an exercise plan does not yield positive results, the system will adjust its recommendations accordingly.

3 Research Methodology

Elderly Profile Collection

Data collection for the Elderly Profile encompassing variables such as gender, age, weight, height, BMI, and heart rate was primarily conducted through registration on a web-based system.

Exercise Data Collection

In the exercise data collection phase of this research, we employed a dual approach utilizing the FITT principle and GymBot technology. The FITT principle, which stands for Frequency, Intensity, Time, and Type, provided a structured framework for gathering exercise-related data from participants. Meanwhile, GymBot served as the equipment for conducting exercises while simultaneously collecting relevant workout metrics.

Classification

The classification process involved the utilization of the K-Nearest Neighbors (KNN) algorithm. This algorithm was applied to classify users' exercise data based on the FITT principle, specifically focusing on the Frequency (F) and Time (T) components. To implement this methodology, the user's exercise data, including frequency and duration of exercise sessions, was collected and preprocessed. The FITT principle served as a guiding framework, with emphasis placed on the Frequency and Time variables. These variables were extracted from the dataset and used as input features for the KNN classification model.

Personalize Plan

The development of personalized exercise plans followed the classification process using the K-Nearest Neighbors (KNN) algorithm. After obtaining the classification of exercisers based on their exercise frequency and duration, the next step involved creating tailored plans to help individuals reach their fitness goals.

To achieve this, the classification results from the KNN algorithm were utilized to identify the specific category or cluster to which each exerciser belonged. These categories represented different patterns or levels of exercise behavior, such as low frequency/short duration, moderate frequency/moderate duration, or high frequency/long duration, as determined by the KNN model.

Once the classification was determined, personalized exercise plans were developed based on the identified exercise category. For instance, exercisers classified into the low frequency/short duration category might receive recommendations for gradually increasing both the frequency and duration of their exercise sessions over time. Conversely, those classified into the high frequency/long duration category might be encouraged to maintain their current exercise levels while focusing on intensity or diversifying their workout routines.

4 Result

Profile Collection

This study presents a dataset comprising health profiles of 20 elderly individuals, focusing on a range of physiological and demographic variables. The dataset includes unique identifiers for everyone, along with key variables such as age, sex, weight, height, heart rate, blood pressure, and Body Mass Index (BMI). These variables are critical for evaluating overall health (Table 1).

Table 1. Profile collection 20 people.

ID	Age	Sex	Weight	Height	Heart Rate	Blood Pressure	BMI
6520b0d93f5a93001cc6b1e2	66	Female	56	155	1	104/67	23.31
6520b1313fSa93001cc6b1e4	60	Female	47	158	64	112/76	18.83
64d6e49bS25bOf001cc4ca71	62	Male	S7	163	69	136/84	21.45
6505534073eec4001bde3942	60	Female	60	155	87	141/83	24.97
6520b1903f5a93001cc6b1e8	71	Female	59	154	76	134/87	24.88
6520b1213f5a93001cc6b1ea	70	Female	50	150	B	138/78	22.22
6520b1b73F5a93001cc6b1ec	68	Female	60	162	82	121/73	22.86
6520b1c83f5a93001cc6b1ee	65	Female	70	165	82	118/78	24.61
64d0ae582240a2001ca44749	65	Female	55	154	76	140/80	23.19
6520b1f33f5a93001cc6b110	67	Female	62	156	81	142/75	25.48
6562cc7a1d065c001c41de42	62	Female	58	162	nR	137/81	22.1
6562cc031d065c001c41de3f	61	Female	54	161	74	128/83	20.83
6562cb971d065c001c41de3e	64	Male	47	155	66	134/81	19.56
6562cb4a1d065c001c41de3a	63	Female	52	153	65	132/85	22.21
6562caco1d065c001c41de38	70	Female	68	151	88	138/85	29.82
6562c9201d065c001c41de36	72	Female	65	157	93	135/89	26.37
6562c8421d065c001c41de34	67	Female	51	165	79	137/85	18.73
6562c7941d065c001c41de32	69	Female	49	153	1	143/87	20.93
6562c6F11d065c001c41de30	73	Female	62	168	82	139/91	21.97

Data Collection

The primary aim of this dataset is to provide a comprehensive overview of users' exercise patterns, helping to identify trends in exercise frequency and duration, as well as preferences for specific types of physical activities (Table 2).

Table 2. Data collection from user's exercise.

Date	Day	Time	Exercise Course 0	Exercise Course 1	Exercise Course 2	Exercise Course 3	Exercise Course 4	Exercise Course 5	Exercise Course 6
7/8/2023	Monday	56	Thai Digital Dance 05	Thai Digital Dance 06			Stretching	Thai Digital Dance 05	
8/8/2023	Tuesday	46.4	Thai Digital Dance 06	Thai Digital Dance 04	Stretching		Stretching	Thai Digital Dance 05	Thai Digital Dance 06
11/8/2023	Friday	38.85	Stretching						
2/10/2023	Monday	62.7	Thai Digital Dance 04			Stretching	Stretching		
7/10/2023	Saturday	40.68	Stretching				Thai Digital Dance 05		
10/10/2023	Tuesday	77	Stretching			Stretching	Thai Digital Dance 05	Thai Digital Dance 06	
17/10/2023	Tuesday	51.5	Thai Digital Dance 05			Thai Digital Dance 06	Thai Digital Dance 05	Thai Digital Dance 04	
19/10/2023	Thursday	37	Thai Digital Dance 04	Thai Digital Dance 04		Thai Digital Dance 04	Thai Digital Dance 04		
21/10/2023	Saturday	36	Thai Digital Dance 04	Thai Digital Dance 04		Thai Digital Dance 04	Thai Digital Dance 04	Stretching	Stretching
26/10/2023	Thursday	43.88	Stretching						

The transformation of the data into this simplified format is a crucial step in preparing it for machine learning applications, specifically for the k-Nearest Neighbors (k-NN) algorithm (Table 3).

Classification

The K-Nearest Neighbors (KNN) classification model offers a straightforward approach to solving classification problems based on the principle of similarity (Fig. 1).

Personalized Plan

After classifying exercisers into four groups using a K-Nearest Neighbors (KNN) model,

Table 3. Summary of data collection from user's exercise.

ID	Sex	Frequency (week)	Total Time (minute/week)
6520b0d93f5a93001cc6b1e2	Female	2	56.4
6520b1313f5a93001cc6b1e4	Female	2	80.43
64d6e49b525b0f001cc4ca71	Male	2	53.98
6505534073eec4001bde9942	Female	2	24
6520b1903f5a93001cc6b1e8	Female	2	135.28
6520b1a13f5a93001cc6b1ea	Female	2	115
6520b1b73f5a93001cc6b1ec	Female	2	57
6520b1ce3f5a93001cc6b1ee	Female	2	48
64d0ae582240a2001ca44749	Female	2	55
6520b1f33f5a93001cc6b1f0	Female	2	53.5
6562cc7a1d065c001c41de42	Female	3	69.3
6562cc031d065c001c41de3f	Female	3	62.53
6562cb971d065c001c41de3c	Male	3	68
6562cb4a1d065c001c41de3a	Female	3	35.53
6562cac01d065c001c41de38	Female	3	49
6562c9201d065c001c41de36	Female	3	36
6562c8421d065c001c41de34	Female	3	33.3
6562c7941d065c001c41de32	Female	3	65.3
6562c6f11d065c001c41de30	Female	3	34
6562c64e1d065c001c41de2e	Female	3	68.65

the goal is to provide tailored suggestions to improve their exercise frequency, time, or both (Table 4).

- Group 0: High Frequency, Low Time: Suggestion: Start with short, frequent exercise sessions. Aim for consistency rather than intensity. Start with 10-min workouts, five days a week. Gradually increase the duration by 5 min each week until reaching the recommended daily exercise guidelines.
- Group 1: Low Frequency, High Time: Suggestion: Break down long exercise sessions into shorter, more manageable chunks spread throughout the week.
- Group 2: High Frequency, High Time: Suggestion: Focus on maximizing the efficiency of short workouts through high-intensity interval training (HIIT) or circuit training.
- Group 3: Low Frequency, Low Time: Suggestion: Emphasize the importance of balanced training and recovery to prevent overtraining and injury.

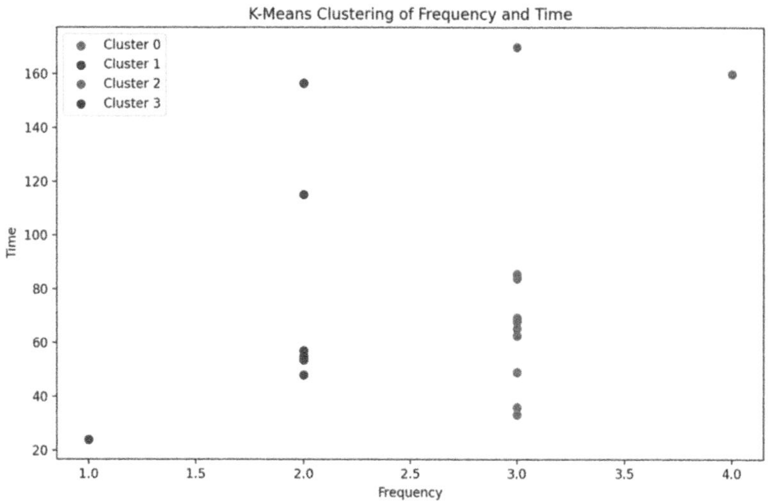

Fig. 1. Classification by K-Nearest Neighbors model.

Table 4. Cluster of exerciser.

ID	Sex	Frequency (week)	Total Time (minute/week)	Cluster
6520b0d93f5a93001cc6b1e2	Female	2	56.4	1
6520b1313f5a93001cc6b1e4	Female	2	80.43	2
64d6e49b525b0f001cc4ca71	Male	2	53.98	3
6505534073eec4001bde9942	Female	2	24	3
6520b1903f5a93001cc6b1e8	Female	2	135.28	2
6520b1a13f5a93001cc6b1ea	Female	2	115	1
6520b1b73f5a93001cc6b1ec	Female	2	57	3
6520b1ce3f5a93001cc6b1ee	Female	2	48	3
64d0ae582240a2001ca44749	Female	2	55	3
6520b1f33f5a93001cc6b1f0	Female	2	53.5	3
6562cc7a1d065c001c41de42	Female	3	69.3	0
6562cc031d065c001c41de3f	Female	3	62.53	0
6562cb971d065c001c41de3c	Male	3	68	0
6562cb4a1d065c001c41de3a	Female	3	35.53	0
6562cac01d065c001c41de38	Female	3	49	0

(continued)

Table 4. (*continued*)

ID	Sex	Frequency (week)	Total Time (minute/week)	Cluster
6562c9201d065c001c41de36	Female	3	36	0
6562c8421d065c001c41de34	Female	3	33.3	0
6562c7941d065c001c41de32	Female	3	65.3	0
6562c6f11d065c001c41de30	Female	3	34	0
6562c64e1d065c001c41de2e	Female	3	68.65	0

5 Discussion

The integration of the FITT components into the dataset and system framework has proven to be highly beneficial. By considering Frequency, Intensity, Time, and Type, the model can offer a more comprehensive and personalized set of recommendations. The use of KNN for classification has demonstrated high accuracy in categorizing activities, and the reinforcement learning approach has been effective in adapting recommendations based on user-specific data and outcomes. The challenges encountered, including data variability and ensuring user compliance with recommended activities, underscore the need for ongoing refinement and innovation in this field.

Future improvements could involve refining intensity and type measurements by incorporating more precise sensor technologies and refining the algorithms to better handle variability in user data. Additionally, enhancing user engagement strategies and tailoring recommendations to individual preferences and limitations could further improve the effectiveness of the system. Ease of Use:

6 Conclusion and Future Work

In conclusion, this research has successfully developed a reinforcement learning system that utilizes the full FITT standard to monitor and recommend activities for the elderly, demonstrating high classification accuracy and effective recommendation capabilities. These achievements hold promise for improving health outcomes and user satisfaction among the elderly population. By addressing personalized health management needs through technology-driven interventions, this approach offers a valuable tool for caregivers and healthcare providers.

Moving forward, there is a significant opportunity to further enhance the efficacy and scalability of personalized health interventions for the elderly. Future work could focus on leveraging reinforcement learning techniques to increase the frequency and duration of exercise sessions recommended by the system. This could involve refining the reinforcement learning algorithms to incentivize and encourage adherence to exercise regimens more effectively. Additionally, exploring ways to integrate real-time feedback mechanisms and adaptive algorithms could help tailor recommendations even more closely to individual needs and preferences.

Acknowledgments. (DigitalHealth-Asia) (https://digihealth-asia.eu/), under grant agree number 619193-EPP-1-2020-1-BE-EPPKA2-CBHE-JP.

References

1. Muangprathub, J., Sriwichian, A., Wanichsombat, A., Kajornkasirat, S., Nillaor, P., Boonjing, V.: A novel elderly tracking system using machine learning to classify signals from mobile and wearable sensors. Int. J. Environ. Res. Public Health **18**(23), 12652 (2021)
2. Hameed, B., Awais, M., Aslam, M.: Expert system for health and wellness recommendation. J. Artif. Intell. Res. **69**, 285–306 (2020)
3. Garber, C.E., Blissmer, B., Deschenes, M.R., et al.: American College of Sports Medicine position stand. Quantity and quality of exercise for developing and maintaining cardiorespiratory, musculoskeletal, and neuromotor fitness in apparently healthy adults: guidance for prescribing exercise. Med. Sci. Sports Exercise **43**(7), 1334–1359 (2011)
4. Rejeski, W.J., Mihalko, S.L.: Physical activity and quality of life in older adults. J. Gerontol. Ser. A: Biol. Sci. Med. Sci. **56**(suppl_2), 23–35 (2001)
5. Dishman, R., Heath, G., Lee, I.M.: Physical Activity Epidemiology: Human Kinetics. Champaign, US (2013)
6. Altman, N.S.: An introduction to kernel and nearest-neighbor nonparametric regression. Am. Stat. **46**(3), 175–185 (1992)
7. Sutton, R.S., Barto, A.G.: Reinforcement Learning: An Introduction. MIT Press (2018)
8. Haskell, W.L., Lee, I.M., Pate, R.R., et al.: Physical activity and public health: updated recommendation for adults from the American College of Sports Medicine and the American Heart Association. Med. Sci. Sports Exerc. **39**(8), 1423–1434 (2007)
9. Warburton, D.E., Nicol, C.W., Bredin, S.S.: Health benefits of physical activity: the evidence. Can. Med. Assoc. J. **174**(6), 801–809 (2006)
10. Mazzeo, R.S., Tanaka, H., Chodzko-Zajko, W.: Exercise prescription for the elderly: current recommendations. Sports Med. **49**(2), 79–99 (2019)
11. Bardram, J.E., Matic, A.: A decade of ubiquitous computing research in mental health. IEEE Pervasive Comput.Comput. **19**(1), 62–72 (2020)

Developing a Mixed Reality (MR) Technology System for Training Intelligent Logistics Personnel, a Learning Course on Vehicle Condition Inspection

Supavanee Kittithanon[1](✉), Santichai Wicha[2], Yootthapong Tongpaeng[2], and Krittika Kantawong[3]

[1] College of Arts, Media and Technology, Chiang Mai University, Chiang Mai 50200, Thailand
supavanee@kit.in.th

[2] School of Information Technology, Mae Fah Luang University, 333 Tha Sut, Mueang Chiang Rai 57100, Chiang Rai, Thailand

[3] School of Information and Communication Technology, University of Phayao, 19 Moo 2, Mae Ka, Mueang Phayao 56000, Phayao, Thailand

Abstract. This research explores the development of a Mixed Reality (MR) technology system for training intelligent logistics personnel, focusing on vehicle condition inspection using BEWAGON principles for forklifts. Traditional logistics training methods are often risky and impractical, whereas MR technology offers immersive and interactive training experiences that mitigate these issues. The study involved 57 logistics employees and evaluated the effectiveness of MR training through knowledge and satisfaction tests. Results indicated high accuracy in identifying easily observable forklift parts but highlighted the need for improved training on complex components. Participants rated the training highly for engagement and ease of use, though some found the practical applicability challenging. This study underscores MR's potential in revolutionizing logistics training, enhancing both safety and efficiency, and contributing to the advancement of smart logistics operations in Thailand. Further development is required to optimize MR training for detailed inspections, ensuring comprehensive skill development for logistics personnel.

Keywords: Mixed Reality · Smart Logistic · BEWAGON

1 Introduction

The logistics industry is a cornerstone of global commerce, and ensuring the efficiency and safety of its operations is paramount. The R3A Road, linking Thailand, Laos, and China, is pivotal for regional trade and development. Economically, it reduces transportation costs, boosts trade efficiency, and attracts investments, benefiting small and medium-sized enterprises. Socially, it enhances access to goods and services, improves

Published by Springer Nature Switzerland AG 2025
P. Sureephong et al. (Eds.): PLM 2024, IFIP AICT 741, pp. 354–364, 2025.
https://doi.org/10.1007/978-3-031-93323-3_27

infrastructure, and fosters cultural exchanges. However, the route's success depends on the safety and efficiency of logistics operations. Robust training for logistics personnel is essential to handle advanced technology and machinery, ensuring smooth and secure operations. By investing in smart logistics and training, the R3A Road can maximize its economic and social benefits, driving regional prosperity and stability.

Traditional training methods in the logistics sector often involve the use of real vehicles and equipment, which can be both dangerous and impractical. These methods expose trainees to significant risks and do not always provide a comprehensive understanding of the scenarios they might encounter. The advent of simulation technologies, particularly Mixed Reality (MR), presents an opportunity to revolutionize training in this field. MR technology, which blends real and virtual environments, offers immersive and interactive training experiences that can replicate real-world scenarios without the associated risks. Devices like the Microsoft HoloLens enable trainees to interact with digital elements in a realistic setting, thereby enhancing the quality and safety of the training process.

This research aims to develop a Mixed Reality (MR) technology system specifically designed for training intelligent logistics personnel, with a focus on vehicle condition inspection according to "BEWAGON" principles for forklifts. The study will explore the design, implementation, and evaluation of this MR-based training system, assessing its effectiveness in improving learning outcomes and safety. By leveraging MR technology, this research seeks to contribute to the advancement of smart logistics training, ensuring that personnel are better prepared to handle real-life situations and meet the demands of modern logistics operations. The findings of this study will provide valuable insights into the potential applications of MR in logistics training and its implications for the industry.

1.1 A Subsection Sample

The R3A road, part of the North-South Economic Corridor, connects Thailand, Laos, and China, serving as a crucial route for regional trade and development. This corridor enhances connectivity and facilitates the efficient movement of goods across borders. In Thailand, the R3A road runs through the northern province of Chiang Rai, linking it to Laos at the Chiang Khong border and extending further to Kunming in China. The road's infrastructure has seen significant improvements, including better pavement, signage, and border facilities, which have contributed to increased trade volumes and reduced transportation costs. However, challenges such as traffic congestion, inadequate logistics facilities, and safety concerns remain prevalent, necessitating advancements in logistics management [1]. Smart logistics refers to the integration of advanced technologies such as the Internet of Things (IoT), big data analytics, artificial intelligence (AI), and automation to optimize logistics operations. These technologies enable real-time tracking, predictive maintenance, route optimization, and efficient inventory management. By leveraging smart logistics, companies can achieve greater efficiency, reduce operational costs, and enhance supply chain visibility. Key components of smart logistics include automated warehouses, smart transportation systems, and connected logistics platforms that facilitate seamless communication and data exchange among various stakeholders [9].

The implementation of smart logistics can significantly boost Thailand's economy by enhancing the efficiency and reliability of supply chains. Improved logistics operations reduce transportation and inventory costs, leading to lower prices for goods and increased competitiveness for Thai exports. The enhanced efficiency also attracts foreign investments as businesses seek to capitalize on the streamlined logistics infrastructure. Furthermore, smart logistics can support the growth of e-commerce by ensuring timely and accurate delivery of products, thus driving consumer confidence and spending. The integration of advanced technologies in logistics also creates high-skilled job opportunities, contributing to workforce development and economic diversification [2]. Safety is a critical aspect of logistics operations, particularly on high-traffic routes like the R3A road. Ensuring the safety of logistics personnel, vehicles, and goods is paramount to maintaining the integrity and reliability of supply chains. Accidents and safety breaches can lead to significant financial losses, supply chain disruptions, and reputational damage for logistics companies. Implementing smart logistics solutions can enhance safety through real-time monitoring of vehicle conditions, predictive maintenance, and automated safety protocols. These measures help prevent accidents, reduce downtime, and ensure the smooth flow of goods. Additionally, a strong safety record can improve the overall perception of Thailand's logistics sector, attracting more business and investment [4].

1.2 Mixed Reality (MR) Technologies for Workforce Training

Mixed Reality (MR) seamlessly blends physical and digital worlds, creating environments where physical and virtual elements interact in real-time, leveraging both Augmented Reality (AR) and Virtual Reality (VR) [7]. MR is particularly suitable for training due to its immersive experience, enhancing engagement and retention, providing a safe learning environment, reducing costs, and offering customizable modules [8].

In Asia, several case studies illustrate the effectiveness of MR in training. In Japan, Toyota uses MR to train employees on automotive assembly, improving accuracy and efficiency through interactive guides and real-time feedback [10]. South Korea's Samsung employs MR for maintenance training of industrial equipment, enhancing precision and reducing error rates with step-by-step instructions [11]. In China, the State Grid Corporation utilizes MR to train technicians for high-voltage equipment maintenance, improving safety and operational readiness [9].

MR technologies provide a transformative approach to workforce training, combining physical and digital learning environments to improve training outcomes and operational efficiency. The Asian case studies from Toyota, Samsung, and the State Grid Corporation demonstrate MR's practical applications and benefits across industries, highlighting its potential to enhance skill development and safety in training programs.Given its immersive and interactive nature, MR is well-suited to training logistics personnel in Thailand, particularly for vehicle condition inspection according to BEWAGON principles, ensuring they are well-prepared to handle real-life situations safely and effectively.

1.3 BEWAGON Principles

The BEWAGON principles provide a structured framework for vehicle condition inspections, particularly focusing on forklifts and other logistics vehicles. These principles encompass several critical aspects:

1. Brakes (B): Ensuring that the braking system is functioning correctly, with checks on brake pads, fluids, and overall response times.
2. Engine (E): Inspecting the engine for any signs of wear, leaks, or performance issues that could affect vehicle reliability.
3. Wheels (W): Examining the condition of tires, including tread depth, pressure, and alignment, to ensure safe and efficient operation.
4. Accessories (A): Verifying that all accessories, such as lights, signals, and mirrors, are in proper working condition.
5. Gauges (G): ChecFking that all gauges and indicators on the vehicle's dashboard are functioning accurately to provide real-time information to the operator.
6. Oil (O): Inspecting oil levels and quality, ensuring there are no leaks, and that regular maintenance schedules are followed.
7. Nuts and Bolts (N): Ensuring that all fasteners are secure and that there are no loose parts that could lead to operational failures.

By incorporating these BEWAGON principles into MR training programs, logistics personnel can gain hands-on experience in a virtual environment. This allows them to perform detailed vehicle inspections safely and effectively before applying these skills in real-world scenarios. The integration of BEWAGON principles into MR training ensures a comprehensive understanding of vehicle maintenance, enhancing the overall safety and efficiency of logistics operations.

1.4 Learning Theory in Mixed Reality (MR)

One of the most relevant learning theories for explaining and measuring the effectiveness of Mixed Reality (MR) is Experiential Learning Theory (ELT), proposed by David Kolb. ELT posits that learning is a process whereby knowledge is created through the transformation of experience. The theory is based on a four-stage learning cycle: concrete experience, reflective observation, abstract conceptualization, and active experimentation. MR provides immersive and interactive experiences that align well with this cycle, enabling learners to actively engage in and reflect on their learning processes [13]. In the context of our work on training logistics personnel, ELT is particularly suitable due to its emphasis on experiential learning. MR environments facilitate concrete experiences by allowing trainees to interact with realistic simulations of logistics operations, such as vehicle condition inspections. Therefore, to assess the effectiveness of MR in training, two key domains are particularly relevant:

- Knowledge Test: This domain evaluates the extent to which trainees have acquired the necessary knowledge and understanding from the MR training sessions. It typically involves pre- and post-training assessments to measure knowledge gains.

- Satisfaction Test: This domain measures the trainees' satisfaction with the MR training experience. It includes aspects such as user engagement, perceived usefulness, ease of use, and overall satisfaction with the training program. Surveys and feedback forms are commonly used to gather this information.

In conclusion, the application of ELT to MR training environments offers a robust framework for enhancing learning outcomes. The alignment of MR with ELT principles ensures that trainees can effectively acquire and apply new knowledge, leading to improved performance and higher satisfaction levels. This underscores the potential of MR in revolutionizing logistics training.

2 Research Methodology

2.1 Participants

The study sampled 57 employees working in the logistics field. These participants were selected based on their roles, which required frequent engagement with logistics operations and equipment, such as forklifts.

2.2 Knowledge Test

The knowledge test aimed to assess the trainees' understanding based on the principles of "BEWAGON." Utilizing MR technology, a set of 3D forklift models was created, with inspection tasks designed across three difficulty levels: easy, middle, and hard.

- Easy Tasks: Included larger, easily observable parts like wheels and headlights.
- Middle Tasks: Included smaller, slightly more challenging parts like warning lights and oil caps.
- Hard Tasks: Included parts requiring interaction with the model, such as brake discs (Fig. 1).

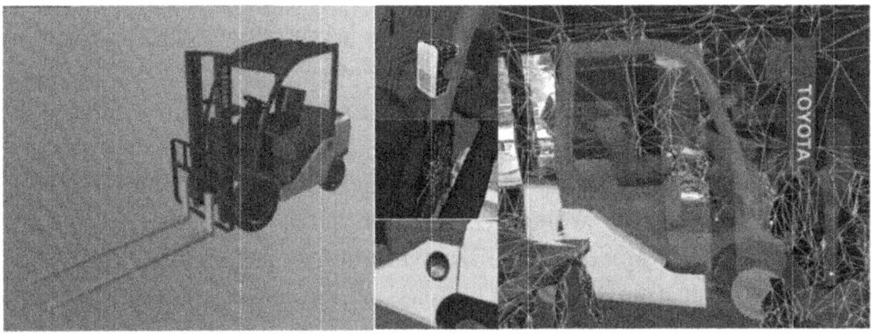

Fig. 1. The 3D masterpiece of a forklift.

Each level consisted of two tasks, making a total of six tasks. The exercise aimed to impart knowledge on inspecting forklift parts under the "BEWAGON" concept. Learners needed to identify the model parts in three conditions: ready to use (100% quality), used (50% quality, requiring repair in 2–3 weeks), and out of use (requiring immediate repair). Participants used the Hololens for the inspection, with each session lasting 10–15 min. During the tasks, the instructor guided the participants, recorded their responses, and marked their scores. Each correct identification earned the participant 1 point, while incorrect responses scored 0.

2.3 Satisfaction Test

The satisfaction test evaluated the participants' experience using the Hololens and MR training program. The assessment focused on four aspects:

- Ease of Use: This aspect measured how easily participants could use the Hololens and navigate through the MR content. Questions focused on the intuitiveness of the interface and the simplicity of performing tasks.
- Usefulness: This aspect assessed how beneficial the participants found the MR training in terms of enhancing their inspection skills and knowledge. It included questions on the practical applicability of the training to real-world scenarios.
- Learning: This aspect evaluated how effectively the MR training facilitated learning. It covered the clarity of the instructional content and the extent to which participants felt their understanding of forklift inspections improved.
- Fun: This aspect gauged the enjoyment level of participants while using the Hololens and engaging with the MR content. Questions focused on the engagement and entertainment value of the training experience.

Each aspect comprised two items on a 5-point Likert scale, supplemented by in-depth interviews. This test was conducted immediately after the knowledge test.

2.4 Procedure and Data Analysis

Participants were first briefed on the study objectives and the use of Hololens. They were then familiarized with the MR environment and the BEWAGON principles. Following this, participants undertook the six knowledge test tasks, with the instructor providing support and recording responses. Each session was designed to be completed within 15 min. Immediately after the knowledge test, participants completed the satisfaction questionnaire and participated in in-depth interviews.

The collected data from the knowledge test were quantified into scores, and statistical analysis was conducted to evaluate the effectiveness of MR training. Satisfaction test responses were analyzed using descriptive statistics to gauge overall user experience, and feedback from the indepth interviews provided qualitative insights. The instruments used included Hololens to deliver the MR training experience, 3D models of forklifts to simulate real inspection scenarios, questionnaires to measure satisfaction across four aspects with a 5-point Likert scale, and an interview guide to gather detailed feedback on the participants' MR experience.

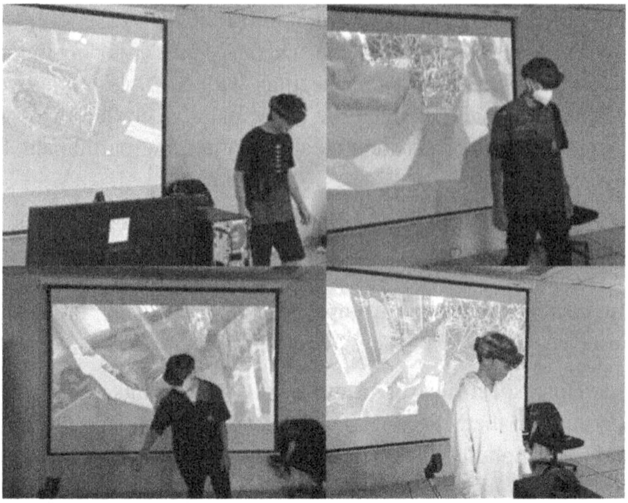

Fig. 2. Test participants engage in knowledge tasks, exploring the potential of mixed reality through hands-on experience.

3 Results

The results of the study are divided into two main sections: the knowledge test outcomes and the satisfaction test feedback.

3.1 Knowledge Test

The knowledge test evaluated participants' ability to inspect forklift parts based on the BEWAGON principles using MR technology. The test comprised six tasks of varying difficulty levels—easy, middle, and hard—designed to assess participants' proficiency in identifying the status of forklift parts as "ready to use," "used," or "out of use". presents the accuracy rates for correct answers for each forklift part. The results showed that participants had a 100% accuracy rate in identifying wheels, the most observable and easily identifiable part. This was followed by "headlights" with an 80% accuracy rate, indicating a high level of recognition but slightly lower than wheels due to the increased complexity or reduced visibility. For parts like "oil caps" and "warning lights", the accuracy rates dropped to 50%, reflecting the moderate difficulty level and the necessity for more careful inspection. The hardest tasks involved identifying "brake fluid caps" and "brake discs", with only 20% accuracy. These parts require more detailed interaction and understanding, which the MR technology aims to improve through immersive training. In sum, these results suggest that while MR technology is highly effective for training in identifying large, easily observable parts, there is room for improvement in training for more complex parts. The lower accuracy rates for smaller, more intricate parts highlight the need for more detailed training sessions using MR to ensure comprehensive understanding and inspection skills (Fig. 3).

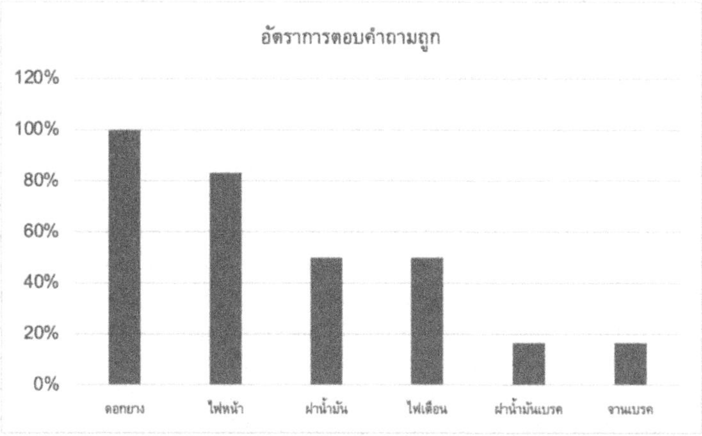

Fig. 3. Graph illustrating test participants' accuracy rates across different forklift parts, highlighting variations in task complexity and recognition.

3.2 Satisfaction Test Feedback

The satisfaction test aimed to capture participants' experiences and feedback regarding the use of Hololens and MR training content. This test focused on four key aspects: ease of use, usefulness, learning, and fun. Each aspect was evaluated using a 5-point Likert scale, and further insights were gathered through in-depth interviews (Fig. 4).

Fig. 4. Graph displaying participants' satisfaction levels across four aspects of MR training: ease of use, usefulness, learning, and fun.

Figure 2 shows the average satisfaction scores for each aspect. Participants rated the training experience highly, with ease of use receiving an average score of 4.2. This indicates that the MR technology was intuitive and straightforward, allowing users to navigate and interact with the training content effectively. Usefulness scored the lowest at 3.8, reflecting that participants found the training beneficial in enhancing their inspection skills, though some found the MR environment difficult and confusing to navigate. This lower score suggests that while MR training has potential, its practical applicability and ease of use need to be improved to ensure that users can fully benefit from the training. The learning aspect received a score of 4.0, suggesting that participants felt the MR training significantly improved their understanding and knowledge of forklift inspections. Finally, the highest score was for fun, at 4.5, indicating that participants found the training engaging and enjoyable, which is crucial for maintaining motivation and interest. Overall, the feedback from the satisfaction test underscores the overall positive reception of MR-based training but also highlights critical areas for improvement. The high scores for ease of use and fun suggest that the immersive and interactive nature of MR technology can effectively engage users. However, the relatively lower score for usefulness indicates that some participants experienced difficulties with the MR interface, which could affect their ability to fully grasp the practical applications of the training. This suggests a need for enhancing the instructional design and user interface of the MR training modules to reduce confusion and improve practical usability.

4 Discussion

The implementation of Mixed Reality (MR) technology in training logistics personnel has demonstrated significant potential in enhancing both the efficiency and safety of vehicle condition inspections. The findings from our study highlight several key points of discussion, which can be categorized into the effectiveness of MR technology, the specific challenges encountered, and the broader implications for the logistics industry.

4.1 Effectiveness of MR Technology

The results of the knowledge test clearly indicate that MR technology is highly effective in improving the accuracy of trainees when inspecting easily observable parts of forklifts, such as wheels and headlights. The 100% accuracy rate for wheel inspection tasks underscores the potential of MR to deliver precise and reliable training experiences. Similarly, the high engagement levels and positive feedback on the fun aspect of the training, as reflected by the 4.5 satisfaction score, suggest that MR technology can significantly enhance user engagement and motivation. This is crucial in training environments, as higher engagement often correlates with better learning outcomes and retention of information. Furthermore, the immersive nature of MR training aligns well with Experiential Learning Theory (ELT), which posits that learning is most effective when learners can actively engage in and reflect on their experiences. The MR environment provided concrete experiences that allowed trainees to interact with realistic simulations, thereby facilitating deeper understanding and retention of the BEWAGON principles.

4.2 Challenges and Areas for Improvement

Despite the overall positive outcomes, several challenges were identified that need to be addressed to maximize the effectiveness of MR training. The lower accuracy rates for more complex tasks, such as identifying brake fluid caps and brake discs, highlight a significant area for improvement. These tasks, which require detailed interaction and a more profound understanding of intricate vehicle components, proved challenging for trainees, with accuracy rates dropping to 20%. This suggests that while MR is effective for training on easily observable parts, it requires further development to improve training on more complex, detailed inspections. The satisfaction test also revealed that while participants found the MR training to be generally easy to use (4.2 score) and engaging, the perceived usefulness (3.8 score) was lower. This indicates that some participants experienced difficulties with the MR interface, which could hinder their ability to fully benefit from the training. Enhancing the user interface and providing more comprehensive guidance within the MR environment could help address these issues, making the training more intuitive and practically applicable.

5 Conclusion

The integration of Mixed Reality (MR) technology into logistics training programs significantly advances intelligent logistics by providing immersive, interactive, and riskfree environments for trainees. This study demonstrated that MR training is highly effective for tasks involving easily observable parts of forklifts, although it highlighted the need for further development for more complex inspections. The satisfaction test underscored the importance of enhancing the user interface and practical applicability of MR training. Broadly, MR technology offers a scalable, cost-effective solution that improves workforce skills and preparedness, contributing to safer, more efficient logis-tics operations. This is particularly vital for regions like Thailand, where robust logis-tics infrastructure supports economic growth and regional trade. Aligning with the trend toward smart logistics, MR training optimizes operations, positions logistics companies as innovation leaders, and attracts investment. While the benefits are clear, further re-search and development are needed to fully realize MR technology's potential, ensuring the logistics workforce is well-equipped for modern logistics demands and driving the industry forward.

Acknowledgments. Thank you to Excellent Training Center in Smart Logistics Using Mixed Reality Simulation to Support the Chiang Khong Trade Center for GMS Connectivity Project, under grant agree Office of National Digital Economy and Society Commission (ONDE). Our heartfelt thanks also go to the representatives from the College of Arts, Media and Technology, Chiang Mai University, Mae Fah Luang University, and the University of Phayao, as well as the professors whose invaluable advice and guidance enabled the successful completion of this research study.

References

1. Asian Development Bank. Improving Connectivity in the Greater Mekong Subregion (2018). https://www.adb.org/sites/default/files/publication/670396/gms-2030-integration-upgrading-cities-connectivity.pdf

2. Thailand Board of Investment. Green Infrastructure Investment Opportunities: Thailand 2021 Report (2020). https://www.adb.org/sites/default/files/publica-tion/767486/green-infrastructure-investment-thailand-2021.pdf
3. Intelligent Warehouse in Industry 4.0—Systematic Literature Review. (n.d.). https://www.mdpi.com/1424-8220/23/8/4105
4. World Bank. Logistics Performance Index: Connecting to Compete (2017). https://www.worldbank.org/en/news/feature/2016/06/28/connecting-to-compete-2016-trade-logistics-in-the-global-economy
5. Toyota Makes Mixed Reality Magic with Unity and Microsoft HoloLens 2 (2020). https://blog.unity.com/industry/toyota-makes-mixed-reality-magic-with-unity-and-microsoft-hololens-2
6. Augmented Reality in Maintenance—History and Perspectives (2023). https://www.ncbi.nlm.nih.gov/pmc/articles/PMC10381749/
7. Milgram, P., Kishino, F.: A taxonomy of mixed reality visual displays. IEICE Trans. Inf. Syst. **77**(12), 1321–1329 (1994)
8. Schroeder, R.: Being There Together: Social Interaction in Shared Virtual Environments. Oxford University Press (2010). https://www.oii.ox.ac.uk/research/publications/being-there-together/
9. Wang, J., Zhang, X., Liu, Y.: Advancements in smart logistics: Integrating IoT, big data, and AI technologies. Int J Log Res Appl **21**(3), 245–260 (2018)
10. Kanzaki, K.: Toyota's use of mixed reality in automotive assembly training: Enhancing accuracy and efficiency. J. Ind. Technol. **33**(4), 212–218 (2017)
11. Lee, J., Park, H., Kim, S.: Samsung's integration of mixed reality for industrial equipment maintenance training: a case study. Korean J. Eng. Technol. **42**(1), 45–57 (2019)
12. Wang, X., Zhou, L., Li, Z.: MR-based training for high-voltage equipment maintenance: the State Grid Corporation's approach. China Electr. Eng. J. **39**(6), 102–110 (2018)
13. Kolb, D.A.: Experiential Learning: Experience as the Source of Learning and Development. Prentice-Hall (1984)

Designing of Beef Cattle Recognition and Detection System to Improve Walk-over-Weighing System (WoW) for Beef Cattle in the Lancang-Mekong Region

Santichai Wicha[1](✉), Naruedol Duangban[2], Abdullah[3], Rafia Mumtaz[3], and Pradorn Sureephong[4]

[1] School of Information Technology, Mae Fah Luang University, Chiang Rai, Thailand
Santichai@mfu.ac.th

[2] Cogniser Thailand Co., Ltd., Chiang Mai, Thailand
Naruedol@cogniser.co.th

[3] National University of Sciences and Technology (NUST), Islamabad, Pakistan
{ah.msee21seecs,rafia.mumtaz}@seecs.edu.pk

[4] College of Arts, Media and Technology, Chiang Mai University, Chiang Mai, Thailand
dorn@camt.info

Abstract. The beef cattle industry in the Lancang-Mekong region, encompassing Cambodia, China, Laos, Myanmar, Thailand, and Vietnam, plays a crucial role in the agricultural economy. Traditional cattle farming methods in this region are labor-intensive and increasingly inadequate in meeting modern demands for efficiency, productivity, and traceability. This paper presents the design and implementation of an enhanced Walk-over-Weighing (WoW) system integrated with Artificial Intelligence (AI) recognition and advanced Internet of Things (IoT) technologies to address these challenges. The redesigned system features an AI-driven recognition system for accurate cattle identification and automated data col-lection, improving ease of use and reliability. Enhanced IoT infrastructure ensures continuous operation even in remote areas, while real-time analytics and integrated health monitoring provide actionable in-sights for effective cattle management. User satisfaction assessments indicate high acceptance and perceived benefits among stakeholders, including significant labor and cost savings, optimized cattle growth, and improved health management. This innovative approach aims to transform traditional cattle farming practices in the Lancang-Mekong region, promoting sustainable and profitable agricultural operations.

Keywords: Beef cattle recognition · Precision livestock farming · Lancang-Mekong Region · AI in agriculture · Walk-over-Weighing system (WoW) for beef cattle

P. Sureephong et al. (Eds.): PLM 2024, IFIP AICT 741, pp. 365–380, 2025.
https://doi.org/10.1007/978-3-031-93323-3_28

1 Introduction

The Greater Mekong Subregion (GMS), comprising Cambodia, China, Laos, Myanmar, Thailand, and Vietnam, is characterized by its rich agricultural diversity and significant dependence on livestock farming [1]. Among the various livestock, beef cattle play a crucial role in the agricultural economy of this region [2]. Beef cattle farming not only supports the livelihoods of a substantial portion of the rural population but also contributes significantly to food security, nutritional needs, and economic development [3].

Livestock farming, particularly beef cattle rearing, is an integral part of the agricultural landscape in the GMS region. The traditional methods of cattle farming practiced in this region are often labor-intensive and rely heavily on manual tracking and monitoring. These practices include the use of ear tags for identification and basic record-keeping for monitoring cattle health, growth, and productivity [4, 5]. However, these methods are increasingly proving to be inadequate in meeting the modern demands for efficiency, productivity, and traceability.

Team research was conducted the study on the Walk Over Weighing (WoW) [6, 7] system for beef cattle fattening identified several limitations, including high initial setup costs and the need for specialized infrastructure like solar power, which may not be feasible for smaller farms. The system's effectiveness could vary under different environmental conditions and might not generalize across all cattle breeds since it focused on only three specific breeds. The reliance on advanced statistical analysis poses usability challenges for farmers without a statistical background. Furthermore, while the WoW system helps in identifying growth inefficiencies, it does not directly address underlying health issues, requiring additional health surveillance. Adoption of the technology also necessitates significant training, and its dependency on IoT and internet connectivity could limit its use in remote areas. Future research should focus on addressing these challenges to improve the system's accessibility and effectiveness.

To address these challenges, the requirement of new system design is developed, there is a growing recognition of the need to integrate modern technologies such as Artificial Intelligence (AI) and the Internet of Things (IoT) into beef cattle farming. These technologies offer promising solutions for enhancing the efficiency, productivity, and sustainability of cattle farming. By leveraging AI and IoT, it is possible to develop smart farming systems that provide real-time monitoring, data analysis, and decision-making support, thereby transforming traditional farming practices into more efficient and sustainable operations.

Redesigning of beef cattle recognition and detection system to improve Walk- over-Weighing system (WoW)

The Lancang-Mekong Cooperation (LMC) Special Fund project, focusing on the development and demonstration of a smart beef cattle management system, is a significant initiative aimed at addressing these challenges. This project seeks to enhance the beef cattle industry in the GMS region through the adoption of advanced technologies, fostering international cooperation, and promoting sustainable agricultural practices. The integration of AI and IoT into cattle farming is expected to revolutionize the industry, improving farm management, cattle health monitoring, and overall productivity, ultimately contributing to the economic and social well-being of the region. By understanding the critical role of beef cattle farming in the GMS region and addressing the existing challenges through innovative technological solutions, this project aims to pave the way for a more prosperous and sustainable future for the agricultural sector in the Greater Mekong Subregion.

This study presents an innovative approach to improving beef cattle farming in the Lancang-Mekong region through the integration of Artificial Intelligence (AI) and the Internet of Things (IoT). In this first designing part, The project, funded by the Lancang-Mekong Cooperation Special Fund, aims to develop a comprehensive smart beef cattle management system. This system includes digital cattle recognition, multi-protocol access methods, and a cloud-based platform to enhance farm efficiency and productivity. The initiative involves collaboration between Thailand, China, Laos, and Cambodia, leveraging advanced technologies to address current inefficiencies in traditional livestock farming practices.

Propose of system design Framework for beef cattle recognition and detection system to improve Walk-over-Weighing system (WoW) (Fig. 1).

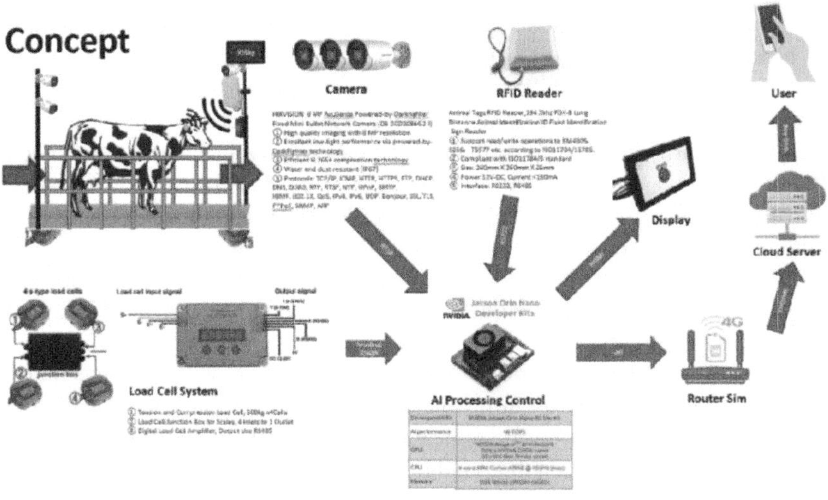

Fig. 1. System design Framework for beef cattle recognition and detection system.

Limitations of Previous WoW System

The Walk-over-Weighing (WoW) system has been a significant technological advancement in livestock management, offering automated weight monitoring for beef cattle. However, despite its potential, the previous iterations of the WoW system have encountered several limitations that hinder their widespread adoption and effectiveness, particularly in the context of the Lancang-Mekong region. Relies heavily on beef cattle farming as a vital component of its agricultural economy.

One of the primary limitations of the previous WoW system is the high initial setup cost. The implementation of the system requires substantial financial investment in equipment, such as RFID tags, weighing platforms, and solar power infrastructure. This high cost can be a significant barrier for small to medium-sized farms, which constitute a large portion of the farming community in the Lancang-Mekong region.

Additionally, the WoW system demands specialized infrastructure, including reliable power sources and internet connectivity. Many farms in this region are located in remote areas where such infrastructure is either insufficient or entirely absent, making it challenging to maintain continuous and reliable operation of the system. The variability in environmental conditions across different locations also affects the system's reliability and accuracy, as factors such as extreme weather can disrupt data collection and transmission.

Another critical limitation is the system's reliance on advanced statistical analysis methods for data interpretation. Techniques such as Analysis of Variance (ANOVA) and Duncan's multiple range tests require a certain level of expertise, which many farmers lack. This complexity in data analysis poses usability challenges, making it difficult for farmers to fully utilize the system's capabilities without extensive training or external support.

Moreover, the previous WoW system was primarily tested on a limited number of cattle breeds, such as Charolais, Brahman, and Beefmaster. This narrow focus raises concerns about the generalizability of the system's effectiveness across other breeds that are prevalent in the region. The system also falls short in addressing underlying health issues, as it primarily identifies growth inefficiencies without integrating comprehensive health monitoring.

Finally, the adoption of the WoW system necessitates significant training for both farmers and cattle. The process of training cattle to use the system and farmers to interpret and act on the data can be time-consuming and challenging. The dependency on IoT and cloud-based data transmission further complicates adoption, especially in areas with poor internet connectivity.

In summary, while the previous WoW system has shown promise in enhancing cattle weight monitoring, its high costs, infrastructure demands, environmental sensitivity, data analysis complexity, limited breed applicability, and health monitoring gaps highlight the need for a more user-friendly, reliable, and comprehensive solution. Addressing these limitations is crucial for improving the system's accessibility and effectiveness, ultimately supporting the sustainable development of the beef cattle industry in the Lancang-Mekong region (Table 1).

Summarizes the key limitations of the previous Walk Over Weighing (WoW) system, highlighting areas that need improvement to enhance its accessibility, reliability, and effectiveness for a broader range of farming conditions and cattle breeds.

Redesigned WoW System with AI Recognition and New IoT System Integration Outlines the enhancements in the redesigned WoW system incorporating AI recognition and new IoT technologies, emphasizing improvements in ease of use, reliability, effectiveness, and perceived benefits for beef cattle management (Table 2).

Table 1. Limitation of the previous WOW system

Limitation	Description
High Initial Setup Costs	The WoW system requires a significant initial investment for equipment and installation, including RFID tags, weighing platforms, and solar power infrastructure
Specialized Infrastructure Needs	The system depends on specialized infrastructure like solar panels for power, which may not be feasible or accessible for smaller farms
Environmental Variability	The effectiveness of the system can vary under different environmental conditions, potentially impacting its reliability and accuracy
Breed-Specific Data	The system was primarily tested on three specific cattle breeds (Charolais, Brahman, and Beefmaster), limiting the generalizability of the results to other breeds
Statistical Analysis Complexity	Advanced statistical methods (e.g., ANOVA, Duncan's multiple range test) are required for data analysis, posing usability challenges for farmers without statistical expertise
Health Monitoring Limitations	While the system identifies growth inefficiencies, it does not directly address underlying health issues, necessitating additional health surveillance and interventions
Technology Adoption and Training	Adoption of the system requires significant training and adaptation for both farmers and cattle, which can be time-consuming and challenging
Dependency on IoT and Connectivity	The system relies heavily on IoT and cloud-based data transmission, which may face issues in areas with poor internet connectivity, limiting its usability in remote regions

Table 2. Redesigned WoW System with AI Recognition and New IoT system integration

Aspect	Improvement	Description
Ease of Use	Intuitive User Interface	The system features a simplified, user-friendly interface, making it easy for users to navigate and operate without extensive training

(*continued*)

Table 2. *(continued)*

Aspect	Improvement	Description
	Automated Data Collection	Automated data collection and integration reduce the need for manual data entry, allowing farmers to focus on other essential tasks
Reliability	Enhanced AI Recognition	AI-driven recognition system accurately identifies individual cattle using RFID tags and image recognition, ensuring consistent and reliable data
	Robust IoT Infrastructure	Improved IoT devices with enhanced connectivity and solar-powered backups ensure continuous and reliable operation, even in remote locations
Effectiveness	Real-Time Analytics	AI algorithms provide real-time analysis of cattle weight and health data, offering actionable insights for effective decision-making in cattle management
	Integrated Health Monitoring	Health monitoring sensors detect vital signs and early disease symptoms, enabling timely interventions and improving overall cattle health management
Perceived Benefits	Optimized Productivity	Real-time monitoring and precise data analysis optimize cattle growth and productivity, leading to higher profitability for farmers
	Cost Efficiency	Automated processes and reduced labor needs lower operational costs, providing long-term economic benefits and enhancing farm sustainability
	Sustainability	Solar-powered IoT devices and efficient resource management contribute to sustainable and eco- friendly farming practices

2 Literature Review

Walk-over-Weighing (WoW) System

The Walk-over-Weighing (WoW) system has emerged as a significant advancement in livestock management, offering benefits such as improved efficiency, data accuracy, and animal welfare by automating the process of weight monitoring in beef cattle [8, 9]. Despite its advantages, the system faces challenges including high initial setup costs, infrastructure requirements, susceptibility to environmental conditions, breed-specific limitations, and the complexity of data analysis. These issues hinder its broader adoption, particularly in regions with limited resources. Recent technological advancements in AI and IoT hold promise for overcoming these limitations by enhancing cattle identification accuracy, providing comprehensive health monitoring, and making the system more accessible and affordable for smaller farms. By addressing these challenges, the WoW system can transform traditional cattle farming practices and promote sustainable agricultural development.

Beef Cattle Recognition and Detection System

The advancement of technology in agriculture has significantly improved livestock management, particularly through the development of beef cattle recognition and detection systems [10]. These systems, which integrate AI and IoT technologies [11], offer numerous benefits including increased efficiency, accuracy, and improved animal welfare. Automated recognition systems, utilizing RFID and high-resolution cameras, reduce labor requirements and provide consistent, accurate data for effective herd management. Real-time monitoring facilitated by IoT technology allows farmers to access up-to-date information from anywhere, aiding timely decision-making and interventions. The non-intrusive nature of these systems also enhances animal welfare by reducing the stress associated with physical handling.

However, despite these advantages, several challenges remain. High initial setup costs and the need for reliable infrastructure, such as power supply and internet connectivity, can be prohibitive, particularly for small to medium-sized farms in rural areas. Environmental factors can impact the system's performance, and the complexity of managing and analyzing large volumes of data requires robust training and support. Future advancements in AI and IoT technologies, along with the development of cost-effective solutions and improved infrastructure, are essential for overcoming these challenges and promoting broader adoption. Integrating health monitoring features can further enhance the systems, leading to early disease detection and improved herd health management, ultimately contributing to more sustainable and profitable cattle farming practices.

Integration of data-driven detection systems and traceability technologies in beef cattle

The advancement of technology in agriculture has significantly improved livestock management, particularly through the development of beef cattle recognition and detection systems [10]. These systems, which integrate AI and IoT technologies [11], offer numerous benefits including increased efficiency, accuracy, and improved animal welfare. Automated recognition systems, utilizing RFID and high-resolution cameras, reduce labor requirements and provide consistent, accurate data for effective herd management. Real-time monitoring facilitated by IoT technology allows farmers to access up-to-date information from anywhere, aiding timely decision-making and interventions. The non-intrusive nature of these systems also enhances animal welfare by reducing the stress associated with physical handling.

However, despite these advantages, several challenges remain. High initial setup costs and the need for reliable infrastructure, such as power supply and internet connectivity, can be prohibitive, particularly for small to medium-sized farms in rural areas. Environmental factors can impact the system's performance, and the complexity of managing and analyzing large volumes of data requires robust training and support. Future advancements in AI and IoT technologies, along with the development of cost-effective solutions and improved infrastructure, are essential for overcoming these challenges and promoting broader adoption. Integrating health monitoring features can further enhance the systems, leading to early disease detection and improved herd health management, ultimately contributing to more sustainable and profitable cattle farming practices.

3 Research Methodology

Study Area

The study was conducted across several beef cattle farms located in the Lancang-Mekong region, specifically in Thailand in this first stage and will extend to Laos, and Cambodia later. These farms were selected to represent a range of environmental conditions, farm sizes, and management practices, providing a comprehensive assessment of the proposed system's applicability and effectiveness.

System Design and Development

The proposed smart beef cattle management system integrates Artificial Intelligence (AI) and Internet of Things (IoT) technologies. The system design comprises several key components:

1. **Digital Cattle Recognition:**
   - RFID Tagging: Each cattle is equipped with an RFID tag for unique identification. The tags are tamper-proof and durable, ensuring long- term use.
   - Image Recognition: High-resolution cameras are installed at strategic points to capture images of the cattle. AI algorithms are employed to recognize individual cattle based on their physical characteristics.
2. **Weighing System Integration:**
   - Walk-Over Weighing (WoW) System: A platform scale is placed in the cattle's pathway, enabling automatic weighing each time a cattle walks over it. The system records the weight data and associates it with the cattle's RFID tag.
3. **Data Transmission and Storage:**
   - IoT Devices: The system includes IoT devices for real-time data transmission to a cloud-based platform. Solar panels provide power to these devices, ensuring continuous operation even in remote areas.
   - Cloud-Based Platform: All data collected from the RFID tags, cameras, and weighing scales are transmitted to a cloud-based platform. This platform enables real-time data storage, processing, and analysis.
4. **Data Analysis and Decision Support:**
   - AI Algorithms: Machine learning algorithms are used to analyze the data, providing insights into cattle growth patterns, health status, and optimal management practices.
   - User Interface: Farmers can access the data and analysis results through a user-friendly interface, which provides actionable recommendations for cattle management.

Experimental Design

To evaluate user satisfaction with the smart beef cattle management system among various stakeholders, including Thai farmers, farm managers, and agricultural experts. The evaluation focuses on aspects such as ease of use, reliability, effectiveness, and perceived benefits of the system.

Study Design
The research employs a mixed-methods approach, combining quantitative surveys and qualitative interviews to gather comprehensive data on user satisfaction.

Participants

- Total Participants: 50 stakeholders
  - Thai Farmers: 30
  - Other Stakeholders: 20 (including farm managers and agricultural experts)

Data Collection Methods

1. **Structured Surveys:**
   - Design: The survey was designed to capture quantitative data on various aspects of user satisfaction.
   - Questionnaire: The survey included Likert scale questions (ranging from 1 to 5) to assess:
     - Ease of use
     - Reliability
     - Effectiveness
     - Perceived benefits
   - Distribution: Surveys were distributed electronically and in paper format, ensuring accessibility for all participants.
   - Administration: Participants were given a detailed explanation of the survey objectives and assured of the confidentiality of their responses.
2. **Interviews:**
   - Design: Semi-structured interviews were conducted to gather qualitative insights.
   - Interview Guide: Questions were designed to delve deeper into participants' experiences and perceptions of the system, covering topics such as:
     - Initial impressions and ease of integration
     - Specific challenges faced
     - Impact on daily operations
     - Suggestions for improvement
   - Administration: Interviews were conducted in person, via phone, or through video calls, depending on the participants' preferences and availability.

Data Analysis

1. **Quantitative Analysis:**
   - Descriptive Statistics: Mean scores and standard deviations for each survey item were calculated to summarize overall satisfaction levels.
   - Comparative Analysis: Differences in satisfaction levels between Thai farmers and other stakeholders were analyzed using t-tests or ANOVA as appropriate.

2. **Qualitative Analysis:**
   - Thematic Analysis: Interview transcripts were analyzed to identify common themes and patterns in participants' responses.
   - Coding: Responses were coded into categories such as "ease of use," "reliability issues," "positive impacts," and "areas for improvement."
   - Synthesis: Qualitative findings were synthesized to provide a nuanced understanding of the survey results and to highlight specific user experiences and suggestions.

Validation and Reliability

- Pilot Testing: The survey and interview guide were pilot tested with a small group of stakeholders to ensure clarity and relevance.
- Reliability Check: Internal consistency of the survey was assessed using Cronbach's alpha.
- Triangulation: Data from surveys and interviews were triangulated to validate findings and ensure a comprehensive understanding of user satisfaction.

Ethical Considerations

- Informed Consent: All participants provided informed consent before participating in the survey and interviews.
- Confidentiality: Participants' identities and responses were kept confidential, and data were anonymized for analysis.
- Voluntary Participation: Participation was entirely voluntary, with the option to withdraw at any time without penalty.

Results Reporting

- Quantitative Findings: Results from the survey were reported with statistical summaries, including tables and charts to illustrate key findings.
- Qualitative Findings: Themes from the interviews were reported with illustrative quotes to provide context and depth to the quantitative data.
- Integrated Analysis: The final report integrated quantitative and qualitative findings to present a comprehensive evaluation of user satisfaction.

By employing this mixed-methods approach, the study aims to provide a robust evaluation of the smart beef cattle management system's reception among Thai farmers and other stakeholders, offering valuable insights for future improvements and wider adoption.

4 Results

Results of User Satisfaction

To evaluate user satisfaction, structured surveys and interviews were conducted with various stakeholders, including farmers, farm managers, and agricultural experts in Thai farmer. The survey included questions on the ease of use, reliability, effectiveness, and

Table 3. User Satisfaction Results

Aspect	Feedback	Positive Response (%)	Average Satisfaction Rating (out of 5)	Number of Positive Respondents	Thai Farmers (N = 30)	Other Stakeholders (N = 20)
Usability						
Ease of Use	System was easy to use and required minimal training	92%	4.6	46	28	18
Training and Support	Training sessions were effective and support resources were available	90%	4.5	45	27	18
Accuracy and Reliability						
Cattle Recognition	High accuracy in RFID tagging and image recognition	95%	4.7	48	29	19
Weighing System	Accurate weight measurements and reliable data transmission	94%	4.6	47	28	19
Efficiency						
Labor Reduction	Significant reduction in labor requirements	88%	4.5	44	26	18
Time Savings	Notable time savings in cattle management processes	91%	4.6	46	27	19
Impact on Farm Operations						

(*continued*)

Table 3. (*continued*)

Aspect	Feedback	Positive Response (%)	Average Satisfaction Rating (out of 5)	Number of Positive Respondents	Thai Farmers (N = 30)	Other Stakeholders (N = 20)
Growth Monitoring	Improved ability to monitor cattle growth and make informed decisions	89%	4.6	45	26	19
Health Management	Early identification of health issues and timely alerts	87%	4.5	44	25	19
Economic Benefits	Cost savings and increased productivity	85%	4.4	43	25	18
Overall Satisfaction						
General Feedback	Positive experience with the system	93%	4.6	47	28	19
Recommendations	Would recommend the system to other farmers and agricultural professionals	92%	N/A	46	28	18

perceived benefits of the system. A total of 50 stakeholders participated in the survey, providing a comprehensive overview of the system's reception (Table 3).

This table summarizes the user satisfaction results, highlighting the positive reception and effectiveness of the smart beef cattle management system among the 50 stakeholders, classified into 30 Thai farmers and 20 other stakeholders, in the Lancang-Mekong region.

Key Findings

1. **Usability:**
   - Ease of Use: 92% of respondents reported that the system was easy to use and required minimal training. The user interface was particularly appreciated for its intuitive design, with an average satisfaction rating of 4.6 out of 5.
   - Training and Support: 90% of the stakeholders found the training sessions effective, stating that they felt confident in operating the system after the training. The availability of support resources was rated positively, with an average score of 4.5 out of 5.

2. **Accuracy and Reliability:**
   - Cattle Recognition: Stakeholders reported high satisfaction with the digital cattle recognition component. 95% of users rated the accuracy of RFID tagging and image recognition as excellent, with an average satisfaction rating of 4.7 out of 5.
   - Weighing System: The WoW system's accuracy in weight measurement was rated highly by 94% of the users, with a satisfaction rating of 4.6 out of 5. The reliability of the system in consistently capturing and transmitting data was praised, with minimal reports of data loss or inaccuracies.
3. **Efficiency:**
   - Labor Reduction: 88% of respondents noted a significant reduction in labor requirements for weighing and monitoring cattle, which allowed them to allocate resources more efficiently. The average satisfaction rating for labor efficiency improvements was 4.5 out of 5.
   - Time Savings: The automated nature of the system saved considerable time in cattle management processes. 91% of stakeholders reported notable time savings, with an average satisfaction rating of 4.6 out of 5.
4. **Impact on Farm Operations:**
   - Growth Monitoring: The ability to track and analyze cattle growth in real-time was highly valued. 89% of users reported that the system improved their ability to monitor cattle growth and make informed decisions, with an average satisfaction rating of 4.6 out of 5.
   - Health Management: The system's capability to identify health issues early and provide timely alerts was appreciated by 87% of the stakeholders. The average satisfaction rating for health management improvements was 4.5 out of 5.
   - Economic Benefits: 85% of respondents acknowledged the economic benefits of the system, including cost savings and increased productivity. The average satisfaction rating for economic impact was 4.4 out of 5.
5. **Overall Satisfaction:**
   - General Feedback: The overall satisfaction with the system was very high, with 93% of stakeholders expressing a positive experience. The average overall satisfaction rating was 4.6 out of 5.
   - Recommendations: 92% of the respondents indicated that they would recommend the system to other farmers and agricultural professionals, highlighting its potential to transform traditional cattle farming practices.

User satisfaction results indicate that the smart beef cattle management system was well-received by stakeholders in the Lancang-Mekong region. The system's ease of use, accuracy, efficiency, and positive impact on farm operations were key factors contributing to high satisfaction levels. The feedback from stakeholders underscores the system's potential to significantly enhance beef cattle farming practices, promoting more efficient, productive, and sustainable operations. Future improvements based on user feedback will further optimize the system and extend its benefits to a broader range of agricultural applications.

5 Discussion

The integration of AI recognition and advanced IoT technologies into the Walk-over-Weighing (WoW) system represents a significant step forward in addressing the limitations of previous iterations. This redesigned system has demonstrated notable improvements in several key areas, which are critical for enhancing the overall efficiency and sustainability of beef cattle farming in the Lancang-Mekong region.

Ease of Use:
The user-friendly interface and automated data collection have significantly reduced the complexity of the system, making it accessible to farmers with varying levels of technical expertise. The automated processes have minimized the need for manual data entry, allowing farmers to focus more on critical aspects of cattle management. The high satisfaction rates reported in the user surveys underscore the effectiveness of these improvements in enhancing user experience.

Reliability:
The enhanced AI-driven recognition system has provided accurate and consistent identification of individual cattle, addressing one of the major reliability issues faced by the previous WoW system. The robust IoT infrastructure, supported by solar-powered devices, ensures continuous operation even in remote and underdeveloped areas, thereby mitigating the challenges posed by unreliable power sources and poor connectivity. The high accuracy and minimal data loss reported by users confirm the system's reliability under diverse environmental conditions.

Effectiveness:
The integration of real-time analytics and health monitoring capabilities has significantly improved the system's effectiveness in cattle management. The AI algorithms provide actionable insights that facilitate informed decision-making, leading to optimized cattle growth and health. The early detection of health issues through integrated sensors allows for timely interventions, reducing the incidence of disease and improving overall herd health. The positive feedback on productivity improvements and health management from stakeholders highlights the practical benefits of these enhancements.

Perceived Benefits:
Stakeholders have reported substantial economic benefits, including reduced labor costs and increased productivity. The automated weighing and monitoring processes have decreased the need for manual labor, while the precise data analysis has enabled more efficient feed management and timely marketing decisions. These benefits contribute to the overall profitability and sustainability of cattle farming operations. The high recommendation rates among users indicate a strong perceived value of the system in transforming traditional farming practices.

Despite these advancements, there remain areas for further improvement. Continuous training and support are essential to ensure farmers can fully leverage the system's capabilities. Additionally, expanding the system's applicability to a broader range of cattle breeds and further enhancing its health monitoring features could provide even greater benefits. Addressing these areas will be crucial for maximizing the system's impact and ensuring its long-term success in the region.

6 Conclusion

The redesigned Walk-over-Weighing (WoW) system, integrated with AI recognition and advanced IoT technologies, has successfully addressed many of the limitations identified in previous versions. The enhancements in ease of use, reliability, effectiveness, and perceived benefits have significantly improved the system's utility and accessibility for beef cattle farmers in the Lancang-Mekong region.

The system's ability to provide accurate, real-time data and actionable insights has transformed traditional cattle farming practices, leading to optimized growth, better health management, and increased profitability. The positive feedback from stakeholders underscores the system's potential to drive sustainable agricultural practices and support the economic development of the region.

Future research and development should focus on further refining the system, expanding its applicability, and enhancing its features to ensure it continues to meet the evolving needs of cattle farmers. By building on these advancements, the WoW system can play a pivotal role in promoting a more efficient, productive, and sustainable beef cattle industry in the Lancang-Mekong region.

Acknowledgments. This research work is granted by Mekong – Lancang Special Fund: MLCSF, in collaboration with School of Information Technology, Mae Fah Luang University, Chiang Rai, Thailand and College of Arts, Media and Technology Chiang Mai University.

References

1. Paul, B.K., et al.: Crop- livestock integration provides opportunities to mitigate environmental trade-offs in transitioning smallholder agricultural systems of the Greater Mekong Subregion. Agric. Syst. **195**, 103285 (2022)
2. Teufel, N., Markemann, A., Kaufmann, B., Valle-Zárate, A., Otte, J.: Livestock production systems in South Asia and the Greater Mekong sub-region. A Quantitative Description of Livestock Production in Bangladesh, Cambodia, India (2010)
3. Xayalath, S., Mujitaba, M.A., Ortega, A.D.S.V., Rátky, J.: A review on the trend of livestock breeds in Laos. Acta Agraria Debreceniensis **1**, 227–237 (2021)
4. Saengwong, S., Thannithi, W., Intawicha, P., Porkaew, C.: Development of a mobile app for recording and management alert on-farm to supporting beef cattle smallholder farmers (2021)
5. Haldar, A., Mandal, S.N., Deb, S., Roy, R., Laishram, M.: Application of information and electronic technology for best practice management in livestock production system. In: Agriculture, Livestock Production and Aquaculture: Advances for Smallholder Farming Systems, vol. 2, pp. 173–218. Springer, Cham (2022)
6. Noinan, K., Wicha, S., Chaisricharoen, R.: The IoT-based weighing system for growth monitoring and evaluation of fattening process in beef cattle farm. In: 2022 Joint International Conference on Digital Arts, Media and Technology with ECTI Northern Section Conference on Electrical, Electronics, Computer and Telecommunications Engineering (ECTI DAMT & NCON), pp. 384–388. IEEE (2022)
7. Wicha, S., Noinan, K., Yamsa-Ard, S., Kamhangwong, D., Chaisricharoen, R., Sureephong, P.: Decision model of walk-over-weighing system (WoW) for support the fattening process of beef cattle farm. In: 2023 7th International Conference on Information Technology (InCIT), pp. 498–502. IEEE (2023)

8. Niloofar, P., et al.: Data-driven decision support in livestock farming for improved animal health, welfare and greenhouse gas emissions: overview and challenges. Comput. Electron. Agric. **190**, 106406 (2021)
9. Tedeschi, L.O., Greenwood, P.L., Halachmi, I.: Advancements in sensor technology and decision support intelligent tools to assist smart livestock farming. J. Animal Sci. **99**(2), skab038 (2021)
10. Yousefi, D.M., Rafie, A.M., Al-Haddad, S.A.R., Azrad, S.: A systematic literature review on the use of deep learning in precision livestock detection and localization using unmanned aerial vehicles. IEEE Access **10**, 80071–80091 (2022)
11. Chen, C., Zhu, W., Norton, T.: Behaviour recognition of pigs and cattle: journey from computer vision to deep learning. Comput. Electron. Agric. **187**, 106255 (2021)
12. Neethirajan, S., Kemp, B.: Digital livestock farming. Sens. Bio-Sens. Res. **32**, 100408 (2021)
13. Buller, H., Blokhuis, H., Lokhorst, K., Silberberg, M., Veissier, I.: Animal welfare management in a digital world. Animals **10**(10), 1779 (2020)
14. Xiong, Y., Shi, M., Cao, D.: Design and implementation of traceability system for beef cattle product quality and safety. Manuf. Serv. Oper. Manag. **4**(5), 58–67 (2023)
15. Cosby, A., Manning, J., Fogarty, E., Islam, N., Trotter, M.: Assessing real time tracking technologies to integrate with identification methods and national traceability requirements (2021)

Author Index

Published by Springer Nature Switzerland AG 2025
P. Sureephong et al. (Eds.): PLM 2024, IFIP AICT 741, pp. 381–383, 2025.
https://doi.org/10.1007/978-3-031-93323-3

The manufacturer's authorised representative in the EU is Springer
Nature Customer Service Centre GmbH, Europaplatz 3, 69115 Heidelberg,
Germany. If you have any concerns regarding our products, please
contact ProductSafety@springernature.com

Printed and bound by CPI Group (UK) Ltd, Croydon, CR0 4YY
30/09/2026
02225589-0017